Polarized Neutrons and Synchrotron X-Rays for Magnetism Conference 2009

(PNSXM 2009)

Satellite Conference of the ICM 2009
Journal of Physics: Conference Series Volume 211

Bonn, Germany
2-5 August 2009

ISBN: 978-1-61738-266-6

Printed from e-media with permission by:

Curran Associates, Inc.
57 Morehouse Lane
Red Hook, NY 12571

Some format issues inherent in the e-media version may also appear in this print version.

Printed by Curran Associates, Inc. (2010)

For permission requests, please contact the Institute of Physics
at the address below.

Institute of Physics
Dirac House, Temple Back
Bristol BS1 6BE UK

Phone: 44 1 17 929 7481
Fax: 44 1 17 920 0979

techtracking@iop.org

Additional copies of this publication are available from:

Curran Associates, Inc.
57 Morehouse Lane
Red Hook, NY 12571 USA
Phone: 845-758-0400
Fax: 845-758-2634
Email: curran@proceedings.com
Web: www.proceedings.com

TABLE OF CONTENTS

COMPLEX BULK MAGNETISM: FROM MULTIFERROICS TO FRUSTRATED MAGNETISM

THIN FILMS AND NANOPARTICLES

NEW METHODS AND INSTRUMENTATION

Journal of Physics: Conference Series **211** (2010) 011001 doi:10.1088/1742-6596/211/1/011001

PNSXM 2009

2nd Workshop on Polarized Neutrons and Synchrotron X-Rays for Magnetism

Thomas Brückel, Werner Schweika and Jean-Marc Tonnerre

PNSXM 2009

The second workshop on Polarized Neutrons and Synchrotron X-Rays for Magnetism PNSXM 2009 took place in Bonn from August 2nd to 5th 2009, as a satellite conference of the ICM 2009 in Karlsruhe. This second edition follows the spirit of the PNSXM 2003 in Venice in bringing together the experts and explorers of these advanced scattering probes to magnetism on atomic and nano-scales. The workshop gathered a large international audience with 110 participants from seventeen countries and five continents in a pleasant enjoyable atmosphere. The program for the three days was organized in plenary sessions with 39 oral presentations including nine invited talks and two topically mixed poster sessions with 74 contributions. Thanks to the international advisory committee many young participants could be given the opportunity for an oral presentation of their research. The topics, with a large span of problems, addressed current challenges in correlated electrons, chirality and complex spin structures, multiferroics, orbital and charge order, thin film and nanomagnetism. Applications to materials and new trends in methods and instrument developments were discussed in the poster sessions. Dedicated to structure, dynamics and kinetics in materials, a wealth of studies were presented utilizing polarized scattering, nuclear resonance spectroscopy, x-ray magnetic circular dichroism, coherence and holography, fast dynamics using the free electron laser and also neutron radiography. Among these studies 32 manuscripts have now been collected for publication in this proceeding band of the conference. The picture shows the participants of the PNSXM 09 gathering near the banks of the Rhine river; a moment just before the conference dinner, which was taken on a boat while cruising up the Rhine river for the setting sun and returning in a full moons light. We express our sincere gratitude to all of you who made this conference a remarkable event and success.

Program Comittee

PNSXM 2009

Prof. Stefan Blügel, Forschungszentrum Jülich, Germany
Prof. Thomas Brückel, Forschungszentrum Jülich, Germany
Dr. Mechthild Enderle, ILL, Grenoble, France
Prof. Christian Pfleiderer, Technische Universität München, Germany
Prof. Hartmut Zabel, Ruhr-Universität Bochum, Germany

International Advisory Committee

Prof. Massimo Altarelli, DESY, Hamburg, Germany
Dr. Andrew Boothroyd, University of Oxford, UK
Prof. Roberto G. M. Caciuffo, European Commission, Karlsruhe, Germany
Prof. Robert Cywinsky, University of Leeds, UK
Prof. Wolfgang Eberhardt, BESSY, Berlin, Germany
Prof. Yasuo Endoh, Kobe, Japan
Dr. José Luis García-Muñoz, CSIC, Bellaterra, Spain
Prof. Alan I. Goldman, Iowa State University, USA
Prof. Bruce Harmon, Iowa State University, USA
Prof. Andrew Harrison, ILL, Grenoble, France
Dr. John P. Hill, BNL, New York, USA
Prof. Kazuhisa Kakurai, JAEA, Tokai, Japan
Dr. Shane Kennedy, ANSTO, Menai, Australia
Prof. Wen-Hsien Li, Taiwan National Central University, Taiwan (R.O.C.)
Dr. Charles Majrkzak, NIST, Gaitersburg, USA
Prof. Des McMorrow, University College London, UK
Prof. Joel Mesot, ETH Zürich & Paul Scherrer Institut, Villigen, Switzerland
Prof. Oscar Moze, Universita di Modena e Reggio Emilia, Modena, Italy
Dr. Stephen E. Nagler, ORNL, Oak Ridge, USA
Prof. Sunil K. Sinha, UCLA, San Diego, USA
Prof. W. G. Stirling, ESRF, Grenoble, France
Prof. Joachim Stöhr, Stanford Linear Accelerator Center, Menlo Park, USA
Prof. Alan Tennant, HMI, Berlin, Germany
Prof. Yoshinori Tokura, University of Tokyo, Japan
Dr. Jean-Marc Tonnerre, CNRS, Grenoble, France
Dr. Boris P. Toperverg, PNPI, St. Petersburg, Russia

Local Organizing Committee

Thomas Brückel
Ursula Funk-Kath
Emmanuel Kentzinger
Barbara Köppchen
Clarissa Rick
Ulrich Rücker
Dirk Schlotmann
Werner Schweika
Jörg Voigt
Angela Wenzik

Journal of Physics: Conference Series **211** (2010) 011001 doi:10.1088/1742-6596/211/1/011001

Sponsors of the PNSXM

All papers published in this volume of *Journal of Physics: Conference Series* have been peer reviewed through processes administered by the proceedings Editors. Reviews were conducted by expert referees to the professional and scientific standards expected of a proceedings journal published by IOP Publishing

Polarized Neutrons and Synchrotron X-rays for Magnetism Conference 2009 IOP Publishing
Journal of Physics: Conference Series **211** (2010) 012001 doi:10.1088/1742-6596/211/1/012001

Polarized-neutron-scattering studies on the chiral magnetism in multiferroic $MnWO_4$

T Finger[1], D Senff[1], K Schmalzl[2], W Schmidt[2], L P Regnault[3], P Becker[4], L Bohatý[4] and M Braden[1]

[1] II. Physikalisches Institut, Universität zu Köln, Zülpicher Str. 77, D-50937 Köln, Germany
[2]Institut für Festkörperforschung, Forschungszentrum Jülich GmbH, JCNS at ILL, 38042 Grenoble Cedex 9, France
[3]Institut Nanosciences et cryogénie, SPSMS-MDN, CEA-Grenoble, DRFMC-SPSMS-MDN, F-38054 Grenoble Cedex 9, France
[4]Institut für Kristallographie, Universität zu Köln, Zülpicher Str. 49b, D-50674 Köln, Germany

E-mail: braden@ph2.uni-koeln.de

Abstract. Neutron diffraction with spherical polarization analysis is a powerful tool for studying the multiferroic materials where the ferroelectric polarization arises from a complex magnetic structure. Analyzing the off-diagonal terms in the polarization matrix one may directly detect the chiral contributions even in a multidomain arrangement. In $MnWO_4$ one can control the chiral magnetism by varying an electric field at constant temperature. The analysis of multiferroic hysteresis cycles at four equivalent magnetic Bragg peaks fully agrees with a nearly monodomain chiral arrangement controlled by the electric field. A pronounced asymmetry of the hysteresis cycles and memory effects point to strong pinning of the chiral magnetism in $MnWO_4$. We find a second-order harmonic modulation which exhibits both magnetic and structural character and which may be related with the domain pinning. The observed interference between the nuclear and the magnetic modulation is another manifestation of the coupling between the crystal structure and the magnetism in the multiferroic oxides.

1. Introduction

The complex magnetism in the recently discovered multiferroic transition-metal oxides attracts considerable attention, as these materials exhibit very large magnetoelectric effects possibly opening a route towards magneto-electric data-storage devices [1, 2, 3, 4, 5, 6]. In most of the newly discovered compounds, the multiferroic or magnetoelectric coupling arises from the asymmetric part of the magnetic interaction. In the same manner, as a low-symmetry crystal structure causes a canting of magnetic moments, the non-collinear magnetic order induces a structural distortion which may sum up into a ferroelectric polarization under the correct symmetry. Following references [7, 8, 9] the induced electric polarization of a single pair of spins S_i, S_j separated by a distance vector $\mathbf{r}_{i,j}$ is given by : $\mathbf{P}_{FE} \propto \mathbf{r_{ij}} \times (\mathbf{S_i} \times \mathbf{S_j})$. For this inverse Dzyaloshinski-Moriya (IDM) mechanism a non-collinear magnetic structure is a necessary condition rendering longitudinal and spherical polarization analysis a powerful and nearly unique method [10].

In the $REMnO_3$ series, depending on temperature and on the type of the RE, several magnetic-field induced effects are described: One may induce a finite ferroelectric polarization out of a paraelectric phase, one may suppress a ferroelectric polarization existing at zero magnetic field, or one may flop a ferroelectric polarization into another direction [3, 6, 11]. However, the opposite sense of the multiferroic coupling is more demanding to study, as the complex magnetic ordering requires a microscopic probe. A circular cycloid magnetic structure yields optimum conditions in view of the generation of ferroelectric polarization. A magnetic cycloid does not possess a scalar chirality or handedness (for example defined as $q_{prop} \cdot (S_1 \times S_2)$), since the propagation vector is parallel to the spiral plane of the magnetic moments. However, the IDM mechanism involves the cross product of neighboring spins and can, therefore, be associated with the chiral term :

$$\mathbf{M}_{ch}(\mathbf{Q}) = i\{\mathbf{M}_{\perp}(\mathbf{Q}) \times \mathbf{M}^{*}_{\perp}(\mathbf{Q})\}, \qquad (1)$$

entering the polarized-neutron-scattering cross sections [10]. By poling the multiferroic compounds in moderate magnetic fields it was possible to prove by polarized neutron scattering, that the sign of the chiral term follows the sign of the poling electric field in several transition-metal oxide multiferroics : in $TbMnO_3$ [12] , in $LiCu_2O_2$ [13] and in $MnWO_4$ [14]. This magnetic poling gives strong support to the IDM multiferroic mechanism [7, 8, 9]. The IDM coupling is, however, a second-order effect, which, in addition to strong frustration as cause of the complex magnetism, yields only a limited amount of ferroelectric polarization. In the $REMnO_3$ [3, 6, 15] series the polarization is about two orders of magnitude smaller than in a typical perovskite ferroelectric [15]. Other materials, like $MnWO_4$ [16, 17, 18] or the pyroxenes [19] exhibit a ferroelectric polarization reduced by more than three orders of magnitude compared to $BaTiO_3$ [3, 15] hindering the observation of electric-field induced effects in the magnetic structure as well as applications.

In our recent work we have studied $MnWO_4$ where the multiferroic poling exhibits a memory effect which is not erased upon heating deeply into the paramagnetic and paraelectric phase [20]. In addition it has been possible to switch the chiral magnetism by varying the applied electric field at constant temperature, which is the relevant effect in view of potential applications. Here we extend our work by further hysteresis-curve measurements and by an analysis of the second-order harmonic modulation, which in $MnWO_4$ has both nuclear and magnetic character.

The crystal structure of $MnWO_4$, see Figure 1, is characterized by zigzag chains formed by the MnO_6 octahedra [21] (space group P2/c, a=4.835Å, b=5.762Å, c=4.992Å and β=91.08^o) suggesting a one-dimensional magnetic character. However, the analysis of the spin-wave dispersion [22] finds strong interchain interactions indicating that the magnetic character in this material is not one-dimensional but frustrated. Upon cooling, $MnWO_4$ exhibits a first transition into an incommensurate magnetic phase labelled AF3 with propagation vector $\mathbf{q}_{ic}$=(-0.214,0.5,0.457) and collinear moments aligned in the a,c-plane, T_{AF3}=13.2 K [21, 22] followed by a second transition to a still incommensurate but non-collinear phase, AF2, this magnetic structure is displayed in Figure 1. In the AF3 phase the magnetic moment aligns along an easy axis, $\mathbf{e_{easy}}$, in the a,c-plane which forms an angle to the propagation vector $\mathbf{q}_{ic}$ of 83°. The transition into the AF2 phase is nearly continuous and can be characterized by the continuous growing of the b-component of the ordered moment, see Figure 1 and reference [20]. The diffraction studies [21] were unable to fix the phase between the easy-axis and the b components of the AF2 structure, but a chiral arrangement was indirectly deduced from the size of the moments. The transition AF3→AF2 is accompanied by the occurrence of the ferroelectric polarization [16, 17, 18]. Finally, the magnetic structure transforms into a commensurate collinear AF1 state with $\mathbf{q}_c$=(-0.25,0.5,0.5) at T_{AF1}=7.0 K, where the ferroelectric polarization disappears and where magnetic moments again exhibit a collinear arrangement aligned along $\mathbf{e_{easy}}$. The AF1 magnetic structure seems to be very anharmonic [21].

Journal of Physics: Conference Series **211** (2010) 012001 doi:10.1088/1742-6596/211/1/012001

2. Polarized-neutron-scattering experiments on $MnWO_4$

2.1. Experimental

We have performed the polarized-neutron-diffraction studies on the IN12 cold triple-axis spectrometer at the Institut Laue-Langevin using the Cryopad-equipment for spherical polarization analysis [10]. A supermirror bender and a (111)-Heusler crystal were used for the polarization control on the k_i and k_f sides, respectively. We use the common cartesian coordinate system with **x** along **Q**, **y** in the scattering plane but perpendicular to **Q**, and **z** vertical. In our notation $\sigma_{ij}^{\downarrow\uparrow}$ denotes the intensity in the channel with the outgoing polarization along j when the incoming polarization is set along i with the arrows indicating the directions of polarizations. With the experimental setup we achieved flipping ratios of 35 and 40 in the two experimental runs, as measured on a nuclear Bragg peak. The precision of the spherical analysis was checked on the purely nuclear (020) Bragg reflection, where no rotation of polarization should occur. At the (020) reflection the rotation of the neutron polarization is very small underlining the precision of the neutron polarimetry. The polarization in the three longitudinal (diagonal) channels amounts to 94.5%; in the other channels we find: xy 4.9%, xz 1.2%, yx 3.8%, yz 0.7%, zx 1.6%, and zy 2.3%. These polarizations of the outgoing beam show that the neutron polarization can be controlled within a precision of about a degree, just in the xy, yx-channels a still small erroneous leakage seems to occur. In order to access the chiral terms in the incommensurate magnetic phases we had to use the unconventional $(0,1,0),(-0.214,0,0.457)$ scattering plane using a synthetic single crystal [16]. At the magnetic Bragg peaks (±0.214,0.5,∓0.457) and (±0.214,-0.5,∓0.457) we find a flipping ratio in the xx channel of 33 to be compared to the value of 35 found at the purely nuclear Bragg peak in the same run. This measurement shows that at the first-order magnetic Bragg peak there is only a negligible nuclear contribution of less than $5 \cdot 10^{-3}$ and that the orientation of the polarimetry coordinate system versus the scattering vector is very precise. The electric field was always applied parallel to the b-direction by placing the plate-shaped crystal of 4mm thickness between Al-plates which are nearly transparent for neutrons.

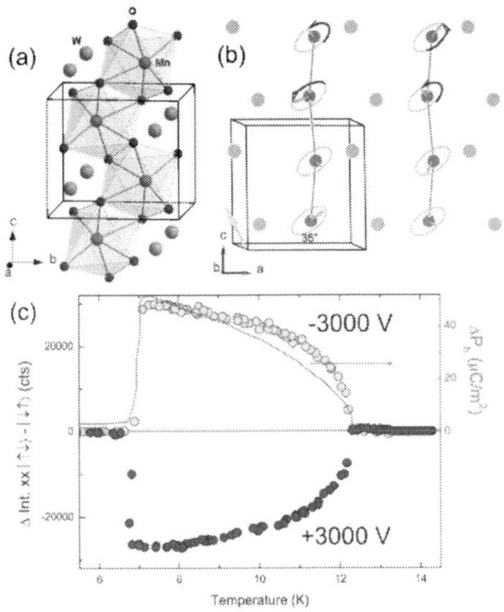

Figure 1. (a) Crystal structure in $MnWO_4$ with chains of MnO_6 octahedra running along the c direction; b) the magnetic structure of $MnWO_4$ in the incommensurate chiral phase AF2; the blue arrow indicates the magnetic propagation vector which is almost perpendicular to $\mathbf{e_{easy}}$. c) temperature dependence of the chiral scattering in $MnWO_4$ obtained in opposed electric poling fields at the first-order magnetic Bragg peak (0.214,0.5,-0.457). The field was applied along the b-direction. The chiral contribution only roughly scales with the electric polarization taken from reference [17] due to a minor amount of misaligned chiral domains, which furthermore can be temperature dependent.

2.2. Poling of chiral magnetism

In the used scattering plane, the easy-axis of $MnWO_4$ points almost perfectly along the **z**-direction. Therefore, the spin-flip scattering in the longitudinal (diagonal) channels along **y** and **z** sense the $\mathbf{e_{easy}}$ and b components of the magnetic order, respectively. Following the temperature dependencies of the $\sigma_{yy}^{\downarrow\uparrow}$ and the $\sigma_{zz}^{\downarrow\uparrow}$ spin-flip scattering one measures the appearance of the magnetic b-component in the AF2 phase and its suppression in the AF1 phase at low temperature [20], note that the propagation vector is almost perpendicular to $\mathbf{e_{easy}}$. In addition, the chiral contribution can be effectively measured at **Q**=(-0.214,0.5,0.457) and (-0.214,1.5,0.457). The chiral term contributes to three channels in spherical polarization. In comparison to the total magnetic scattering it is measured in the xx channel as $\sigma_{xx}^{\downarrow\uparrow} = [M_y(\mathbf{Q}) \cdot M_y^*(\mathbf{Q}) + M_z(\mathbf{Q}) \cdot M_z^*(\mathbf{Q})] - M_{ch}(\mathbf{Q})$ and $\sigma_{xx}^{\uparrow\downarrow} = [M_y(\mathbf{Q}) \cdot M_y^*(\mathbf{Q}) + M_z(\mathbf{Q}) \cdot M_z^*(\mathbf{Q})] + M_{ch}(\mathbf{Q})$. Therefore it is useful to analyze the chiral ratio at a certain scattering vector **Q**:

$$r_{chir} = \frac{M_{ch}(\mathbf{Q})}{M_y(\mathbf{Q}) \cdot M_y^*(\mathbf{Q}) + M_z(\mathbf{Q}) \cdot M_z^*(\mathbf{Q})} = \frac{\sigma_{xx}^{\uparrow\downarrow} - \sigma_{xx}^{\downarrow\uparrow}}{\sigma_{xx}^{\uparrow\downarrow} + \sigma_{xx}^{\downarrow\uparrow}}. \tag{2}$$

The chiral contribution, $M_{ch}(\mathbf{Q})$, and the chiral ratio, r_{chir}, can also be measured at the non-diagonal components of the polarization tensor :

$$r_{chir} = \frac{\sigma_{yx}^{\uparrow\downarrow} - \sigma_{yx}^{\uparrow\uparrow}}{\sigma_{yx}^{\uparrow\downarrow} + \sigma_{yx}^{\uparrow\uparrow}} = \frac{\sigma_{yx}^{\downarrow\downarrow} - \sigma_{yx}^{\downarrow\uparrow}}{\sigma_{yx}^{\downarrow\uparrow} + \sigma_{yx}^{\downarrow\downarrow}}, \tag{3}$$

in the yx and similarly in the zx channels [10]. Since the studied magnetic Bragg peaks (besides the second-order peaks, see below) are purely magnetic we may fully neglect any nuclear term and its interference with the magnetism. The polarization matrices measured at the purely magnetic Bragg peaks further document the high polarization precision, as the xy and xz channels exhibit polarizations of only 3%.

Similar to earlier observations on this and other chiral multiferroics it is possible to pole the chiral component with an electric field [12, 13, 14, 20]. In $MnWO_4$ we observe, after cooling in an electric field of 3500V/4mm, a chiral ratio of 81% at **Q**=(-0.214,0.5,0.457) and T=7.7 K [20] which is slightly below the expected value. Supposing that the phase between the $\mathbf{e_{easy}}$ and the b components is 90° and that the spiral is perfectly circular, one calculates an expected chiral ratio of 96%. This value still deviates from one due to the projections of the magnetic moment to the (**x**,**y**,**z**) coordinate system defined above. There are several reasons which may explain the reduced experimental chiral ratio. First the imperfect polarization reduces the measured chiral ratio. In addition, we find that the neutron beam becomes slightly depolarized for initial polarization along **y** and **z**. From that and the polarization corrected chiral ratio we may conclude that about 5% of the sample volume remain in the *wrong* chiral arrangement [20]. Taking further into account that the two magnetic components are of equal size [20], we may further conclude that the chiral structure of $MnWO_4$ is close to ideal, i.e. circular and with a 90° phase shift between the components. The relative amount of the misaligned domains may vary as function of temperature in the used large sample crystal explaining the limited scaling between the chiral contribution and the ferroelectric polarization, shown in Figure 1. Indeed a better description of the ferroelectric polarization is obtained for plotting the product of the two magnetic components, see reference [20], indicating that the phase of the magnetic structure amounts to about 90° throughout the range of the AF2 phase.

2.3. Electric-field driven hysteresis curves

The control of magnetism by an external field at constant temperature is most important in view of the desired future applications, but only very little experiments of this type have been reported

Journal of Physics: Conference Series **211** (2010) 012001 doi:10.1088/1742-6596/211/1/012001

in general [23, 24, 25, 26, 27, 28]. In particular the electric-field control of the chiral magnetism in the multiferroics driven through the IDM mechanism has not been studied until recently. In $MnWO_4$ it has been possible to switch the electric polarization in thin crystals by varying the electric field [29]. From this effect one may indirectly deduce via the IDM mechanism a switching of the chiral domains. In addition, studies of the second-harmonic generation find a switching of magnetic domains induced by an electric field [30]. A direct observation of multiferroic hysteresis cycles was obtained by the polarized neutron diffraction experiments [20], where it was possible to control the chiral ratio defined in equation (2,3) by varying the electric field. After completion of this work we learned that a similar hysteresis curve was also reported in [31].

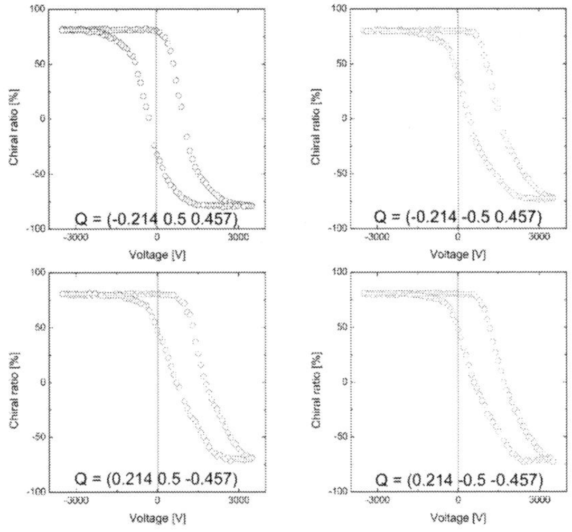

Figure 2. Hysteresis curves obtained by measuring the chiral ratio as a function of external electric field at a constant temperature of T=10.5 K; nearly identical hysteresis cycles were obtained at the four equivalent **Q**-positions ($\pm$0.214,0.5,$\mp$0.457) and ($\pm$0.214,-0.5,$\mp$0.457) studied.

All the hysteresis cycles reported in reference [20] were recorded at **Q**=(-0.214,+0.5,+0.457). Here we present in addition the cycles at the symmetrically equivalent **Q**-positions. Between these measurements the sample was always heated to 15 K and cooled with a voltage of -3500V. The hysteresis cycles shown in Figure 2 were then recorded by varying the voltage from -3500V to +3500V and back to -3500V. Due to the monoclinic symmetry one may only invert the signs of the q_h and q_l components simultaneously in order to reach an equivalent position in reciprocal space, corresponding to an inversion of the incommensurate part of the propagation vector. The inversion of q_k is irrelevant, as k=$\pm$0.5 generates the same magnetic structure with antiferromagnetic coupling of neighboring chains along the b direction (note that the difference $k = \pm 0.5$ is just a reciprocal lattice vector). The real crystal consists of domains corresponding to ($\pm$0.214,0.5,$\mp$0.457) modulation vectors which are associated with different signs of the chiral component at a given scattering vector, see equation (1). Alternatively, one may consider the chiral domains as arising from an inversion of one of the magnetic components, either that along b or that along $\mathbf{e}_{easy}$. One of these chiral arrangements is shown in Figure 1. The real domain structure of this magnetic configuration may be more complex due to phase shifts of the magnetic moments. The four hysteresis-cycles recorded at T=10.5 K at the four **Q**-positions ($\pm$0.214,0.5,$\mp$0.457) and ($\pm$0.214,-0.5,$\mp$0.457) are almost perfectly identical thereby documenting the reproducibility of the setup. When analyzing the inverted **Q**-position in this nearly perfect chiral mono-domain structure, one might expect to find the scattering in the reverted channel, $\sigma_{xx}^{\uparrow\downarrow}$ versus $\sigma_{xx}^{\downarrow\uparrow}$. However, one has to keep in mind that the (**x**,**y**,**z**) coordinate system is fixed to the scattering vector, and that it is rotated by 180° in respect with the crystal coordinate system by the inversion of the scattering vector. For a given chiral domain, we measure thus the identical chiral ratio at the ($\pm$0.214,$\pm$0.5,$\mp$0.457) scattering vectors. The simple inversion of q_k does not change neither the incommensurate part of the propagation vector nor the orientation of the (**x**,**y**,**z**) coordinate system with respect to the magnetic components projected perpendicular to the scattering vector. Therefore, there is again no reversal of the measured chiral ratios, so that all four hysteresis curves are almost identical. Minor differences arise from the difficulty to fully reproduce the initial conditions and the different orientation of the large crystal in the neutron beam. We may conclude that the hysteresis cycles at the four equivalent magnetic Bragg positions perfectly agree with a nearly monodomain chiral structure, which can be inverted by the application of an external electric field.

2.4. Second-order anharmonic modulation

The poling of the chiral component exhibits a complex memory effect in $MnWO_4$ [20]. The crystal remembers a preference for a certain chiral arrangement even when heating deeply into the paramagnetic and paraelectric phase. A related memory effect has also been observed when cycling the sample across the paraelectric AF1 phase [32] and when suppressing the ferroelectric AF2 phase by a magnetic field [30]. The latter study of the second-harmonic generation nicely documents that the identical domain structure is recovered when suppressing the AF2 phase by a magnetic field and setting the magnetic field back to zero. In particular our observation that the memory is not erased upon heating up to 30 K renders the given interpretation [32] in terms of multiferroic embryos very unlikely. We interpret the memory due to the pinning of the ferroelectric and magnetic domain walls by defects. Furthermore, relaxation processes similar to those reported for $REMnO_3$ might be involved in the memory effects [33].

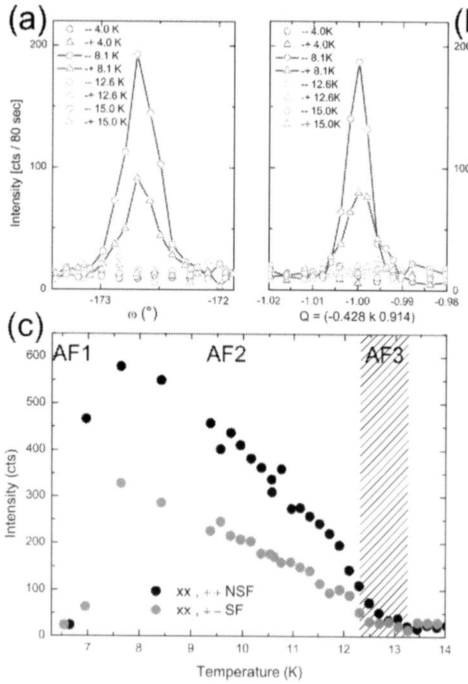

Figure 3. Analysis of the scattering at the second-order harmonic reflection (-0.428,-1,0.914); part a) and b) show scans across the reflection in transversal (ω-scan) and [010] direction, respectively for spin-flip and non-spin-flip scattering in the xx channel; part c) presents the temperature dependencies of the magnetic and nuclear contributions across the magnetic phases of $MnWO_4$.

If the ferroelectric structure is purely electronic in origin, it appears difficult to explain a strong magneto-elastic pinning. Due to the weakness of the ferroelectric polarization, even a nuclear ferroelectric origin might be too weak in order to cause an efficient pinning. We have, therefore, searched for other processes pinning the multiferroic phase by analyzing the scattering at twice the propagation vector in the AF2 and AF3 phases. The results are given in Figure 3. We have studied the second-order modulation at $\mathbf{Q}$ =(-0.428,1.0,0.914) finding a sizeable signal. Already in the collinear AF3 phase we find intensity in the non-spin-flip channels σ_{xx}. An incommensurate collinear magnetic structure is always coupled to a structural modulation of half the period in real space (and twice the propagation vector) due to the exchange striction modulating the magnetic interaction. In the AF2 phase this second-order harmonic scattering seems to increase much more than one might expect from the total increase in the ordered moment. Furthermore, the usual exchange-striction mechanism should not apply in an ideal chiral magnetic structure as the scalar product of coupled spins, $S_i \cdot S_j$, along the propagation vector does not vary along the propagation vector. Most interestingly we also find a magnetic component at the position of the second-order harmonics, which is limited to the AF2 phase. The magnetic second-order scattering is, however, three orders of magnitude smaller than the first-order magnetic scattering. Therefore, the anharmonic modulation of the magnetic structure in $MnWO_4$ is only a minor distortion. The ferroelectric chiral AF2 phase exhibits thus a $2 \cdot \mathbf{q}_{ic}$ modulation of both the magnetic and the crystal structure. These anharmonic modulations might be key elements for the understanding of the anomalous memory and asymmetry effects in $MnWO_4$ and in other multiferroics.

The measurement of the full polarization matrix at $\mathbf{Q}$ =(-0.428,1.0,0.914) gives further insight into the nature of the second-order modulation: There is a strong nuclear-magnetic interference term indicating that the magnetic and nuclear modulations are coexisting and intimately coupled. The real part of the nuclear-magnetic interference R_i is determined in the longitudinal (diagonal) channels through [10]: $R_i = \frac{\sigma_{ii}^{\uparrow\uparrow} - \sigma_{ii}^{\downarrow\downarrow}}{2}$. We find R_y=1738-226 and R_z=378-1263 counts, far beyond statistical errors. Similarly significant evidence is also found in the non-diagonal channels xz, yz and xy, zy indicating a dominating real part of the nuclear magnetic interference. The nuclear-magnetic interference clearly documents that the second-order harmonic distortions are closely coupled. Thereby the chiral magnetism gets fixed to an incommensurate lattice modulation. Since the latter can be effectively pinned by defects or by an extrinsic distortions, the entire multiferroic domains will be pinned much more effectively than one what one would expect for just a ferroelectric structural modulation.

3. Conclusion

Neutron diffraction with spherical polarization analysis is a very powerful tool for studying the multiferroic transition-metal oxides, where ferroelectric coupling arises from a chiral magnetic structure. Analyzing the diagonal and off-diagonal terms one may directly access the chiral arrangement in a mono-domain sample, and the depolarization of the beam can point to strong chiral components even in a chiral multi-domain arrangement [10].

In $MnWO_4$ we find the close to perfect poling of the chiral components in large single-crystals, which exhibits an anomalous memory effect. Detailed analysis of the spherical polarization indicates that $MnWO_4$ at low temperature in the AF2 phase exhibits a nearly perfect circular chiral state. The product of the magnetic components scales very well with the ferroelectric polarization giving further support to the IDM mechanism.

The chiral arrangement can be controlled at constant temperature by varying the electric field [20], which corresponds to the effect desired for applications. However, the hysteresis cycles are broad and asymmetric due to an effective pinning of the domains in the chiral ferroelectric phase. Hysteresis cycles at four equivalent $\mathbf{Q}$-positions agree perfectly to each other and to the expectations for a nearly monodomain chiral phase, which can be inverted by the application of an external electric field.

As a possible origin of the pinning of the multiferroic phase, we may identify a structural and magnetic $2 \cdot \mathbf{q}_{ic}$ modulation. In the chiral ferroelectric AF2 phase, $MnWO_4$ exhibits an anharmonic magnetic modulation detected at the second-order scattering position. This magnetic anharmonicity seems to disappear in the collinear AF3 phase. In the AF3 and in the AF2 phase there is in addition a structural modulation characterized by the doubled modulation vector, which however, is much stronger in the ferroelectric AF2 phase. The structural and the second-order harmonic magnetic modulations in the multiferroic AF2 phase seem to be relevant for the understanding of the pinning of magnetic and ferroelectric domains. The observation of strong nuclear-magnetic interference terms at the second-order peak is another manifestation of the strong coupling between nuclear structure and magnetism in the multiferroic materials.

Acknowledgments

This work was supported by the Deutsche Forschungsgemeinschaft in the Sonderforschungsbereich 608.

References

[1] Eerenstein W, Mathur N D and Scott J F 2006 *Nature* **442** 759
[2] Cheong S-W and Mostovoy M 2007 *Nature Materials* **6** 13
[3] Kimura T, Ishihara S, Shintani H, Arima T, Takahashi K, Ishizaka K and Tokura Y 2003 *Nature* **426** 55
[4] Hur N, Park S, Sharma P, Ahn J, Guha S and Cheong S-W 2004 *Nature* **429** 392
[5] Lawes G, Harris A B, Kimura T, Rogado N, Cava R J, Aharony A, Entin- Wohlman O, Yildirim T, Kenzelmann M, Broholm C and Ramirez A P 2005 *Phys. Rev. Lett.* **95** 087205
[6] Goto T, Kimura T, Lawes G, Ramirez A and Tokura Y 2004 *Phys. Rev. Lett.* **92** 257201
[7] Katsura H, Nagaosa N and Balatsky A 2005 *Phys. Rev. Lett.* **95** 057205
[8] Mostovoy M 2006 *Phys. Rev. Lett.* **96** 067601
[9] Sergienko I and Dagotto E 2006 *Phys. Rev.* B **73** 094434
[10] Chatterji T 2006 *Neutron Scattering from Magnetic Materials*, Elsevier, Amsterdam
[11] Aliouane N, Schmalzl K, Senff D, Maljuk A, Prokes K, Braden M and Argyriou D N 2009 *Phys. Rev. Lett.* **102** 207205
[12] Yamasaki Y, Miyasaka S, Kaneko Y, He J-P, Arima T and Tokura Y 2007 *Phys. Rev. Lett.* **98**, 147204
[13] Seki S, Yamasaki Y, Soda M, Matsuura M, Hirota K and Tokura Y 2008 *Phys. Rev. Lett.* **100** 127201
[14] Sagayama H, Taniguchi K, Abe N, Arima T, Soda M, Matsuura M and Hirota K 2008 *Phys. Rev.* B **77** 220407
[15] Kimura T, Lawes G, Goto T, Tokura Y, and Ramirez A 2005 *Phys. Rev.* B **71** 224425
[16] Heyer O, Hollmann N, Klassen I, Jodlauk S, Bohatý L, Becker P, Mydosh J A, Lorenz T and Khomskii D 2006 *J. Phys. Condens. Matter* **18** L471
[17] Taniguchi K, Abe N, Takenobu T, Iwasa Y and Arima T 2006 *Phys. Rev. Lett.* **97** 097203
[18] Arkenbout A H, Palstra T T M, Siegrist T and Kimura T 2006 *Phys. Rev.* B **74** 184431
[19] Jodlauk S, Becker P, Mydosh J A, Khomskii D I, Lorenz T, Streltsov S V, Hezel D C and Bohatý L 2007 *J. Phys. Condens. Matter* **19** 432201
[20] Finger T, Senff D, Schmalzl K, Schmidt W, Becker P, Bohatý L and Braden M *arXiv:0907.5319*
[21] Lautenschläger G, Weitzel H, Vogt T, Hock R, Bonnet M and Fuess H 1993 *Phys. Rev.* B **48** 6087
[22] Ehrenberg H, Weitzel H, Fuess H and Hennion B 1999 *J. Phys. Condens. Matter* **11** 2649
[23] Asher E, Rieder H, Schmid H and Stössel H 1966 *J. of Appl. Physics* **37** 1404
[24] Lottermoser T, Lonkai Th, Amann U, Hohlwein D, Ihringer J and Fiebig M 2004 *Nature* **430** 541
[25] Bodenthin Y, Staub U, García-Fernández M, Janoschek M, Schlappa J, Golovenchits E I, Sanina V A and Lushnikov S G 2008 *Phys. Rev. Lett.* **100** 027201
[26] Zhao T, Scholl A, Zavaliche F, Lee K, Barry M, Doran A, Cruz M P, Chu Y H, Ederer C, Spaldin N A, Das R R, Kim D M, Baek S H, Eom C B and Ramesh R 2006 *Nature Materials* **5** 823
[27] Radaelli P G, Chapon L C, Daoud-Aladine A, Vecchini C, Brown P J, Chatterji T, Park s and Cheong S W 2008 *Phys. Rev. Lett.* **101** 067205
[28] Saito M, Ishikawa K, Konno S, Taniguchi K and Arima T 2009 *Nature Materials* **8** 634
[29] Kundys B, Simon C and Martin C 2008 *Phys. Rev.* B **77** 172402
[30] Meier D, Maringer M, Lottermoser Th, Becker P, Bohatý L and Fiebig M 2009 *Phys. Rev. Lett.* **102** 107202
[31] Poole A, Brown P J and Wills A S 2009 *J. Phys. Conf. Series* **145** 012074
[32] Taniguchi K, Abe N, Ohtani S and Arima T 2009 *Phys. Rev. Lett.* **102** 147201
[33] Schrettle F, Lunkenheimer P, Hemberger J, Ivanov V Yu, Mukhin A A, Balbashov A M and Loidl A 2009 *Phys. Rev. Lett.* **102** 207208

$FeTe_2O_5Br$ system: new ferroelectric with an incommensurate spin modulation

O Zaharko[1], M Pregelj[2], D Arčon[2], P J Brown[3], D Chernyshov[4], U Stuhr[1] and H Berger[5]

[1]Laboratory for Neutron Scattering, ETHZ & PSI, CH-5232 Villigen, Switzerland
[2]Department for Condensed Matter Physics, Jozef Stefan Institute, 1000 Ljubljana, Slovenia
[3]Institut Laue-Langevin, 156X, F-38042 Grenoble Cedex 9, France
[4]Swiss-Norwegian Beamlines, ESRF, F-38042 Grenoble Cedex 9, France
[5]Institute of Physics of Complex Matter, EPFL, 1015 Lausanne, Switzerland

E-mail: Oksana.Zaharko@psi.ch

Abstract. The magnetic structure of the geometrically frustrated layered cluster compound $FeTe_2O_5Br$ has been determined by single crystal neutron diffraction and neutron spherical polarimetry. Incommensurate magnetic ordering with wave vector $\mathbf{k}=(\frac{1}{2}\ 0.463\ 0)$ develops below $T_N=10.6$ K. It is accompanied by ferroelectric ordering. X-ray synchrotron single crystal diffraction confirms that the observed magnetoelectric coupling can arise from exchange striction involving frustrated tetramer clusters and intercluster exchange bridges with polarizable lone-pair electrons on Te^{4+} ions.

1. Introduction

The new class of multiferroic materials in which magnetism induces electric polarization introduces new mechanisms by which strong magnetoelectric coupling may be achieved For example in $TbMnO_3$ [1] competition between different frustrated magnetic interactions results in a transverse incommensurate spiral [2] which breaks the inversion symmetry and gives rise to electric polarization. It has been shown[4, 5] that the inverse Dzyaloshinski-Morya interaction driven by spin-orbit coupling is the dominant mechanism coupling the two order parameters in such noncollinear systems. The same mechanism is active in other spiral arrangements such as the kagomé staircase compound $Ni_3V_2O_8$[3, 6], in the triangular lattices $Ba_3NbFe_3Si_2O_{14}$[7, 8], $RbFe(MnO_4)_2$[9] and in $CuFeO_2$[10]. Strong magnetoelectric coupling also can be induced in collinear or canted magnetic arrangements, such as YMn_2O_5[11] and $TmMnO_3$[12]; in these multiferroic behaviour arises from exchange striction related to the magnetic Jahn-Teller effect. In this paper we present yet another system, the layered compound $FeTe_2O_5Br$, in which incommensurate magnetic order gives rise to a ferroelectric polarization. The spontaneous electric polarization arises from exchange striction in frustrated tetramer clusters and in intercluster exchange paths involving Te^{4+} ions with polarizable lone-pair electrons. It is the first example of a new mechanism for coupling magnetism and ferroelectricity.

Journal of Physics: Conference Series **211** (2010) 012002 doi:10.1088/1742-6596/211/1/012002

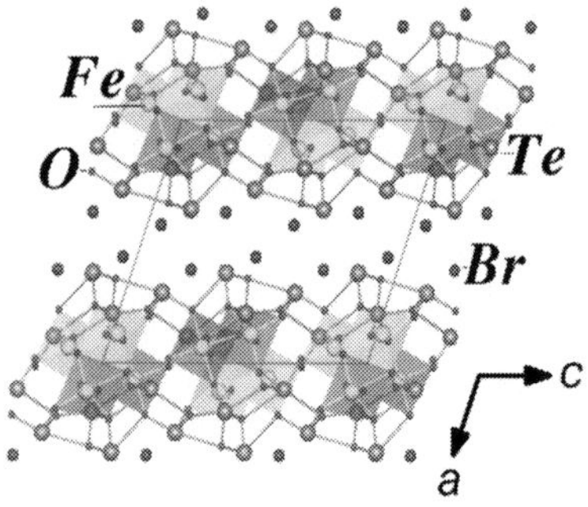

Figure 1. Crystal structure of $FeTe_2O_5Br$ viewed along the *b*-axis.

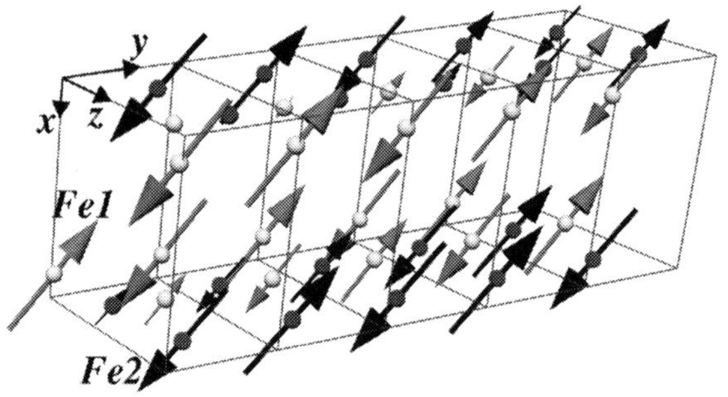

Figure 2. Incommensurate amplitude modulated magnetic structure. The moments of the Fe1 and Fe2 sites are denoted by brown and black arrows.

2. Neutron single crystal diffraction: spherical polarimetry and integrated intensity measurements

$FeTe_2O_5Br$ has a monoclinic layered structure [13] (space group $P2_1/c$, a=13.346, b=6.590, c=14.242 Å, β=108.204°). The layers consist of $[Fe_4O_{16}]^{20-}$ tetramer clusters linked by $[Te_4O_{10}Br_2]^{6-}$ bridges and are stacked along the a^* axis (Figure 1). The four Fe^{3+} (S=3/2) ions of each tetramer occupy two different 4(c) sites, Fe1 and Fe2, which have distorted $[FeO_6]$ octahedral coordination. All the Fe-O-Fe bond angles and distances within a tetramer are very similar which suggests that the Fe1-Fe2 and Fe1-Fe1 antiferromagnetic exchange interactions should also be similar. Competition between antiferromagnetic interactions within tetramers leads to magnetic frustration and, though the Curie-Weiss temperature T_{CW}=-110 K, long-range magnetic order only sets in below the Néel temperature, T_N=10.6 K.

Below T_N $FeTe_2O_5Br$ was found to order magnetically with an incommensurate wave vector $\mathbf{k}=(\frac{1}{2}\ 0.463\ 0)$ in a neutron diffraction experiment using the single crystal diffractometer TriCS (λ=2.32 Å) at SINQ (Villigen, Switzerland). Further details of the magnetic structure were obtained from neutron spherical polarimetry data collected at 1.8 K with CRYOPAD II installed on the IN20 spectrometer (λ=2.34 Å, polarization efficiency 95%) at ILL (Grenoble, France). Neutron polarimetry has proved to be a very sensitive technique for determining a zero field magnetic structure [15, 16, 17]; it can distinguish a collinear moment arrangement from a helical one, can determine the direction of magnetic moments and can detect the presence of magentic domains. The relation between the polarization of the incident neutrons $\mathbf{P}$ and the polarisation $\mathbf{P}'$ of those scattered by a single domain is given by the equation $\mathbf{P}' = P^t\mathbf{P} + \mathbf{P}''$ [17]. Here the tensor P^t (the transpose of matrix P given in the Table 1) describes rotation and $\mathbf{P}''$ - creation of the polarization during the scattering process and both depend on the magnetic interaction vector $\mathbf{M}^{\perp}_{\mathbf{hkl}}$, the projection of the magnetic structure factor perpendicular to the scattering vector $\mathbf{Q}_{\mathbf{hkl}}$. The components of the polarization matrix P_{ij}, which are the measured quantities, are commonly expressed in the polarization reference frame: the scattering vector $\mathbf{Q}_{\mathbf{hkl}}$ points along the x-axis, the z-axis is chosen perpendicular to the horizontal scattering plane (in our case it is the c-axis of the crystal) and y completes the right-handed Cartesian coordinate system. With this frame the P_{yx} and P_{zx} matrix elements can only be non-zero for spiral, cycloid or canted arrangements. In our experiment these elements were not observed (Table 1). Also, the scattered beam retains almost full polarization, showing that the absence of the P_{yx} and P_{zx} components is not due to the presence of chiral domains of opposite handedness which, although they could eliminate P_{yx} and P_{zx}, would depolarize the scattered beam.

The combined refinement [18] of polarization (25 $hk0$ reflections) and integrated intensities (41

Table 1. Measured and calculated representative polarization matrices P_{ij} at 1.8 K (i - incident and j - scattered components of polarization). Measured matrices have been corrected for incomplete neutron polarization.

$j \rightarrow$	x	y	z	x	y	z
		measured		calculated		
$i \downarrow$		-0.5 -1.537 0				
x	-1.000(7)	-0.01(2)	-0.00(2)	-1.00	0.00	0.00
y	0.03(2)	0.87(1)	-0.47(2)	0.00	0.89	-0.45
z	0.00(2)	-0.46(2)	-0.88(1)	0.00	-0.45	-0.89
		0.5 2.463 0				
x	-0.999(6)	0.00(2)	-0.04(2)	-1.00	0.00	0.00
y	0.00(2)	0.889(9)	0.48(2)	0.00	0.89	0.46
z	-0.04(2)	0.50(1)	-0.894(9)	0.00	0.46	-0.89
		2.5 -1.463 0				
x	-0.992(7)	-0.08(2)	-0.10(2)	-1.00	0.00	0.00
y	0.02(2)	0.83(1)	-0.43(2)	-0.01	0.87	-0.40
z	-0.00(2)	-0.39(2)	-0.82(2)	-0.01	-0.40	-0.87
		1.5 -2.537 0				
x	-1.00(1)	-0.03(2)	-0.00(2)	-1.00	0.00	0.00
y	0.04(2)	0.86(1)	-0.50(2)	0.01	0.89	-0.44
z	-0.01(2)	-0.53(2)	-0.87(1)	0.01	-0.44	-0.89
		4.5 0.463 0				
x	-0.99(1)	-0.07(3)	-0.11(2)	-1.00	0.00	0.00
y	0.02(2)	0.91(1)	0.20(2)	0.00	0.87	0.12
z	0.04(2)	0.25(2)	-0.92(1)	0.00	0.12	-0.87
		4.5 0.537 0				
x	-0.992(7)	0.10(2)	0.07(2)	-1.00	0.00	0.00
y	-0.03(2)	0.90(1)	0.26(2)	0.00	0.85	0.17
z	-0.02(2)	0.28(2)	-0.89(1)	0.00	0.17	-0.85

hkl reflections) revealed two equally populated domains related by the 2_{1y} axis. The best fit was obtained for the amplitude modulated arrangement presented in Figure 2. The magnetic moments are almost collinear, with a small 7(3)° canting between the Fe1 and Fe2 sites. They lie close to the a^*b plane inclined at 100(1)° to the c-axis and are orthogonal to the wave vector $\mathbf{k}$. Table 2 gives the polar angles θ and ϕ used to define the orientation of magnetic moment S_{0k} relative to the a^*bc coordinate system and the phases ψ_{kl} of eight individual amplitude modulation waves $S(i,k,l) = S_{0k}\cos(\mathbf{k}\cdot\mathbf{r}_i + \psi_{kl})$. The vector $\mathbf{r}_i$ defines the origin of the i-th cell, k counts the sites and l - the iron atoms within the site.
This magnetic arrangement reflects the strong anisotropy of the crystal structure. Presumably amplitude modulation arises from competing interactions within the geometrically frustrated iron tetramers and between the tetramers. The result of the unconstrained refinement do not conform to models derived from representation symmetry analysis. Two irreducible representations Γ_1 and Γ_2 are possible and both must be invoked to account for the proposed structure. Moreover, our model includes a phase difference between the waves on the atoms related by the center of symmetry. This allows ferroelectric order and indeed spontaneous electric polarization appears at T_N with the largest component along the c-axis, $P(c) = 8.5(2)\mu C/m^2$

(see [14] for details). Surprisingly the electric polarization is linearly proportional to $\mathbf{M}^{\perp}_{\mathbf{hkl}}$ in contrast to the $P \propto (\mathbf{M}^{\perp}_{\mathbf{hkl}})^2$ dependence observed in the majority of multiferroics with incommensurate magnetism.

To obtain a better understanding of the coupling between the magnetic and ferroelectric order [14] we have examined the evolution of the magnetic structure with temperature. Figure 3 shows the temperature dependence of magnetic intensity at the ($\frac{1}{2}$ 0.537 -1) position. Two breaks in the curve can be distinguished which may indicate two transitions separated by less than 0.2 K. The electric polarization only appears at the second, lower temperature, transition. In the higher temperature phase (HT) the magnetic peaks are at the same positions as in the low temperature phase (LT), but their intensities are significantly smaller. We therefore measured their integrated intensities with the three-axis neutron spectrometer TASP at SINQ (λ=2.36 Å) using the analyzer to reduce the background. 40 reflections were measured at 10.7 K and for systematic comparison the same reflections were also collected at 1.5 and 9.5 K in the LT phase. The results of the structure refinements listed in Table 2 suggest that the main difference between the LT and HT arrangements is in the direction of the moments: the ϕ angle changes from almost -45° in the LT phase (1.5 K and 9.5 K) to -75° in the HT phase (10.7 K). With this limited experimental data the differences between the phases of the amplitude modulated waves could not be properly determined. However we speculate that, since the HT phase paraelectric, $\psi_{k1} = \psi_{k2}$ and $\psi_{k3} = \pi + \psi_{k4}$ at 10.7 K, so that the inversion operation is restored. With the large estimated standard deviations in the phases such a scenario is plausible. The ordered magnetic moment determined, which approaches 5.44(9) μ_B/atom at 1.5 K, is probably overestimated as only strong nuclear reflections affected by extinction were available for scaling.

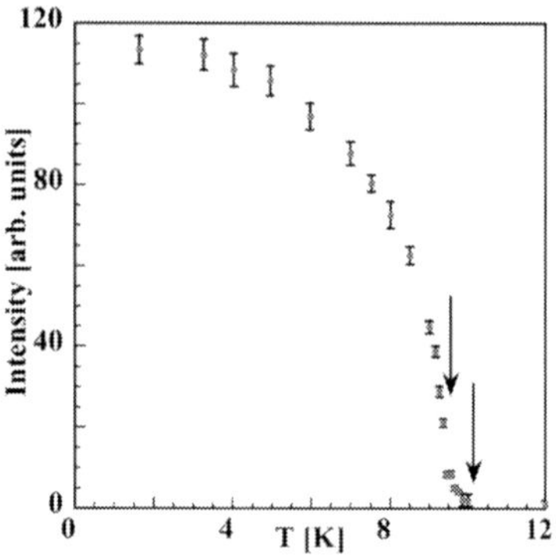

Figure 3. Temperature dependence of the ($\frac{1}{2}$ 0.537 -1) peak.

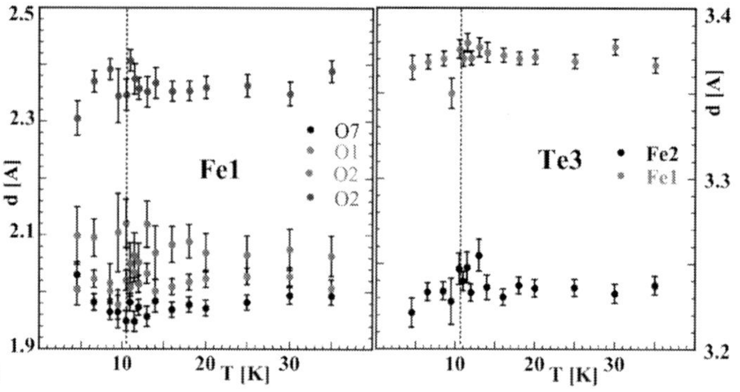

Figure 4. Variation of the selected Fe-O and Te-Fe interatomic distances. The atoms are labeled as in [13], T_N is marked by dashed lines.

3. X-ray synchrotron single crystal diffraction

Since the magnetic structure is almost collinear, the observed magnetoelectric coupling is expected to involve exchange striction. To assess the magnitude of this effect we have made low temperature single crystal X-ray diffraction measurements at the BM01A Swiss-Norwegian Beamline of ESRF (Grenoble, France). Data sets (typically 780 reflections per temperature point) were collected in the temperature range 4.5 K to 35 K at a wavelength of 0.64 Å using a closed-cycle He cryostat mounted on a six-circle kappa diffractometer KUMA. No structural transitions leading to measurable changes in the lattice symmetry or cell metrics were found near T_N. However the high quality of the data allowed us to refine [19] the atomic positions and to deduce the interatomic distances at each measured temperature. The Fe-O and Fe-Te distances shown in Figure 4 show significant changes below T_N which confirm that the crystal lattice responds to the onset of magnetic order. The changes are tiny, in the range of 0.03 Å, as expected from the small magnitude of electric polarization. Apparently there is striction not only in the tetramers, but also in the Fe-O-Te-O-Fe intercluster exchange bridges containing the Te^{4+} ions with the lone electron pairs.

Table 2. Parameters of the magnetic structure. The atoms labeled by l=2...4 are generated from the Fe_{11} (0.1184(6) -0.001(1) -0.0243(7)) and Fe_{21} (0.9386(6) 0.296(1) 0.8568(6)) atoms by the symmetry elements i, 2_{1y} and $2_{1y}i$. Top: refinement from combined polarized (1.8 K) and integrated intensity (5 K) data. Bottom: refinement from integrated intensity data collected with an analyzer at 1.5, 9.5 and 10.7 K. Here the tilt to the c-axis θ has been fixed to 100°.

	θ[°]	ϕ[°]	μ [μ_B]	ψ_{k1} 1	ψ_{k2} i	ψ_{k3} 2_{1y}	ψ_{k4}[°] $2_{1y}i$
Fe_{1l}	100(1)	-52(3)	4.02(9)	0	17(4)	55(5)	260(10)
Fe_{2l}	100(1)	-45(3)	4.02(9)	10(5)	-10(11)	113(5)	274(10)
Fe_{1l}	1.5 K	-44(2)	5.36(9)	0	13(3)	65(6)	272(12)
Fe_{2l}		-47(2)	5.36(9)	26(8)	-4(14)	82(7)	251(14)
Fe_{1l}	9.5 K	-42(2)	3.99(5)	0	16(3)	70(5)	280(9)
Fe_{2l}		-50(2)	3.99(5)	38(9)	8(17)	92(10)	269(19)
Fe_{1l}	10.7 K	-81(4)	1.75(6)	0	11(7)	66(14)	265(27)
Fe_{2l}		-70(3)	1.75(6)	28(15)	-12(25)	81(20)	232(34)

4. Conclusions

Neutron and x-ray single crystal diffraction experiments have been carried out on the new multiferroic system $FeTe_2O_5Br$. Combination of neutron polarized and nonpolarized diffraction allows to suggest the amplitude modulated magnetic arrangement of the Fe^{3+} moments as the best model. Just below T_N the magnetic moments are nearly parallel to the b axis and most probably the inversion symmetry relations between the amplitude modulated waves are maintained. At lower temperatures, with increase of the ordered moment, the moments become oriented normal to the wave vector **k** and the inversion symmetry is lost allowing emergence of spontaneous electric polarization. The exchange striction in the $[Fe_4O_{16}]^{20-}$ tetramer clusters and in the $[Te_4O_{10}Br_2]^{6-}$ exchange bridges, associated with the spontaineous polarization has been determined using x-ray synchrotron single crystal diffraction. The magnetoelastic coupling arises from the strong anisotropy of the crystal structure and from geometrical frustration of antiferromagnetic exchange in the tetramer clusters. Importantly, the exchange bridges contain polarizable lone-pair electrons which assist in propagating the ferroelectric order within the layers. The unique interplay between magnetic frustration and Te^{4+} lone-pairs bridging magnetic intercluster exchange appears vital for the occurrence of the magnetoelectric effect. We plan to investigate whether other $3d$ oxohalides containing lone-pair cations display similar ferroelectric properties.

Acknowledgments

We thank Ya. Filinchuk for his help in x-ray diffraction experiment. The sample preparation was supported by the NCCR research pool MaNEP of the Swiss NSF.

References

[1] Kimura T, Goto T, Shintani H, Ishizaka K, Arima T and Tokura Y 2003 *Nature* **426** 55
[2] Kenzelmann M, Harris A B, Jonas S, Broholm C, Schefer J, Kim S B, Zhang C L, Cheong S-W, Vajk O P and Lynn J W *Phys. Rev. Lett.* 2005 **95** 087206
[3] Lawes G, Harris A B, Kimura T, Rogado N, Cava R J, Aharony A, Entin-Wohlman O, Yildirim T, Kenzelmann M, Broholm C and Ramirez A P 2005 *Phys. Rev. Lett.* **95** 087206
[4] Katsura H, Nagaosa N and Balatsky A 2005 *Phys. Rev. Lett.* **95** 057205
[5] Mostovoy M 2006 *Phys. Rev. Lett.* **96** 067601
[6] Kenzelmann M, Harris A B, Aharony A, Entin-Wohlman O, Yildirim T, Huang Q, Park S, Lawes G, Broholm C, Rogado N, Cava R J, Kim K H, Jorge G and Ramirez A P 2006 *Phys. Rev.* B **74** 014429
[7] Marty K, Simonet V, Ressouche E, Ballou R, Lejay P and Bordet P 2008 *Phys. Rev. Lett.* **101** 247201
[8] Zhou H D, Lumata L L, Kuhns P L, Reyes A P, Choi E S, Dalal N S, Lu J, Jo Y J, Balicas L, Brooks J S and Wiebe C R 2009 *Chem. mater.* **21** 156
[9] Kenzelmann M, Lawes G, Harris A B, Gasparovic G, Broholm C, Ramirez A P, Jorge G A, Jaime M, Park S, Huang Q, Shapiro A Ya and Demianets L A 2007 *Phys. Rev. Lett.* **98** 267205
[10] Kimura T, Lashley J C and Ramirez A P 2006 *Phys. Rev.* B **73** 220401(R)
[11] Radaelli P G, Vecchini C, Chapon L C, Brown P J, Park S and Cheong S-W 2009 *Phys. Rev. Lett.* **79** 020404(R)
[12] Pomjakushin V Yu, Kenzelmann M, Dönni A, Harris A B, Nakajima T, Mitsuda S, Tachibana M, Keller L, Mesot J, Kitazawa H and Takayama-Muromachi E 2009 *New Journal of Physics* **11** 043019
[13] Becker R, Johnsson M, Kremer R K, Klauss H H and Lemmens P 2006 *J. Am. Chem. Soc.* **128** 15469
[14] Pregelj M, Zaharko O, Zorko A, Kutnjak Z, Jeglič P, Brown P J, Jagodič M, Jagličić Z, Berger H and Arčon D 2009 *Phys. Rev. Lett.* **103** 147202
[15] Blume M 1963 *Phys. Rev.* B **130** 1670
[16] Maleev S V, Bar'yaktar V G and Suris P A 1963 *Sov. Phys. - Solid State* **4** 2533
[17] Brown P J 2001 *Physica* B **297** 198
[18] Brown P J and Matthewman J C computer program CCSL 2008
[19] Sheldrick G M, computer program SHELXL97, University of Göttingen: Göttingen Germany 1997

Advanced resonant soft x-ray diffraction to study ordering phenomena in magnetic materials

Urs Staub

Swiss Light Source, Paul Scherrer Insitut, 5232 Villigen PSI, Switzerland

Urs.Staub@psi.ch

Abstract. Here we present recent advances in the resonant soft x-ray diffraction technique and present a review of recent experiments in the field. Our review is focused on the study of charge and orbital phenomena occurring in manganites systems. It is shown how resonant diffraction can access directly the 3d shell of Mn by tuning the x-ray energy to the Mn $L_{2,3}$ edges. Similarly, by tuning the energy to the oxygen K edge, empty oxygen 2p states can be probed. It is shown how the technique probes corresponding atomic multipoles, the monopole (charge), the magnetic dipole, and the orbital (electronic quadrupole) of the open valence shell.

1. Introduction

The interaction between electronic and magnetic degrees of freedom has been intensively studied in oxides in recent years. These interactions result in interesting macroscopic effects such as high-temperature superconductivity, colossal magneto-resistance and multiferroicity. In some of these systems, the charge, magnetic, orbital and the structural degrees of freedom are strongly coupled. X-ray probes have strongly contributed to the understanding of such couplings in recent years. One of the evolving techniques is resonant diffraction which provides direct access to ordering phenomena of the electronic and magnetic sublattices. Since the early work on Ho metal,[1] the technique rapidly developed into a powerful tool in the study of charge, magnetic and orbital ordering phenomena in materials. In particular, resonant magnetic scattering is widely used in studies of multiferroic materials, such as the $RMnO_3$ [2-7] and RMn_2O_5 [8-12] families.

Here we are concerned mainly with the ordering of charge and orbitals. Resonant diffraction studies started with the claim of direct observation of orbital order at the Mn K edge about a decade ago [13,14]. Subsequent theoretical calculations [15] challenged this interpretation. These calculations predict that at the K edge, the probed splitting of 4p states is dominated by the effect of the oxygen 2p Mn-4p overlap caused by the Jahn-Teller distortion. These experimental studies represent the first measurement of an electric quadrupole moment of Mn 4p type states in this strongly correlated electron system. It has soon after been pointed out by theoretical calculations[16], that the Mn $L_{2,3}$ edge contains a much more detailed view of the orbital order, which was soon confirmed by first experiments. [17,18] Here we review more recent work, made possible by the use of different incoming x-ray polarizations and analysis of the polarization of the scattered x-rays as well as by rotation of the sample around the Bragg wave vector (azimuthal angle scans). Moreover, an example is given showing what type of information can be retrieved by the study of the energy dependence of reflections, even on polycrystalline samples.

2. Concepts of resonant diffraction

In this report we concentrate of the electric-electric dipole transition, which is in many cases a good approximation for resonant diffraction in the soft x-ray regime. A theoretical review can be found in [19], and extension to other transitions have recently been discussed elsewhere [20]. The structure factor for resonant diffraction can be written in the form

$$F(E) \propto \sum_{K,Q}\sum_{q}(-1)^{Q} H^{K}_{-Q} D^{K}_{Qq} \sum_{\mathbf{d}} \langle \mathbf{T}^{k}_{q} \rangle_{\mathbf{d}} e^{i\mathbf{d}\cdot\tau} \quad (1)$$

where **d** is the position of the resonant ions in the unit cell, τ the studied reciprocal lattice vector and $\langle \mathbf{T}^{k}_{q} \rangle$ represents an atomic multipole moment of rank k and projection q in the local axis system of the resonant ion. D^{K}_{Qq} represents an axis transformation of the operators from the local site system to the experimental setup and H^{K}_{-Q} contains the dependence on the x-ray polarization. The resulting tensor of rank k represents now atomic multipoles of the probed valence shell with k=0 charge (monopole), k=1 the magnetic dipole and k=2 the electronic quadrupole. The resonance is reflected by the energy dependence of $\langle \mathbf{T}^{k}_{q} \rangle$

$$\langle \mathbf{T}^{k}_{q} \rangle \propto \sum_{n} \frac{\langle g|O|n\rangle\langle n|O^{*}|g\rangle}{E_{n} - E_{g} - \hbar\omega + i\Gamma} \quad (2)$$

where the sum over n reflects the sum over all probed intermediate states in the resonant process. The resonant enhancement of the magnetic and electronic contrast is directly related to the denominator, which makes the expectation values for the atomic multipoles large when the energy of the x-rays approach the absorption edges. However, as the intermediate states contain a hole in the core state, the intermediate states have to be considered with this hole, making them more atomic like. The presence of the core hole leads to a complex energy spectra even for a single electronic quadrupole of the Mn 3d shell due to the occurrence of multiplet structure (see fig. 1).

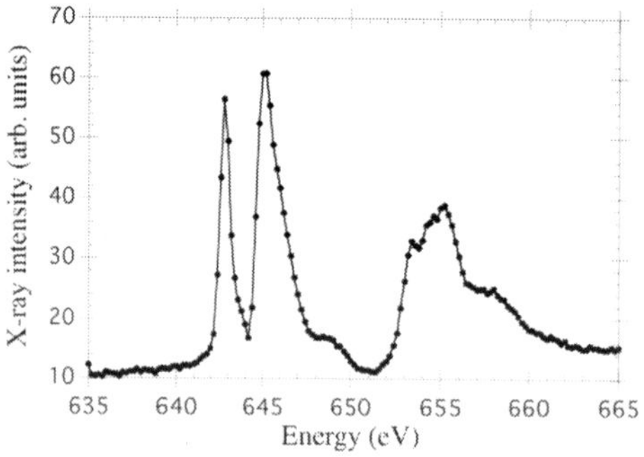

Figure 1. Orbital (¼ ¼ 0) reflection of $La_{0.5}Sr_{1.5}MnO_4$ taken at 160K at the Mn $L_{2,3}$ edges in π incident polarization in the vicinity of the Mn $L_{2,3}$ edges, representing the energy splitting of the multiplets.

3. Polarization (analysis) and azimuthal angle dependence

Most soft x-ray diffraction studies are concerned primarily with the measurement of a Bragg reflection and its energy, temperature, and (in the case of incommensurate structures) momentum dependence. In some cases, x-rays with two different linear incoming polarizations were used. Very few studies used the freedom to rotate the sample around the scattering wave vector (azimuthal scan) to explore the behavior of the internal moments and their symmetry properties [21-24], as is now standard for resonant hard x-ray Bragg diffraction [25]. Studies using polarization analysis of scattered radiation are even more rare [21,23].

Polarization of the scattered radiation is performed using a graded C/W multilayer, with details in Ref. [26] The information obtainable with polarized radiation contains the orientation of the orbitals at the resonant sites. Resonant diffraction with linear polarized x-rays tests directly the local symmetry of the resonant ion sampled by the structure factor of the probed reflection. For a simple checker-board type orbital order, as is commonly assumed to occur in half doped manganites, the azimuth angle (Ψ) dependence of the orbital reflection is simply two fold symmetric and is proportional to a $\sin^2(\Psi)$ term with a single electric quadrupole $\langle T_1^2 \rangle$. This symmetry of the azimuthal angle dependence, as shown in Figure 2, is simple because the probed orbital reflection (1/4 1/4 0) is along a high-symmetry direction. Note that for these long x-ray wave lengths, there is only one orbital reflection accessible at the Mn $L_{2,3}$ edges. In addition, due to the 2 fold local axis at the Mn for this orbital ordering, we expect scattering only in the rotated polarization channels, as experimentally verified in Figure 3.

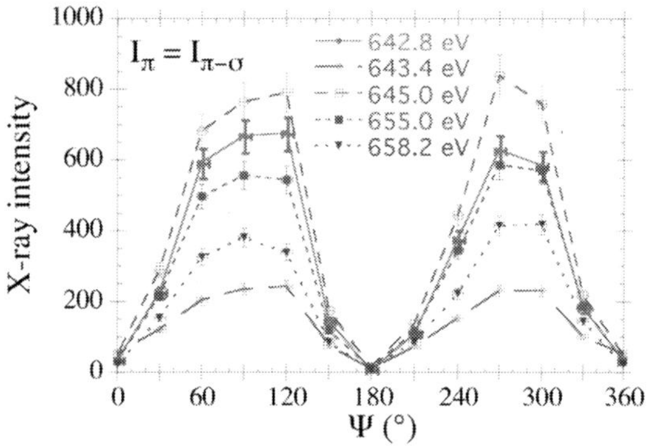

Figure 2. Azimuthal angle dependence of the orbital (¼ ¼ 0) reflection taken at 160K with π incident radiation of $La_{0.5}Sr_{1.5}MnO_4$,[21] above the antiferromagnetic transition temperature. The individual energy positions has been chosen to reflect different spectral features shown in Figure 1.

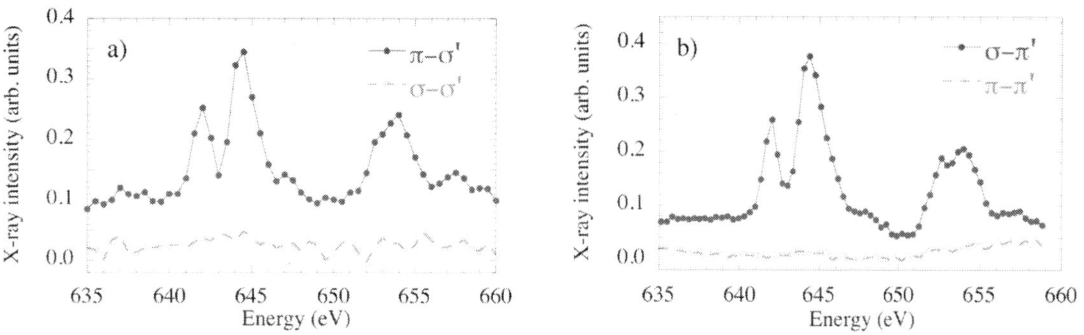

Figure 3. Energy dependence of the orbital (¼ ¼ 0) reflection in the four standard polarization channels of $La_{0.5}Sr_{1.5}MnO_4$, [26] in the antiferromagnetic transition temperature (left 40K and right 60K).

Testing the local symmetry of the Mn ions has significant impact on the possible ordering schemes in manganites such as the Zener-polaron or ferroelectric phases as well as for possible magnetic structures, which may break the symmetry of the Mn 3d states. Theoretical calculations have shown that depending on the magnetic moment direction of the Mn^{3+} ions, there might be a considerable contribution to scattering in the non-rotated polarization channels σ-σ and π-π, [27] where π and σ are vectors that define x-ray polarization as in and perpendicular to the scattering plane, respectively. The absence of scattering in these channels has therefore impact on the magnetic moment orientation of the Mn ions. More recently, the polarization dependence of the models such as the Zener-polaron has been calculated,[28] which would lead again to significant deviations in both the azimuthal angle dependence as well as to substantial contributions in the unrotated polarization channels. This would also be reflected by unequal intensities of the orbital reflection for π and σ incident radiation, without an analysis of the polarization of the outgoing radiation. Such a change in intensity is caused by the loss of inversion symmetry at the Mn sites, as the electron is shared by two Mn ions forming a dimer (Zener-polaron). This is neither observed in simple perovskite $(R/T)MnO_3$ (R=Bi, Nd, Pr) and (T=Sr, Ca) systems [29,30] nor in A-site ordered manganites for all studied dopings.[28]

4. Oxygen K edge

To gain more information of the interaction of the Mn with the oxygen, the x-ray energy can be tuned to the oxygen K edge. Providing that the doping level (Ca concentration) is significantly large (and the ordering wave vector sufficiently small), the orbital order reflection will lie in the Ewald sphere and can be studied. This has been performed for 2/3 doping in the $Bi/CaMnO_3$ system, where strong reflection was observed at 528 eV. [31] This reflection has a single peak (single harmonic oscillator) shape as a function of energy, indicative of a significant fraction of unoccupied oxygen 2p states. Previously, the occurrence of such a narrow spectral shape of a superlattice reflection has also been found in cuprates [32] and magnetite, [33] where it was interpreted as either caused by a charge order at the oxygen or a charge and orbital order of the Fe-O subsystem. However, the required dependence on the polarization with respect to a given setting has not been performed to establish an orbital or charge scattering origin of the reflection. For the case of the manganites discussed here, the reflection had a strong azimuthal dependence and had equal intensities for both incident polarization channels. This is only compatible with an orbital origin. It shows that the reflection is caused by unoccupied oxygen 2p states which are aspheric, reflecting an electric quadrupole moment at the oxygen. This can be understood in terms of overlap of Mn 3d states with the oxygen 2p states inducing an orbital order at the oxygen sites as illustrated by band structure calculation presented in Ref. [34].

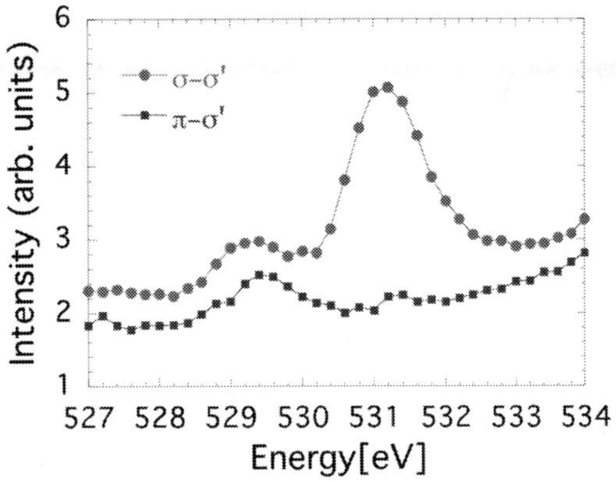

Figure 4. Polarization dependence of the (0 0 ½) charge allowed reflection taken at the O K-edge in $La_{7/8}Sr_{1/8}MnO_3$ for fixed σ scattered radiation. [26,31]

For the occurrence of charge order with the periodicity of a probed superlattice reflection, the reflection must usually fulfill three minimal requirements: A) The reflection must be space group allowed, indicated by non zero intensity away from the absorption edge. Note that the space group is often unknown. B) Despite resonant charge reflection is often accompanied by an orbital (aspherical) contribution, the intensity should be considerably stronger in the σ than in the π incidence channels, when no polarization analysis is used. C) When an azimuthal angle dependence is observed, there should be a clear non-zero intensity at all azimuthal angles. This has been observed in 1/8 doped $LaMnO_3$,[31] where the data has been interpreted as caused by a charge order of different planes along the c-axis, leading also to a charge difference at the oxygen. An example of a stronger signal in the σ–σ channel is exemplified in figure 4 for this system. [26] We would like to note that the direct measurement of the charge order in the soft x-ray regime remains a challenge, because it is a difficult task to separate the aspheric orbital contributions from the spherical charge.

5. Experiments on polycrystalline materials

Very recently, resonant soft x-ray diffraction has been successfully applied to study polycrystalline materials. Due to the very large enhancement of the signals at the $L_{2,3}$ edges for 3d ions and the K edge of oxygen, magnetic and orbital order reflection have been observed in nickelates and manganites, respectively.[28,30,35,36] These observations have become possible due to the use of an in vacuum CCD camera with high sensitivity and the possibility of collecting a larger part of the Debye Scherrer ring of a resonant powder reflection, as shown in Fig. 5. These experiments have been performed out of focus, with a spot size of approximately 2 x 2 mm, to obtain a reasonable averaging over powder grains. The probed volume is very small, in particular due to the very limited penetration depth at resonance in the order of only 1000 Å. This limitation has the advantage that the resonant reflection is significantly broadened along **q**, allowing to get intensity on the CCD camera also from grains slightly misaligned in theta (rocking curve) angle. This effect leads to a broadening of intensity on the camera, but helps to obtain a reasonable powder averaging.

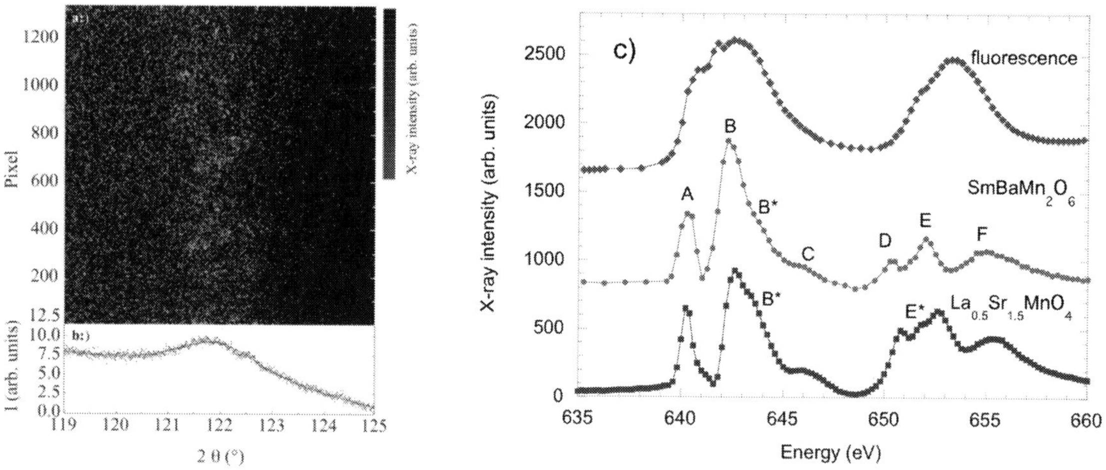

Figure 5. left a) Image of the cut of the Debye-Scherrer ring of the orbital reflection of $SmBaMn_2O_6$.[36] b) vertical integration of the image with the orbital reflection and a Lorentzian fit. Right c) Energy dependence of the orbital reflections of $La_{0.5}Sr_{1.5}MnO_4$ [21] compared with that of $SmBaMn_2O_6$ [36] and its fluorescence.

The type of information which can be retrieved for these experiments are based on the energy, polarization, momentum transfer **q** and temperature dependence. From the energy dependence of the orbital reflection of A-site ordered half doped $SmBaMn_2O_6$ (see fig 5a) it was concluded that the orbital order is of x^2-y^2 type. This is in contrast to that found for the layered $La_{0.5}Sr_{1.5}MnO_4$, which would be of $3z^2$-r^2 type. Moreover, an unusual temperature dependence of the orbital reflection was found in $La_{0.5}Sr_{1.5}MnO_4$, which was interpreted in terms of a different temperature evolution of Jahn-Teller distortion and the electronic quadrupoles (orbitals). [37] The independence of the intensity of the orbital reflection on the incoming x-ray polarization has been used to argue for the absence of Zener-Polaron formation for the $RBaMn_2O_6$ system for a large region of the doping range. [28] Such arguments excludes also other models based on a charge localization on the Mn-O bonds. Moreover, the momentum dependence of the orbital reflection was found to be linear as a function of doping. The temperature dependence at half doping could resolve the origin of the orbital restacking transition, whereas for higher doping it shows an anomalous melting transition of the orbitals, not associated with the onset of magnetic order.

Conclusions

Here we have presented recent progress in the use and applicability of the resonant soft x-ray diffraction technique. The examples given here on the studies of charge and orbital ordering in manganites show that the technique can lead to important conclusions on the electronic ground states properties of the Mn-O network. It shows that both the incoming x-ray polarization and the polarization of the scattered radiation imply stringent constrains on the symmetry of the Mn 3d and oxygen 2p electronic states. It also shows that experiments can be performed on polycrystalline materials, allowing easier access to many interesting materials, with order in the charge, magnetic or orbital degrees of freedom.

References

[1] Gibbs D, Grübel G, Harshman D R, Isaacs E D, McWhan D B, Mills D, and Vettier C 1988 *Phys. Rev. Lett.* **61** 1241

[2] Strempfer J, Bohnenbuck B, Zegkinoglou I, Alinouane N, Landsgesell S, Zimmermann H v and Argyriou D N 2008 *Phys. Rev.* B **78** 024429

[3] Nandi S, Kereyssig A, Tan L, Kim J W, Yan J Q, Lang J C, Haskel D, McQueeney R J and Goldman A I 2008 *Phys. Rev. Lett.* **100** 217201
[4] Mannix D, McMorrow D F, Ewings R A, Boothroyd A T, Prabhakaran D, Joly Y, Janousova B, Mazzoli C, Paolasini L and Wilkins S 2007 *Phys. Rev.* B **76** 184420
[5] Voigt J, Persson J, Kim J, W Bihlmayer G and Brückel T 2007 *Phys. Rev.* B **76** 104431
[6] Prokhnenko O, Feyerherm R, Mostovoy M, Alinouane N, Dudzik E, Wolter A U B, Maljuk A and Argyriou D N 2007 *Phys. Rev. Lett.* **99** 177206
[7] Forrest T R, Bland S R, Wilkins S, Walker H C, Beale T A, W Hatton P D, Prabhakaran D, Boothroyd A T, Mannix D, Yakhou F and McMorrow D F 2008 *J. Phys.: Condens. Matter* **20** 422205
[8] Beutier G, Bombardi A, Vecchini C, Radaelli P G, Park S, Cheong S-W and Chapon L C 2008 *Phys. Rev.* B **77** 172408
[9] Ewings R A, Boothroyd A T, McMorrow D F, Mannix D, Walker H C and Wanklyn B M R 2008 *Phys. Rev.* B **77** 104415
[10] Okamoto J, Huang D J, Mou C-Y, Chao K S, Lin H-J, Park S, Cheong S-W and Chen C T 2007 *Phys. Rev. Lett.* **98** 157202
[11] Koo J, Song C, Ji S, Lee J S, Park J, Jang T-H, Yang C-H, Park J-H, Jeong Y H, Lee K-B, Koo T K, Park Y J, Kim J-Y, Wermeille D, Goldman A I, Srajer G, Park S and Cheong S-W 2007 *Phys. Rev. Lett.* **99** 197601
[12] Bodenthin Y, Staub U, García-Fernández M, Janoschek M, Schlappa J, Golovenchits E I, Sanina V A and Lushnikov S G 2008 *Phys. Rev. Lett.* **100** 027201
[13] Murakami Y, Hill J P, Gibbs D, Blume M, Koyama I, Tanaka M, Kawata H, Arima T, Tokura Y, Hirota K and Endoh Y 1998 *Phys. Rev. Lett.* **81** 582
[14] Murakami Y, Kawada H, Kawata H, Tanaka M, Arima T, Moritomo Y and Tokura Y 1998 *Phys. Rev. Lett.* **80** 1932
[15] Benfatto M, Joly Y and Natoli C R 1999 *Phys. Rev. Lett.* **83** 636
[16] Castleton C W M and Altarelli M 2000 *Phys. Rev.* B **62** 1033
[17] Wilkins S B, Spencer P D, Hatton P D, Collins S P, Roper M D, Prabhakaran D and Boothroyd A T 2003 *Phys. Rev. Lett.* **91** 167205
[18] Dhesi S S, Mirone A, De Nadai C, Ohresser P, Bencok P, Brookes N B, Reutler P, Revcolevschi A Tagliaferri A, Toulemonde O and van der Laan G 2004 *Phys. Rev. Lett.* **92** 056403
[19] Lovesey S W, Balcar E, Knight K S and Fernández-Rodríguez J 2005 *Phys. Rep.* **411** 233
[20] Lovesey S W and Scagnoli V 2009 *J. Phys.: Condens. Matter* **21** 474214
[21] Staub U, Scagnoli V, Mulders A M, Katsumata K, Honda Z, Grimmer H, Horisberger M and Tonnerre J M 2005 *Phys. Rev.* B **71** 214421
[22] Wilkins S B, Stojic N, Beale T A W, Binggeli N, Castleton C W M, Bencok P Prabhakaran, D Boothroyd A T, Hatton P D and Altarelli M 2005 *Phys. Rev.* B **71** 245102
[23] Scagnoli V, Staub U, Mulders A M, Janousch M, Meijer G I, Hammerl G, Tonnerre J M and Stojic N 2006 *Phys. Rev.* B **73** 100409(R)
[24] Herrero-Martín J, García J, Subías G, Blasco J, Concepción-Sánchez M and Stanescu S 2006 *Phys. Rev.* B **73** 224407
[25] Paolasini L, Detlefs C, Mazzoli C, Wilkins S, Deen P P, Bombardi A, Kernavaois N, Bergevin F d, Yakhou F, Valade J P, Breslavetz I, Fondacaro A, Pepellin G and Bernard P 2007 *J. Syn. Rad.* **14** 301
[26] Staub U, Scagnoli V, Bodenthin Y, García-Fernández M, Wetter R, Mulders A M ,Grimmer H and Horisberger M 2008 *J. Syn. Rad.* **15** 469
[27] Stojic N, Binggeli N and Altarelli M 2005 *Phys. Rev.* B **72** 104108
[28] García-Fernández M, Staub U, Bodenthin Y, Scagnoli V, Pomjakushina V, Lovesey S W, Mirone A, Herrero-Martín J, Piamonteze C and Pomjakushina E 2009 *Phys. Rev. Lett.* **103** 097205

[29] Grenier S, Kiryukhin V, Cheong S-W, Kim B G, Hill J P, Thomas K J, Tonnerre J M, Joly Y Staub U and Scagnoli V 2007 *Phys. Rev.* B **75** 085101
[30] Staub U, García-Fernández M, Bodenthin Y, Scagnoli V, De Souza R A, Garganourakis M, Pomjakushina E and Conder K 2009 *Phys. Rev.* B **79** 224419
[31] Grenier S, Thomas K J, Hill J P, Staub U, Bodenthin Y, García-Fernández M, Scagnoli V, Kiryukhin V, Cheong S-W, Kim B G and Tonnerre J M 2007 *Phys. Rev. Lett.* **99** 206403
[32] Abbamonte P, Blumberg G, Rusydi A, Gozar A, Evans P G, Siegrist T, Venema L, Eisaki H, Isaacs E D, and Sawatzky G A 2004 *Nature* **431** 1078
[33] Huang D J, Lin H-J, Okamoto J, Chao K S, Jeng H-T, Guo G Y H, Hsu C-H, Huang C-M, Ling D C, Wu W B, Yang C S and Chen C T 2006 *Phys. Rev. Lett.* **96** 96401
[34] Grenier S, Hill J P, Kiryukhin V, Ku W, Kim Y-J, Thomas K J, Cheong S-W, Tokura Y, Tomioka Y, Casa D and Gog T 2005 *Phys. Rev. Lett.* **94** 047203
[35] Staub U, García-Fernández M, Mulders A M, Bodenthin Y, Martínez-Lope M J and Alonso J A 2007 *J. Phys.: Condens. Matter* **19** 092201
[36] García-Fernández M, Staub U, Bodenthin Y, Lawrence S M, Mulders A M, Buckley C E, Weyeneth S, Pomjakushina E and Conder K 2008 *Phys. Rev.* B **77** 060402(R)
[37] Staub U, Scagnoli V, Mulders A M, Janousch M, Honda Z and Tonnerre J M 2006 *Europhys. Lett.* **76** 926

Little volume change in orbital ordering transition in manganite thin films

Yusuke Wakabayashi[1], Hiroshi Sawa[2], Naoko Takubo[3,5], Masao Nakamura[3,6], Yasushi Ogimoto[3,4] and Kenjiro Miyano[3,4]

[1]Graduate School of Engineering Science, Osaka University, Toyonaka 560-8531, Japan
[2]Department of Applied Physics, Nagoya University, Nagoya 464-8603, Japan
[3]Research Center for Advanced Science and Technology, University of Tokyo, Tokyo 153-8904, Japan
[4]CREST, Japan Science and Technology Agency, Honcho 4-1-8, Kawaguchi 332-0012, Japan

E-mail: wakabayashi@mp.es.osaka-u.ac.jp

Abstract. We have investigated the structural feature of orbital-ordering manganite thin films fabricated on the perovskite (011) substrates by means of synchrotron x-ray diffraction technique. Like bulk manganites, these films show large jumps in lattice parameters at the orbital ordering transition temperature, which clearly show the transitions are first order. However, the magnitude of the volume change at the transition temperature for the films is much smaller than that for the orbital ordering transition in bulk manganite. This difference is understood by taking into account the stress from the substrate.

1. Introduction

Many transition metal oxides show various properties that can be used for devices, and thin films made of them are therefore extensively studied with a view to device application. Among the oxides, manganites are one of the most promising materials because of their metal-insulator switching controlled by temperature, magnetic field[1, 2, 3], and photo-irradiation[4, 5]. These properties are known to be related to the orbital occupancy of the $3d$ electrons, and in manganites, the orbital occupancy is closely related to the lattice strain. Therefore, the property of the epitaxial film is different than that of bulk crystals. The most well studied system is the films fabricated on perovskite (001) substrates[6, 7, 8, 9, 10]. In these systems, metal-insulator transitions are suppressed and, in many cases, the transitions are reported in films whose in-plane lattice parameters are relaxed from the substrates. When we change the orientation of the substrate from (001) to (011), the property changes drastically[11]. The film shows clear first-order metal-insulator transition at almost the same temperature with bulk system. Such orientation dependence of the property is reported in ref.[12]. Only films made on (011) substrate show clear metal-insulator transition, and no clear transitions are observed in films fabricated on (001) or (111) substrates.

The orbital occupancy of the manganites characterized by a periodicity longer than 11Å is best studied by the soft x-ray resonant scattering technique using Mn L-absorption edge[13],

[5] present address: Condensed Molecular Materials Lab, RIKEN, 2-1 Hirosawa, Wako 351-0198, Japan
[6] present address: Advanced Science Institute (ASI), RIKEN, 2-1 Hirosawa, Wako, 351-0198, Japan

and ferro-ordering of the orbital occupancy is studied by linear dichroism around Mn L-edge, while ordering characterized by an intermediate wavelength cannot be observed through L-edge resonance. Because of this limitation in periodicity, an arrangement of the orbital occupancy (orbital ordering) with unknown periodicity is often studied through Jahn-Teller (JT) mode lattice distortion. Nevertheless, detailed structure refinement of the film system is, unlike bulk systems[14, 15], not very simple[16]. Instead of making structure analyses, one can estimate the magnitude of the JT distortion based on the position of the Bragg reflections and characteristic superlattice reflections, together with utilizing the K-edge resonant x-ray scattering[17], which is known to reflect the local structure around the Mn ions[16]. Based on such structural studies, it has been clarified that the origin of the substrate-orientation dependent property is the fact that the JT distortion in films on (001) substrate is the elongation mode while that in films on (011) substrate is the shear mode [18], as shown in Fig.1. In the films on (001) substrates (panel (a)), the JT mode distortion for e_g electrons involves a large volume change, while that in films on (011) substrates does not (panel (b)). In this paper, we present the little volume change in the orbital ordering transition in three manganite films fabricated on (011) surfaces of the $[(LaAlO_3)_{0.3}(SrAl_{0.5}Ta_{0.5}O_3)_{0.7}]$ (LSAT) substrate. LSAT has the cubic perovskite structure with the lattice parameter of 3.87Å, and all the films we used in this study have in-plane lattice parameters locked into the substrate.

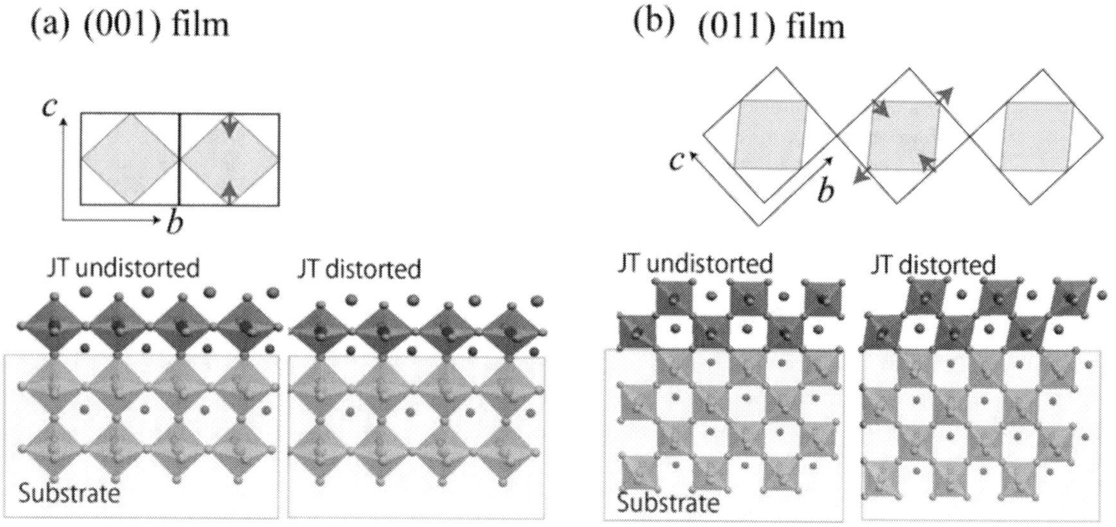

Figure 1. Schematic view of the JT mode lattice distortion in films on (a) (001)-substrate and (b) (011)-substrate. Shear mode strain in films on (011) substrate is a JT mode, while that in films on (001) substrate is not.

2. Experiment

Epitaxial films of $Pr_{0.5}Sr_{0.5}MnO_3$ (PSMO), $Nd_{0.5}Sr_{0.5}MnO_3$ (NSMO), and $Pr_{0.55}(Ca_{0.8}Sr_{0.2})_{0.45}MnO_3$ (PCSMO) were grown on (011) surfaces of LSAT substrates using the pulsed laser deposition method[11, 12]. The typical thickness of the samples was 80 nm. X-ray diffraction experiments were carried out on the BL-4C at the Photon Factory, KEK, Japan and on the X22C at the National Synchrotron Light Source, Brookhaven National Laboratory, USA. The beamlines are equipped with four-circle diffractometers connected to closed-cycle refrigerators. The temperature dependence of the lattice parameters was measured for all the samples with x-rays having energies of 9.5keV (Photon Factory) or 6.5keV (NSLS).

3. Results

Temperature dependences of the lattice parameters a, b and c in pseudo-cubic notation of the films are shown in Fig. 2. For comparison, those of bulk manganite $Nd_{0.5}Sr_{0.5}MnO_3$ (bulk NSMO) are also shown in panel (d).

The lattice parameters of PSMO shown in panel (a) were obtained from 002, 102, and 012 Bragg reflections in a heating run. This figure shows sudden jumps in b and c lattice parameters around 125 K. Since there was no superlattice reflection, this transition was found to be a ferro-orbital ordering transition[18]. Similarly, there is a clear transition in the lattice parameter of NSMO shown in panel (b). This transition is also a ferro-orbital ordering transition, while the ground state of this film is CE-type orbital ordered structure[18]. The lattice parameters of PCSMO are shown in panel (c). Superlattice reflections corresponding to $\sqrt{2} \times 2\sqrt{2}$ structure indicates that this transition is antiferro-orbital ordering transition[19]. Panel (d) shows the temperature dependence of the lattice parameters of another antiferro-orbital ordering material, bulk NSMO reported in ref. [20]. Note that in the cubic notation, the lattice parameters a and b are the same because they are the length of the diagonal line for the rectangle defined by a and b for the orthorhombic $Pbnm$ setting. All of the films as well as the bulk NSMO show clear jumps in their lattice parameters by 0.6% to 2% at the orbital ordering temperatures.

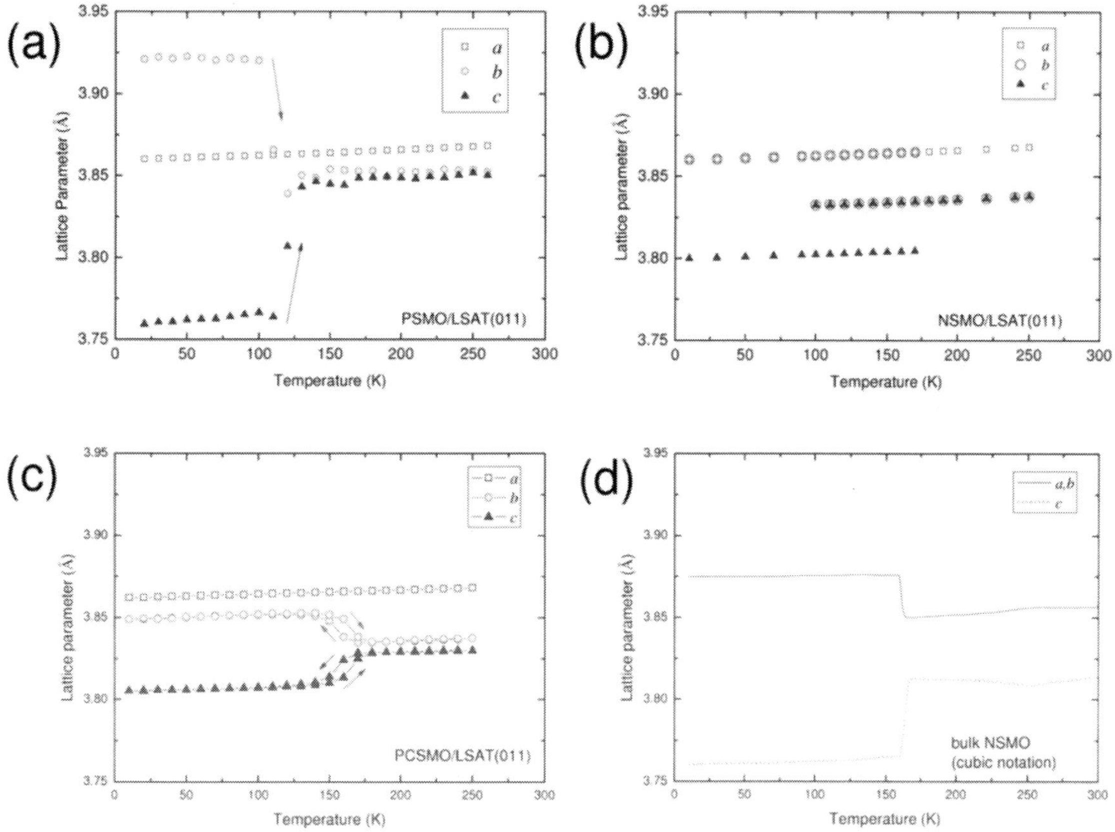

Figure 2. Temperature dependence of the lattice parameters of (a) PSMO, (b) NSMO, (c) PCSMO films as well as (d) bulk NSMO[20]. NSMO film has no clear hysteresis, and orbital ordered and disordered phases coexist in this film within a wide temperature range.

Figure 3 shows the temperature dependence of the cell volumes of the films and bulk NSMO in cubic notation. One can obtain extremely precise cell volume for the film system from the peak position of 011 Bragg reflection, because the in-plane lattice parameters are locked into the substrate. The experimental error of the peak position for PCSMO is the smallest, because the distance between the 011 Bragg reflections for the film and the substrate is the largest, which makes the analysis easier. Only the bulk NSMO shows a clear volume jump at the transition temperature, and the amount is 0.1%. Since all the transitions are first order, finite volume jump is expected at the transition points. However, no noticeable amount of the volume jumps were observed in the three films, while the bulk crystal system show finite volume jump as expected. Therefore, there must be a mechanism to suppress the volume jump in the film system. We discuss the origin of this lack of the volume jump in the film systems in the next section.

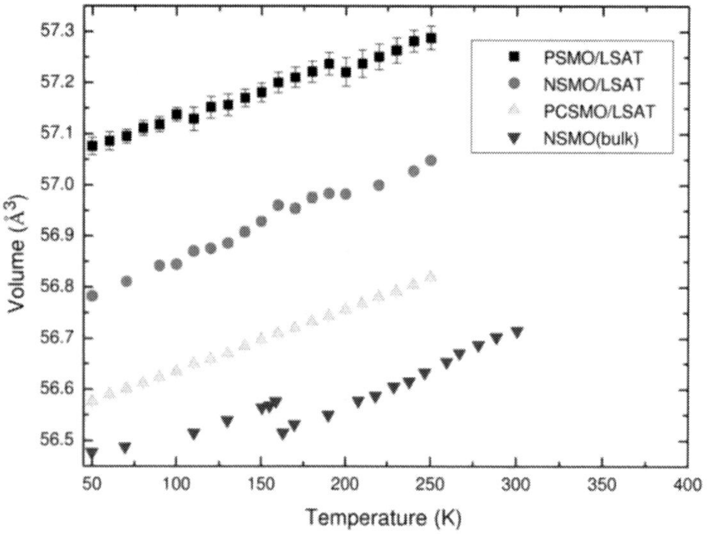

Figure 3. Temperature dependence of the cell volume of PSMO, NSMO, PCSMO films as well as bulk NSMO[20].

4. Discussion

As mentioned above, the origin of the orientation dependent properties is that the JT distortion is the shear mode in films on (011) substrate[18]. The orbital ordered state is stabilized by a large JT mode distortion, which are called q_2 or q_3 modes presented in Fig.4. Other modes of distortion are, at least to the first order, decoupled from the orbital occupancy; From energetic point of view, they only contribute to an elastic energy. The volume change is the distortion having A_{1g} symmetry, in other words, breathing mode or q_1 mode distortion. The q_1 mode distortion decouples from JT mode distortion in films on (011) substrates. This makes the transition easier. By contrast, the q_1 mode distortion couples with JT mode in the films on (001) substrate in order to lock the in-plane lattice parameter to the substrate; The same amount of q_1 distortion with q_3 mode is required to keep the in-plane lattice parameters constant. This strong coupling makes the transition very hard. Despite this difficulty, one exception is recently reported: narrow-bandwidth material $Pr_{0.5}Ca_{0.5}MnO_3$ film on LSAT(001) substrate[21]. This transition was attributed to the similarity of the LSAT lattice parameter with a and b of $Pr_{0.5}Ca_{0.5}MnO_3$ in ordered phases, and treated as an example of a strain assisted ordered structure.

Next, we focus on the little volume change in the orbital ordering transition. The decoupling of the q_1 and JT mode distortions in films on (011) substrate itself does not explain the little volume change. The strain energy ΔE is written as $1/2 \cdot \sum_i KQ_i^2$, where K is the elastic constant

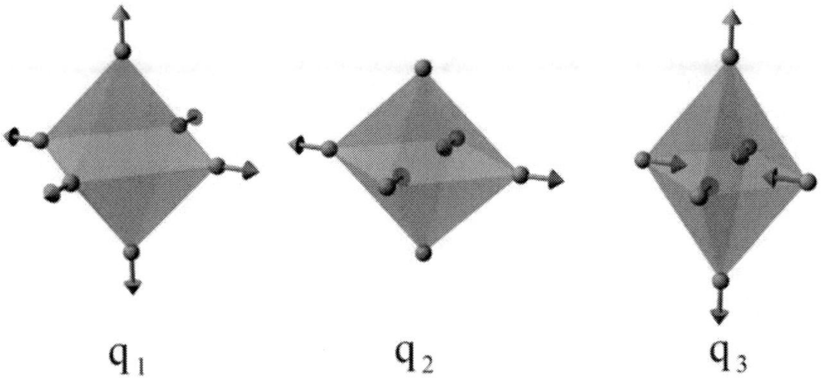

Figure 4. Schematic view of the distortion modes q_1, q_2, and q_3.

and Q_i is the magnitude of the lattice distortion corresponding to the mode q_i. Although Q_2 and Q_3 are very small in the orbital disordered phase, Q_1 is large in the orbital disordered phase in the film system. For example, film NSMO has larger volume than bulk NSMO by 0.6%. This volume difference is a good measure of Q_1 in the orbital disordered phase. At the transition temperature, the volume can be changed by the first order transition. The ΔE caused by this volume change is related to the specific latent heat of this transition. The amount of the volume change of the bulk NSMO is 0.1% from zero strain. The magnitude of ΔE corresponding to this strain is comparable to the ΔE caused by the volume change from 0.6% to 0.61%, which is too small to be observed by our experiment. This is the reason why the film systems do not show any volume jump in the transition. In conclusion, the little volume change at the first order transition is caused by the large lattice strain in the high temperature phase.

Acknowledgements

This work was supported by Murata Science Foundation and by KAKENHI (21740274). Use of the National Synchrotron Light Source, Brookhaven National Laboratory, was supported by the U.S. Department of Energy, Office of Science, Office of Basic Energy Sciences, under Contract No. DE-AC02-98CH10886.

References

[1] Tokura Y, Urushibara A, Moritomo Y, Arima T, Asamitsu A, Kido G and Furukawa N 1994 *J. Phys. Soc. Jpn.* **63** 3931
[2] Tokura Y and Nagaosa N 2000 *Science* **288** 462
[3] Dagotto E 2005 *New J. of Phys.* **7** 67
[4] Miyano K, Tanaka T, Tomioka Y and Tokura Y 1997 *Phys. Rev. Lett.* **78** 4257
[5] Kiryukhin V, Casa D, Hill J P, Keimer B, Vlgliante A, Tomioka Y and Tokura Y 1997 *Nature* **386** 813
[6] Prellier W, Simon C, Haghiri-Gosnet A M, Mercey B and Raveau B 2000 *Phys. Rev.* B **62** R16337
[7] Prellier W, Haghiri-Gosnet A M, Mercey B, Lecoeur P, Hervieu M, Simon C and Raveau B 2000 *Appl. Phys. Lett.* **77** 1023
[8] Ogimoto Y, Izumi M, Manako T, Kimura T, Tomioka Y, Kawasaki M and Tokura Y 2001 *Appl. Phys. Lett.* **78** 3505
[9] Biswas A, Rajeswari M, Srivastava R C, Venkatesan T, Greene R L, Lu Q, de Lozanne A L and Millis A J 2001 *Phys. Rev.* B **63** (2001) 184424
[10] Buzin E R, Prellier W, Simon C, Mercone S, Mercey B, Raveau B, Sebek J and Hejtmanek J 2001 *Appl. Phys. Lett.* **79** 647
[11] Ogimoto Y, Nakamura M, Takubo N, Tamaru H, Izumi M and Miyano K 2005 *Phys. Rev.* B **71** 060403(R)
[12] Nakamura M, Ogimoto Y, Tamaru H, Izumi M and Miyano K 2005 *Appl. Phys. Lett.* **86** 182504

[13] Wilkins S B, Spencer P D, Hatton P D, Collins S P, Roper M D, Prabhakaran D and Boothroyd A T 2003 *Phys. Rev. Lett.* **91** 167205
[14] Radaelli P G, Cox D E, Marezio M and Cheong S-W 1997 *Phys. Rev.* B **55** 3015
[15] Okuyama D, Tokunaga Y, Kumai R, Taguchi Y, Arima T and Tokura Y 2009 *Phys. Rev.* B **80**, 064402
[16] Wakabayashi Y, Sawa H, Nakamura M, Izumi M and Miyano K 2004 *Phys. Rev.* B **69** 144414
[17] Materlik G, Sparks C J and Fischer K 1994 *Resonant anomalous x-ray scattering: theory and applications* (Amsterdam)
[18] Wakabayashi Y, Bizen D, Kubo Y, Nakao H, Murakami Y, Nakamura M, Ogimoto Y, Miyano K and Sawa H 2008 *J. Phys. Soc. Jpn.* **77**, 014712
[19] Wakabayashi Y, Takubo N, Miyano K and Sawa H 2009 *Eur. Phys. J. Special Topics* **167** 67
[20] Shimomura S, Tajima K, Wakabayashi N, Kobayashi S, Kuwahara H and Tokura Y 1999 *J. Phys. Soc. Jpn.* **68** 1943
[21] Okuyama D, Nakamura M, Wakabayashi Y, Itoh H, Kumai R, Yamada H, Taguchi Y, Arima T, Kawasaki M and Tokura Y 2009 *Appl. Phys. Lett.* **95**, 152502

Exchange-striction associated with the elliptical proper helical magnetic structure in the ferroelectric phase of $CuFe_{1-y}Ga_yO_2$

N Terada[1,3,4], T Nakajima[2], S Mitsuda[2], M Matsuda[3], K Kakurai[3], Y Tanaka[4] and H Kitazawa[1]

[1]National Institute for Materials Science (NIMS), Tsukuba, Ibaraki, 305-0047, Japan
[2]Department of Physics, Tokyo University of Science, Tokyo 162-8601, Japan
[3]Japan Atomic Energy Agency, Tokai, Ibaraki, 319-1195, Japan
[4]RIKEN SPring-8 Center, Harima Institute, Sayo, Hyogo 679-5148, Japan

E-mail: terada.noriki@nims.go.jp

Abstract. Recently, the detail magnetic structure in the ferroelectric phase of $CuFe_{1-y}Ga_yO_2$ has been determined to be the proper helical structure with the ellipticity of ~ 0.9 by the comprehensive neutron diffraction study.[Nakajima et al. Phys. Rev. B **79** 214423 (2009)] In the present study, we have investigated the lattice modulation associated with the elliptical magnetic structure, using the resonant X-ray diffraction measurements on $CuFe_{0.965}Ga_{0.035}O_2$. We observed the superlattice reflection indexed as $(0\ 1+2q_{\rm m}\ 0)$, where q_m is the wave number of the magnetic modulation. The energy spectrum near the Fe-K absorption edge for the $(0\ 1+2q_{\rm m}\ 0)$ intensity conforms with that for the fundamental (0 2 0) intensity within experimental accuracy, suggesting that the superlattice modulation is at least associated with displacements of the Fe ions. We discuss the origin of the superlattice modulation with the exchange striction model.

1. Introduction

In recent years, magneto-electric multiferroic materials have attracted much attention with the aim of clarifying the novel microscopic coupling between spin and electric dipole moment.[1] The theoretical study has presented three possible mechanisms regarding coupling between spin and electric dipole moment: (i) exchange-striction, (ii) spin-current[2] and (iii) *d-p* hybridization mechanisms.[3] In a majority of multiferroic materials with a non-collinear magnetic ordering and a ferroelectric polarization, the second mechanism has succeeded in explaining the appearance of their ferroelectricity. In $CuFeO_2$, however, the ferroelectricity can be explained with the third mechanism, which was argued by Arima.[4] He also predicted that the electric dipole moments are modulated with twice larger wave number, $2q_{\rm m}$, than the magnetic one, $q_{\rm m}$ in the ferroelectric phase of $CuFeO_2$. Although the superlattice reflection with $2q_{\rm m}$ was observed in the previous non-resonant X-ray diffraction measurements,[5] the origin of the superlattice modulation has not been revealed so far.

$CuFeO_2$ shows both magnetic field-induced[6] and nonmagnetic (Al or Ga) impurity-substitution-induced[7, 8] ferroelectric phase transitions. The magnetic phase diagram of $CuFe_{1-y}Ga_yO_2$ is shown in figure 1.[9] Recently, Nakajima *et al.* have demonstrated the

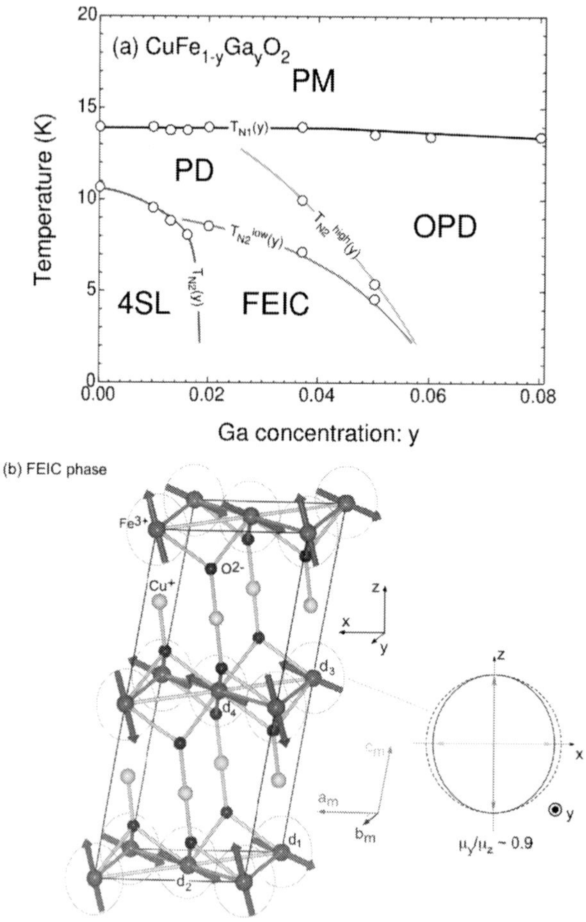

Figure 1. (a) Magnetic phase diagram of $CuFe_{1-y}Ga_yO_2$. The data were taken from Ref. [9]. (b) Elliptical proper helical magnetic structure in the FEIC phase with the monoclinic basis, a_m, b_m and c_m. Cartesian coordinate, x, y and z, is also drawn. The x and y axes are parallel to a_m and b_m axes, respectively. The z axis is perpendicular to the $a_m b_m$ plane.

comprehensive study using both the spherical neutron polarimetry and the unpolarized neutron diffraction with the four-circle diffractometer, and have determined the elliptical proper helical magnetic structure with the ellipticity of ~ 0.9 in $CuFe_{0.965}Ga_{0.035}O_2$.[10] There are two possible mechanisms regarding the $2q_m$ modulation, which are the exchange-striction and the $d-p$ hybridization mechanisms. In the present study, in order to investigate the origin of the $2q_m$ superlattice modulation in the ferroelectric phase, we have performed the resonant X-ray diffraction (RXD) experiments on the single crystal of $CuFe_{0.965}Ga_{0.035}O_2$.

2. Experimental detail

The single crystal of $CuFe_{0.965}Ga_{0.035}O_2$, which was prepared by the floating zone technique,[11] was cut into a disk with a thickness of 2 mm and subsequently polished to remove the surface roughness. The surface is perpendicular to the monoclinic b axis. In this paper, we use the monoclinic notation with the unique axis b and the cell choice 1 in the space group $C2/m$. While the lattice parameters were not refined in $CuFe_{0.965}Ga_{0.035}O_2$, those of $CuFeO_2$ were

Journal of Physics: Conference Series **211** (2010) 012005 doi:10.1088/1742-6596/211/1/012005

reported in the previous paper.[12] The monoclinic unit cell is illustrated in figure 1(b). The RXD experiments were carried out with a standard four-circle diffractometer in BL29XU at SPring-8. The cut crystal was mounted in a liquid ^{4}He refrigerator that can cool a sample down to 3 K. The incident X-ray beam had almost perfect horizontal polarization and its energy was tuned to near the Fe-K absorption edge. ($E \approx 7.112$ keV) The diffracted X-rays were analyzed using the 006 reflection of a pyrolytic graphite crystal.

3. Experimental results

As shown in figure 2(a), the energy dependence of the superlattice reflection at $(0\ 1+2q_{\rm m}\ 0)$ conforms with that of the fundamental 020 reflection, within the experimental accuracy. No resonant enhancement that corresponds to the $2q_{\rm m}$ modulation of the $d-p$ hybridization was observed near the Fe-K edge. The energy dependence of the structure factor of the fundamental 020 reflection near the Fe-K absorption edge can be written as

$$F_{020} \propto f'_{\rm Fe} + i f''_{\rm Fe} + C \tag{1}$$

where $f'_{\rm Fe}$ and $f''_{\rm Fe}$ are real and imaginary parts of anomalous scattering factor of Fe atom, respectively. C is independent of energy. Note that the energy spectrum near the edge depends only on the anomalous scattering factors of Fe atom. Considering that the spectrum of the fundamental reflection conforms with that of the superlattice reflection at $(0\ 1+2q_{\rm m}\ 0)$, we thus conclude that the superlattice reflection at least originates from atomic displacements of Fe atoms.

The spectra of the superlattice reflection at $(0\ 1+2q_{\rm m}\ 0)$ with $\pi-\pi'$ and $\pi-\sigma'$ channels are shown in figure 2(b). While the spectrum of intensity in $\pi-\pi'$ channel is identical to the case without analyzer (figure 2(a)), the intensity in $\pi-\sigma'$ channel is zero in every energy. These results infer that the superlattice reflection originates from the charge scattering and support our conclusion mentioned above.

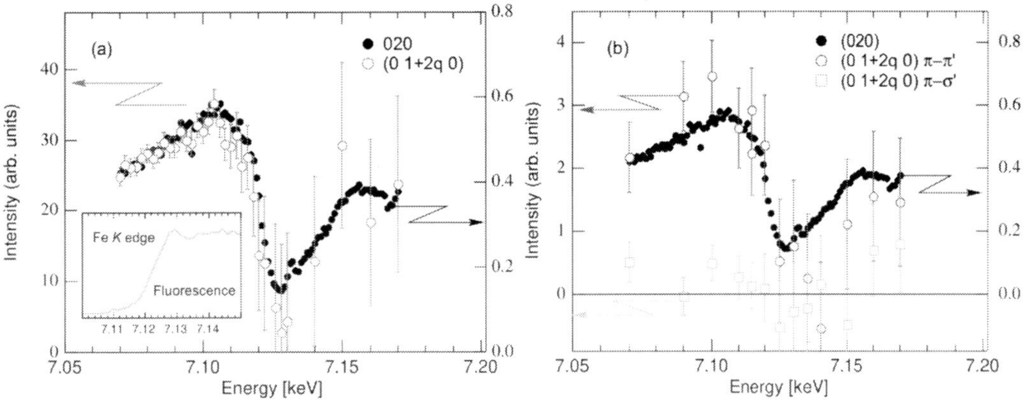

Figure 2. Energy dependence of the intensity of the fundamental reflection at (020) and the superlattice reflection at $(0\ 1+2q_{\rm m}\ 0)$, which are represented by closed and open symbols, respectively. As for the superlattice reflection, the diffracted X-ray beams were not analyzed in (a) and analyzed into $\pi-\pi'$ and $\pi-\sigma'$ channels in (b). The fluorescence spectrum is shown in the inset on (a).

4. Discussion

As mentioned above, we concluded that the superlattice reflection with $2q_{\rm m}$ wave number at least originates from the atomic displacement of Fe atom. In the present work, we focus upon discussing whether the displacement is explained by the exchange-striction mechanism.

4.1. Exchange-striction model

We calculate atomic displacement through the exchange-striction for Fe atoms, associated with the elliptical proper helical magnetic ordering, in the ferroelectric phase of $CuFe_{1-y}Ga_yO_2$. Here, we consider the exchange-striction for two Fe sites, Fe1 and Fe2, which is illustrated in figure 3. Assuming that only nearest neighbor exchange interactions contribute the exchange-striction, the atomic displacement for Fe1(Fe2) is proportional to sum of the magnetoelastic energy between spin on Fe1(Fe2) site and its nearest neighbor spins. The general form is written as

$$\boldsymbol{L}_{\mathrm{Fe}_n} = \sum_i^{NN} \alpha_i (\boldsymbol{r}_n - \boldsymbol{r}_i)(\boldsymbol{S}_n \cdot \boldsymbol{S}_i), \tag{2}$$

where α_i is magnetoelastic constant. In the present calculation, α is assumed to be identical for the six nearest neighbor exchange bonds.

When we define the Cartesian coordinate illustrated in figure 1(b), the spin components[10] are written as

$$\begin{cases} S^x_{l,d_i} &= \mu_x \cos[\boldsymbol{Q}\cdot(\boldsymbol{l}+\boldsymbol{d}_i) - \phi_i] \\ S^y_{l,d_i} &= 0 \\ S^z_{l,d_i} &= \mu_z \sin[\boldsymbol{Q}\cdot(\boldsymbol{l}+\boldsymbol{d}_i) - \phi_i]. \end{cases} \tag{3}$$

$\boldsymbol{Q}$ is the magnetic propagation vector of $(0\ q_{\rm m}\ \frac{1}{2})$. $\boldsymbol{d}_i$ denotes the factional coordinates represented by the monoclinic basis described in figure 1(b). ϕ_i is the relative phase shift at the $\boldsymbol{d}_i$, $\phi_{1,3,4,7,8} = 0$ and $\phi_{2,5,6,9,10} = \delta$.

Using the eqs. (2) and (3), we calculate the atomic displacements along the b axis for Fe1 and Fe2,

$$L^b_{\rm Fe1} = A(\mu_z^2 + \mu_x^2)\sin(\delta) + C(\mu_z^2 - \mu_x^2)\sin(2\pi(2q_{\rm m})y) \tag{4}$$

$$L^b_{\rm Fe2} = -A(\mu_z^2 + \mu_x^2)\sin(\delta) + C(\mu_z^2 - \mu_x^2)\sin(2\pi(2q_{\rm m})y + \Delta), \tag{5}$$

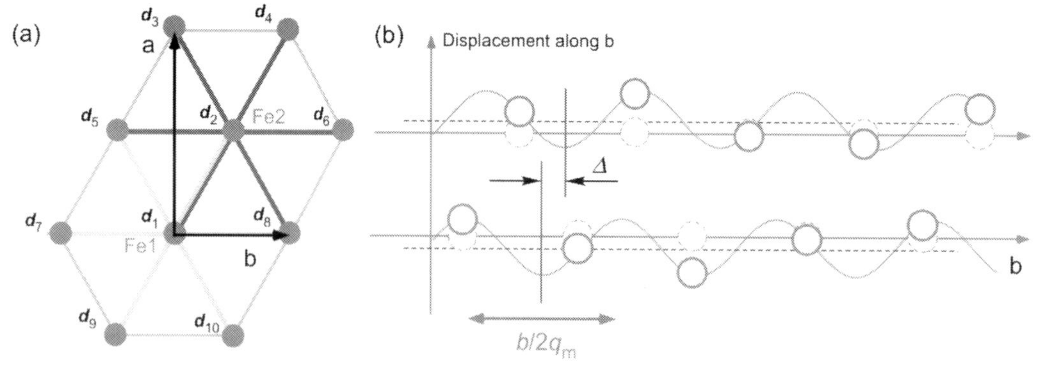

Figure 3. (a) Schematic drawing of the nearest neighbor Fe ions for Fe1 and Fe2 sites in the ab plane. (b) Illustration of the displacement along the b direction. The phase difference between the lattice modulations through Fe1 and Fe2 sites is represented by Δ. Note that the vertical axis denotes the displacements along the b axis.

where $A = \sin(\pi q_m)$, $B = \sin(2\pi q_m)$ and $C = \sqrt{A^2 + B^2 + 2AB\cos(\delta)}$. Δ is the phase difference for the lattice modulation,

$$\Delta = 2\arctan\left(\frac{B\sin(\delta)}{A + B\cos(\delta)}\right). \tag{6}$$

The first terms in eqs. (4) and (5) are constants in space, which reflects uniform displacements. These uniform terms are identical in their absolute value and their signs are opposite. Therefore, the first terms give antiferroic displacements along the b axis on Fe1 and Fe2 sites, when δ is finite. This calculation is consistent with the observation of the $0\ k\ 0$ superlattice reflection with $k =$ odd, which was reported in the previous X-ray diffraction measurements.[5]

The second terms correspond to the displacements modulated with $2q_{\rm m}$. These terms are proportional to $(\mu_z^2 - \mu_x^2)$ that is not zero; because, the ellipticity was determined to be $\mu_x/\mu_z \simeq 0.9$ in the recent neutron diffraction study.[10] Note that the oscillation terms for Fe1 and Fe2 sites are different in their initial phase by the phase difference Δ. When the values $q_m = 0.406$ and $\delta = q_{\rm m}\pi$, which were determined in the neutron diffraction study,[10] are substituted in the eq. (6), we obtain $\Delta = 0.292\pi$. The structure factor of the oscillation terms can be written as

$$F_{2q_{\rm m}} \propto (\mu_z^2 - \mu_x^2)\sum_y \Big[\sin(2\pi(2q_{\rm m})y) + e^{i\pi k}\sin(2\pi(2q_{\rm m})y + \Delta)\Big]e^{2\pi iky}, \tag{7}$$

where k is y component of the index. Writing a delta function as $\delta(\cdots)$ at the moment, eq. (7) is rewritten as

$$(\mu_z^2 - \mu_x^2)\Big[\big\{1 + e^{i(\pi k - \Delta)}\big\}\sum_\tau \delta(k + 2q_{\rm m} - \tau) - \big\{1 + e^{i(\pi k + \Delta)}\big\}\sum_\tau \delta(k - 2q_{\rm m} - \tau)\Big]. \tag{8}$$

Consequently, the satellite reflections indexed as $(0\ k \pm 2q_{\rm m}\ 0)$ are predicted in the present calculation. These results are in agreement with the observation of the superlattice reflection at $(0\ 1 + 2q_{\rm m}\ 0)$, qualitatively.

4.2. Comparison between experimental results and the calculation

We discuss the oscillation terms calculated above, quantitatively. In the previous non-resonant X-ray diffraction measurements on $CuFeO_2$,[13] the reflections at $(0\ 4 - 2q_{\rm m}\ 0)$ and $(0\ 3 + 2q_{\rm m}\ 0)$ are observed in the ferroelectric phase. The intensity ratio between the two reflections observed in the experiments[13] is

$$\frac{I(0\ 3 + 2q_{\rm m}\ 0)}{I(0\ 4 - 2q_{\rm m}\ 0)} \simeq 10^2. \tag{9}$$

On the other hand, in the exchange-striction model above mentioned, the ratio of the intensity in the case k is even or odd is derived using eq. (8) with $\Delta = 0.292\pi$,

$$\frac{|F_{2q_{\rm m}}^{\rm odd}|^2}{|F_{2q_{\rm m}}^{\rm even}|^2} \propto \frac{|1 - e^{i\Delta}|^2}{|1 + e^{i\Delta}|^2} \simeq \frac{1}{4}. \tag{10}$$

This calculated value is not coincident with the experimental value at all. Although the observation of the reflections indexed as $(0\ k_{\rm even} \pm 2q_{\rm m}\ 0)$ with $(0\ k_{\rm odd} \pm 2q_{\rm m}\ 0)$ can be qualitatively explained by the exchange-striction model, the intensity ratio between them cannot be quantitatively done at all. The discrepancy between the experimental facts and the model calculation might be caused by the oversimplification of the exchange-striction model.

In this calculation, the magnetoelastic constants α in eq. (2) are assumed to be identical for the six nearest neighbor bonds, though these bonds are not equivalent owing to the symmetry lowering from rhombohedral to monoclinic below the Neel temperature. In fact, when α of the bonds along the b axis is much larger than the others, the ratio in eq. (10) increases about 10. This explains the tendency of $I(0\ 3+2q_{\mathrm{m}}\ 0) > I(0\ 4-2q_{\mathrm{m}}\ 0)$ observed experimentally.

We should also mention the temperature dependence of the ellipticity that increases with increasing temperature in the ferroelectric phase, which was reported in the recent neutron diffraction study.[10] The intensity of the $2q_{\mathrm{m}}$ reflection is also enhanced with increasing temperature near the phase transition temperature, which is reported in the previous non-resonant X-ray diffraction study.[5] These experimental facts support the exchange-striction mechanism.

5. Summary

We have investigated the lattice modulation associated with the elliptical magnetic structure that has been determined in the recent neutron diffraction experiments,[10] using the resonant X-ray diffraction measurements on $CuFe_{0.965}Ga_{0.035}O_2$. We observed the superlattice reflection indexed as $(0\ 1+2q_{\mathrm{m}}\ 0)$, where q_m is the wave number of the magnetic modulation. The energy spectrum near the Fe-K absorption edge for the $(0\ 1+2q_{\mathrm{m}}\ 0)$ intensity conforms with that for the fundamental (0 2 0) intensity within experimental accuracy, suggesting that the superlattice modulation is at least associated with displacements of the Fe ions. We also calculated the lattice modulation associated with the elliptical proper helical magnetic structure, using the exchange-striction model involving the only nearest neighbor exchange interactions. The exchange-striction model succeeds in explaining the observation of the reflections indexed as $(0\ k_{\mathrm{even}} \pm 2q_{\mathrm{m}}\ 0)$ with $(0\ k_{\mathrm{odd}} \pm 2q_{\mathrm{m}}\ 0)$ qualitatively. However, the intensity ratio between these two reflections cannot be explained quantitatively, which might be caused by anisotropic magnetoelastic constants.

Acknowledgments

This work is partly supported by Grants-in-Aid for Scientific Research "Young Scientists (B), No. 20740209" and "Scientific Research (A), No. 21244049" from the Japan Society for the Promotion of Science. This work is also supported by Grants-in-Aid for Scientific Research on Priority Areas (No. 19052004).

References

[1] Cheong S-W and Mostovoy M 2007 *Nature Mater.* **6** 13
[2] Katsura H, Nagaosa N and Balatsky A V 2005 *Phys. Rev. Lett.* **95** 057205
[3] Jia C, Onoda S, Nagaosa N and Han J H 2007 *Phys. Rev.* B **76** 144424
[4] Arima T 2007 *J. Phys. Soc. Jpn.* **76** 073702
[5] Nakajima T, Mitsuda S, Inami T, Terada N, Ohsumi H, Prokes K and Podlesnyak A 2008 *Phys. Rev.* B **78** 024106
[6] Kimura T, Lashley J C and Ramirez A P 2006 *Phys. Rev.* B **73** 220401(R)
[7] Seki S, Yamasaki Y, Shiomi Y, Iguchi S, Onose Y and Tokura Y 2007 *Phys. Rev.* B **75** 100403(R)
[8] Terada N, Nakajima T, Mitsuda S, Kitazawa H, Kaneko K and Metoki N 2008 *Phys. Rev.* B **78** 014101
[9] Terada N, Nakajima T, Mitsuda S and Kitazawa H 2009 *J. Phys. Conference Series* **145** 012071
[10] Nakajima T, Mitsuda S, Takahashi K, Yamano M, Masuda K, Yamazaki H, Prokes K, Kiefer K, Gerischer S, Terada N, Kitazawa H, Matauda M, Kakurai K, Kimura H, Noda Y, Soda M, Matsuura M and Hirota K 2009 *Phys. Rev.* B **79** 214423
[11] Zhao T R, Hasegawa M and Takei H 1996 *J. Cryst. Growth* **166** 408
[12] Ye F, Ren Y, Huang Q, Fernandez-Baca J A, Dai P, Lynn J W and Kimura T 2006 *Phys. Rev.* B **73**, 220404(R)
[13] Terada N, Mitsuda S, Tanaka Y, Tabata Y, Katsumata K and Kikkawa A 2008 *J. Phys. Soc. Jpn.* **77** 054701

Journal of Physics: Conference Series **211** (2010) 012006 doi:10.1088/1742-6596/211/1/012006

X-ray resonant scattering study of the incommensurate charge-orbital density wave in $La_{2-2x}Sr_{1+2x}Mn_2O_7$ ($x = 0.7$)

T A W Beale[1], S R Bland[1], P D Hatton[1], P Thompson[2], D Prabhakaran[3] and A T Boothroyd[3]

[1] Department of Physics, University of Durham, Rochester Building, South Road, Durham, DH1 3LE, United Kingdom

[2] European Synchrotron Radiation Facility, Boîte Postal 220, F-38043 Grenoble Cedex, France

[4] Department of Physics, University of Oxford, Clarendon Laboratory, Parks Road, Oxford, OX1 3PU, United Kingdom

E-mail: p.d.hatton@durham.ac.uk

Abstract.
Orbital order has been proposed theoretically for more than half a century, and has subsequently been observed in a huge range of materials. Charge ordering is generally accompanied by a transition from a metal to an insulator. In manganites, colossal magneto-resistance (CMR) is associated with competition between charge and orbital antiferromagnetic insulating phases and ferromagnetic metallic phases. Initial models of charge and orbital order in the manganites concentrated on the chequerboard charge order structure, observed in many half doped systems. Incommensurate charge ordering and structural distortions have been identified in $La_{0.6}Sr_{2.4}Mn_2O_7$ through the observation of superlattice reflections using resonant x-ray scattering. Such structural distortions are indicative of concomitant orbital order, leading to our proposal of an incommensurate charge-orbital density wave. Superlattice structural distortion reflections were observed with a modulation vector (0.178, 0.178, 0), and charge order reflections with a modulation vector (0.356, 0.356, 0). Low temperature charge order melting, observed in the half doped bilayer manganite, is not present due to the absence of long range magnetic order.

1. Introduction

Transition metal oxides have been extensively studied due to the complex interactions between the different order parameters. The layered manganites have shown long-range magnetic, charge, orbital and correlated structural distortions. Such order parameters in the Ruddleston-Popper group, $(La,Sr)_{n+1}Mn_nO_{3n+1}$ have been studied with a variety of techniques[1, 2, 3]. In this paper we present resonant x-ray scattering results for the bilayer $n = 2$ member with $x = 0.7$ (Fig. 1). This overdoped system has a hole density provided by the doped strontium ions sufficient for the usual half doped CE chequerboard to become unstable. Indeed, this breakdown of the 1:1 chequerboard pattern and associated herringbone orbital order has been observed to collapse with $x \geq 0.575$[4]. Further investigations have suggested that the orbital and charge order are caused to melt through the onset of antiferromagnetic order[5]; in contrast no such long range magnetic order has been observed in $La_{0.6}Sr_{2.4}Mn_2O_7$[6].

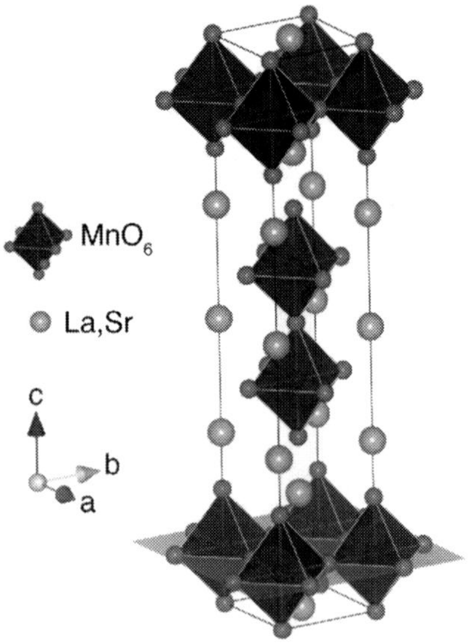

Figure 1. The crystal structure of $La_{2-2x}Sr_{1+2x}Mn_2O_7$. The shaded *ab* plane indicates the plane in which the charge-orbital density wave propagates.

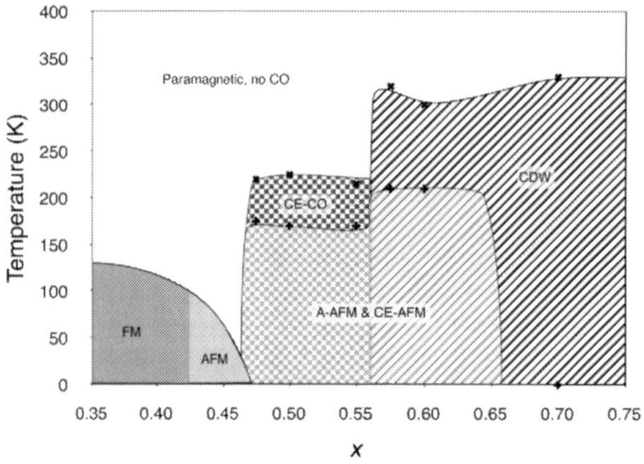

Figure 2. Phase diagram for the bilayer manganite series $La_{2-2x}Sr_{1+2x}Mn_2O_7$ ($0.35 < x < 0.75$). Data points marked by × show the onset of charge order (and JT) through the appearance of superlattice reflections, + data points mark the maxima in the intensity of the CO (or JT) reflections, interpreted as the onset of A-type AFM order. The data points marked by symbols are taken from this and previous publications[5, 4], and the areas of the phase diagram not marked by data symbols are generated from previous phase diagrams[7, 6].

Figure 2 displays the phase diagram for the bilayer manganites, showing a general transition from a ferromagnetic phase in the underdoped system, to an antiferromagnetic phase around half doping, to a paramagnetic overdoped system as a function of increasing doping. Charge order and co-operative Jahn-Teller distortions are observed in the half doped and overdoped systems. We have previously shown that the CE chequer-board charge order breaks down at $x \geq 0.575$[4], and through resonant x-ray diffraction we now extend this study and suggest the formation of a charge density wave and associated orbital density wave in the $x = 0.7$ system.

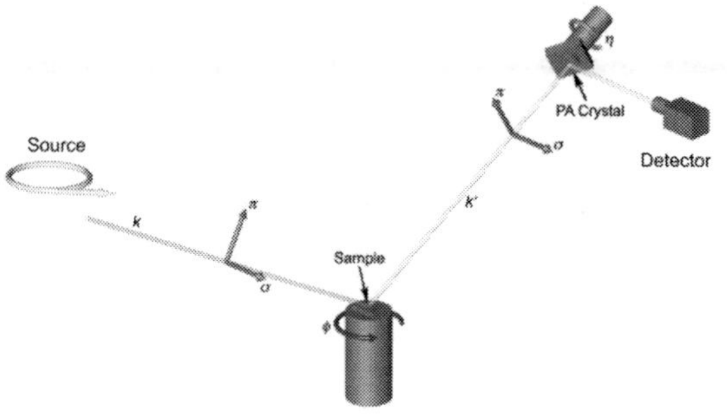

Figure 3. Schematic of the experimental setup, showing the two linear polarisation channels. Polarisation analysis is enabled through the rotation of the PA crystal (η), here shown at $\eta = 90°$.

2. Experimental Details

High quality single crystals of $La_{0.6}Sr_{2.4}Mn_2O_7$ were grown at the University of Oxford using the floating zone method[8, 9]. Due to the large c/a lattice parameter ratio, the samples naturally cleave in the ab plane, providing a suitable scattering surface. A selection of crystals were pre-aligned using a rotating anode diffractometer, with a single domain crystal of approximately $3 \times 4 \times 0.5$ mm selected for experiments using synchrotron radiation.

Resonant x-ray diffraction experiments were performed on BM28 at the ESRF[10]. The beamline is situated on a bending magnet, providing a focussed monochromatic x-ray beam in the energy range 2.4 - 15 keV. For this experiment, the sample was illuminated by an x-ray beam with an incident energy of 6.555 keV, provided by a double bounce Si (111) monochromator with the higher energy x-ray harmonics rejected by a set of rhodium coated mirrors. The sample was mounted on a copper puck in a closed cycle cryostat held in a Eulerian cradle, with the cleaved (ab) plane perpendicular to the scattering plane. Polarisation analysis of the scattered beam was undertaken using a copper (220) polarisation analyser (PA) crystal (Fig. 3). At the manganese K-edge the (220) Bragg reflection of copper gives a leakthrough between the channels of less than 3%.

3. Results and Discussion

Superlattice reflections were observed below 330 K at (0.178, 0.178, 10) and (0.356, 0.356, 10) wavevectors (Fig. 4). Both these reflections showed a resonant enhancement at the manganese K edge, with the (0.178, 0.178, 10) reflection observed in both polarisation channels and the (0.356, 0.356, 10) reflection only present in the unrotated $\sigma - \sigma$ polarisation channel. In addition, further reflections were observed around the (005) Bragg reflection, at (1.178, 0.178, 5). These reflections displayed no resonant enhancement, and reduced in intensity through the absorption edge congruent to the Bragg reflection.

The manganese K edge probes the 4p electron band through the 1s-4p virtual transition, and is thus only sensitive to the symmetry of the 4p band and not directly sensitive to the orbital occupation of the 3d electron orbital occupation. However we are able to observe reflections arising from the anisotropic tensor of susceptibility (ATS)[11]. These ATS reflections arise through the anisotropic charge density in the crystal probed in the resonant condition. In this system the anisotropy of the Jahn-Teller distortion, allows the ATS reflections that we observe at (0.178, 0.178, 10).

In the bilayer manganite systems the orbital order is concomitant and possesses the same symmetry as the Jahn-Teller distortions, and thus by determining the nature of the Jahn-Teller distortion we can deduce that of the orbital order. In contrast to the ATS scattering, the charge order satellites are observed with finite intensity away from the resonant condition, however there is an enhancement in the signal through the absorption edge. This enhancement is caused by the charge order existing primarily on the manganese ions, which is selectively enhanced by

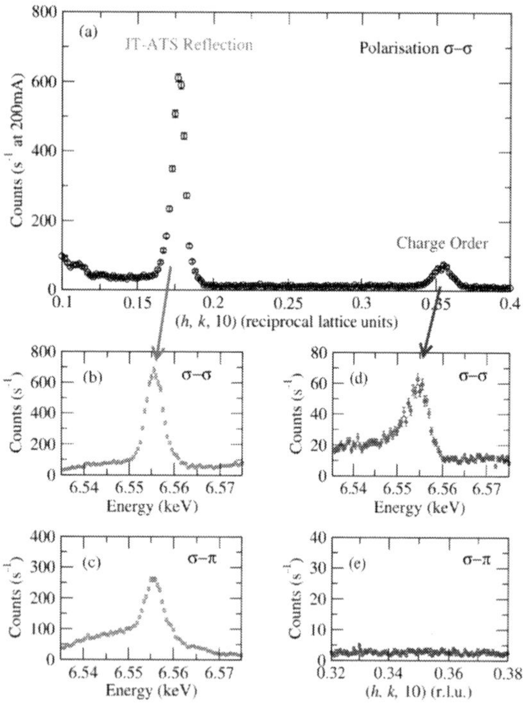

Figure 4. (a) [110] scan in reciprocal space away from the (0,0,10) Bragg peak at 10 K. Two superlattice reflections are observed in the $\sigma \to \sigma$ polarisation channel corresponding to the JT-ATS reflection (0.178, 0.178, 10) and charge order reflection (0.356, 0.356, 10). Energy scans at fixed $\vec{Q}$ of the ATS reflection are shown in the unrotated (b) and rotated (c) polarisation channels. Energy scans at fixed $\vec{Q}$ of the charge order reflection show resonance in the unrotated channel (d). A scan along [110] through the charge order reflection in the rotated channel shows no scattered intensity(e).

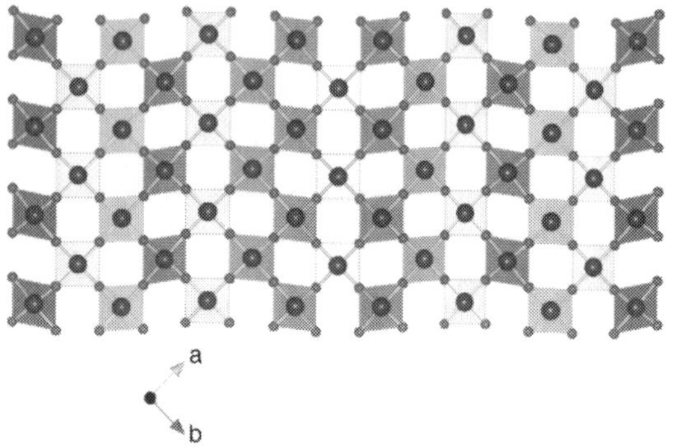

Figure 5. Schematic of the incommensurate Jahn-Teller distortions in $La_{0.6}Sr_{2.4}Mn_2O_7$. The shading represents the charge ordering, with heavily shaded MnO_6 octahedra representing those with more electron density (*i.e.* nearer to Mn^{3+}), and consequently show a bigger distortion. The light MnO_6 represent Mn^{4+} with no distortion.

the resonant condition.

The wavevector of the JT-ATS and charge order reflections reveals the periodicity of the crystal distortion and concomitant orbital order. This incommensurate value, suggests a decoupling of the charge and crystal distortions from the crystal lattice. This has been observed for lower doped systems ($x = 0.6$), however in that case the wavevector was close to a commensurate value and varied with temperature. In the $x = 0.7$ bilayer manganite, we found the wavevector varied between $0.17 \leq \delta \leq 0.178$ where the ATS reflection exists at (δ,δ,10). An incommensurate charge order is by no means unique to LaSrMn2O7, for example, a linear relationship between the doping and charge order wavevector was seen in $Eu(Ba_{1-y})La_y)Mn_2O_6$ and $(Eu_{1-x}Ca_x)BaMn2O6$[12]. Figure 5 shows a schematic of a charge density wave and incommensurate Jahn-Teller distortions with the periodicity derived from the wavevector of the superlattice reflections. Although a sinusoidal charge density wave, as shown in this figure represents the most compelling model, with this data we are unable to distinguish between this and an incommensurate square-wave modulation. A sinusoidal charge-orbital density wave has been proposed in the single layered manganites $Nd_{1-x}Ca_{1+x}MnO_4$ and $Nd_{1-x}Sr_{1+x}MnO_4$, supported by high resolution electron microscopy data[13, 14], and also in cubic $La_{1-x}Ca_xMnO_3$[15]. The surprising aspect of the CDW in the cubic manganite is that this incommensurate wave is observed with a doping of $x = 0.52$, unlike the bilayer manganite series where the charge order is locked into the chequer-board pattern below $x = 0.575$.

4. Conclusion

Resonant x-ray scattering measurements at the manganese K edge in $La_{0.6}Sr_{2.4}Mn_2O_7$ show evidence of charge order and Jahn-Teller distortions with the observation of superlattice reflections. Multiple superlattice reflections corresponding to charge order, Jahn-Teller distortions, and reflections arising from the anisotropic tensor of susceptibility were observed. These show an incommensurate order, that we have suggested form a charge-orbital density wave. Such a wave occurs when the doping level in the bilayer manganite is greater than $x = 0.575$, however it is more stable as $x \to 0.7$.

Acknowledgments

This work was performed on the EPSRC-funded XMaS beam line at the ESRF, directed by M.J. Cooper and C. Lucas. We are grateful to the beam line team of S.D. Brown, L. Bouchenoire, D. Mannix and D.F. Paul for their invaluable assistance, and to S. Beaufoy and J. Kervin for additional support. TAWB, SRB and PDH wish to thank EPSRC and STFC for support.

References

[1] Sternlieb B J, Hill J P, Wildgruber U C, Luke G M, Nachumi B, Moritomo Y and Tokura Y 1996 *Phys. Rev. Lett.* **76** 2169
[2] Murakami Y, Kawada H, Kawata H, Tanaka M, Arima T, Moritomo Y and Tokura Y 1998 *Phys. Rev. Lett.* **80** 1932
[3] Wakabayashi Y, Murakami Y, Koyama I, Kimura T, Tokura Y, Moritomo Y, Endoh Y and Hirota K 2003 *J. Phys. Soc. Japan* **72** 618
[4] Beale T A W, Spencer P D, Hatton P D, Wilkins S B, v Zimmerman M, Brown S D, Prabhakaran D and Boothroyd A T 2005 *Phys. Rev.* B **72** 064432
[5] Wilkins S B, Spencer P D, Beale T A W, Hatton P D, v Zimmermann M, Brown S D, Prabhakaran D and Boothroyd A T 2003 *Phys. Rev.* B **67** 205110
[6] Ling C D, Millburn J E, Mitchell J F, Argyriou D N, Linton J and Bordallo H N 2000 *Phys. Rev.* B **62** 15096
[7] Zheng H, Li Q A, Gray K E and Mitchell J F 2008 *Phys. Rev.* B **78** 155103
[8] Prabhakaran D and Boothroyd A T 2003 *J. Mater. Sci.: Materials in Electronics* **14** 587
[9] Prabhakaran D and Boothroyd A T 2005 *Frontiers in Magnetic Materials* ed Narlikar A (Spinger-Verlag, Berlin) pp 97–115
[10] Brown S D, Bouchenoire L, Bowyer D, Kervin J, Laundy D, Longfield M G, Mannix D, Paul D F, Stunault A, Thompson P, Cooper M J, Lucas C A and Sterling W G 2001 *J. Sync. Rad.* **8** 1172
[11] Dmitrienko V E 1983 *Acta Crystallographica* **39** 29–35
[12] Akahoshi D, Uchida M, Arima T H, Tomioka Y and Tokura Y 2006 *Phys. Rev.* B **74** 012402
[13] Nagai T, Kimura T, Yamazaki A, Tomioka Y, Kimoto K, Tokura Y and Matsui Y 2003 *Phys. Rev.* B **68** 092405
[14] Nagai T, Kimura T, Yamazaki A, Asaka T, Kimoto K, Tokura Y and Matsui Y 2002 *Phys. Rev.* B **65** 060405(R)
[15] Loudon J C, Cox S, Williams A J, Attfield J P, Littlewood P B, Midgley P A and Mathur N D 2005 *Phys. Rev. Lett.* **94** 097202

Polarized Neutrons and Synchrotron X-rays for Magnetism Conference 2009 IOP Publishing
Journal of Physics: Conference Series **211** (2010) 012007 doi:10.1088/1742-6596/211/1/012007

Soft x-ray diffraction from lattice constrained orbital order in $Pr(Sr_{0.1}Ca_{0.9})_2Mn_2O_7$

T A W Beale[1], S R Bland[1], R D Johnson[1], P D Hatton[1], J C Cezar[2], S S Dhesi[3], D Prabhakaran[4], A T Boothroyd[4]

[1] Department of Physics, University of Durham, Rochester Building, South Road, Durham, DH1 3LE, United Kingdom
[2] European Synchrotron Radiation Facility, Boîte Postal 220, F-38043 Grenoble Cedex, France
[3] Diamond Light Source Ltd, Harwell Science and Innovation Campus, Diamond House, Chilton, Didcot, Oxfordshire, OX11 0DE, United Kingdom
[4] Department of Physics, University of Oxford, Clarendon Laboratory, Parks Road, Oxford, OX1 3PU, United Kingdom

E-mail: p.d.hatton@durham.ac.uk

Abstract. Controlling orbital occupancy is a fundamental prerequisite for orbitronics. It has been shown in the orthorhombic bilayer manganite $Pr(Sr_{0.1}Ca_{0.9})_2Mn_2O_7$ that the direction of orbital order stripes can be influenced by controlling temperature or through inducing strain in the material. In this paper we have used resonant soft x-ray diffraction at the Mn L-edge to confirm the rotation of the orbital direction T_{OO2} and furthermore prove that there is no change in the occupied orbital type, however the orbital rotation causes a switch from $3_{x^2-r^2}$ to $3_{y^2-r^2}$ on a single site. We find that unlike the tetragonal bilayer manganites, where an onset of A-type AFM quenches the orbital order, no such effect is found on the orbital order below T_N.

1. Introduction

Orbital order is a key ingredient in the transition metal oxides[1, 2, 3], however the control or even measurement of this order parameter has been elusive. Crystallographic distortions, in particular Jahn-Teller distortions appear to be ubiquitous in orbitally ordered materials, and many systems also contain magnetic and even charge order. This makes direct measurements of orbital order particularly difficult. Despite this, there is considerable interest in measuring, and possibly controlling, the orbital occupancy in a material[4].

In the material $Pr(Sr_{0.1}Ca_{0.9})_2Mn_2O_7$, Tokunaga and co-workers have recently shown that the direction of orbital chains can be switched through a thermal transition[5], and more recently through applying uniaxial stress[6]. Like many layered manganites, $Pr(Sr_{0.1}Ca_{0.9})_2Mn_2O_7$ possesses both long range magnetic and charge order, in addition to orbital order. Diffraction techniques are inherently wavevector sensitive and so, in principle, can selectively probe individual order parameters with unique periodicities. This is complicated by the concomitant nature of the Jahn-Teller distortions and orbital order[7]. Not only do these phenomena both occur with the same wavevector, but they also have the same transition temperatures. However, it was conclusively demonstrated by Castleton and Altarelli that orbital order and Jahn-Teller distortions can be disentangled through the resonant diffraction spectrum at the Mn L edges[8, 9].

$Pr(Sr_{0.1}Ca_{0.9})_2Mn_2O_7$ crystallises in a orthorhombic lattice, and so unlike other tetragonal bilayer manganites[10], the a and b axes can be easily separated through diffraction, even on a twinned crystal. The crystal can be described in the *Amam* spacegroup, with lattice parameters $a = 5.410$ Å, $b = 5.462$ Å and $c = 19.277$ Å at 405 K. Through the anisotropy of optical conductivity, Tokunaga and co-workers concluded that the orbital order formed with chains running along the b axis below 350 K[5]. The orbital chains were constrained by the orthorhombic crystal lattice, whereas tetragonal bilayer manganites have shown orbital chains simultaneously along both a and b axes. Intriguingly below $\sim$ 325 K the stripe direction of the orbital chains flips to lie along the a axis. There is simultaneously a small crystallographic change, however the b axis remains longer than the a axis.

In this paper we describe the use of soft x-ray diffraction to observe the rotation of orbital stripes in $Pr(Sr_{0.1}Ca_{0.9})_2Mn_2O_7$ using soft x-ray resonant diffraction at the Mn L edges. The anisotropy of the stripes can be determined in each phase through azimuthal measurements, and the complex temperature dependence of the orbital order signals are explained.

2. Experimental Method

Soft x-ray diffraction measurements were undertaken at beamlines ID08 at the ESRF and I06 at Diamond Light Source (DLS). A sample of $Pr(Sr_{0.1}Ca_{0.9})_2Mn_2O_7$ was prepared with the [100] face surface normal. Due to the similarity of the a and b lattice parameters of the orthorhombic crystal, a large degree of twinning was observed. Thus it was possible to observe scattering reflections in both the [100] and [010] directions simultaneously. The experiments were conducted under high vacuum conditions, using the soft x-ray diffraction endstation at ID08, and using the Daresbury soft x-ray diffractometer at DLS. A new sample mount was designed for use at DLS incorporating a piezo rotation device to enable azimuthal measurements.

3. Results and Discussion

At 330 K a diffraction signal was observed at a 2θ angle of 121° corresponding to the $(0,\frac{1}{2},0)$ orbital order reflection (Fig. 1). This displayed a complex energy spectrum with split resonances at both the manganese L_3 and L_2 edges[11]. Reducing the temperature of the sample to 315 K revealed a double reflection, observed by two peaks in the θ-2θ scan. The orthorhombic nature of the sample causes the 2θ value of the $(\frac{1}{2},0,0)$ reflection to be slightly higher than that of the $(0,\frac{1}{2},0)$ reflection, however it is clear from the scan at 315 K that these two reflections can be resolved. The individual crystallographic domains of the twinned were sufficiently small such that no observable relative change in intensity of either reflection was observed when raster scanning the synchrotron x-ray beam over the sample. At DLS the x-ray beam was only of order 100 μm horizontally and 20 μm vertically. As the sample was cooled further, the original high temperature $(0,\frac{1}{2},0)$ disappears, and the 2θ value of the $(\frac{1}{2},0,0)$ reflection increased due to thermal contraction of the sample. At 90 K there was a further transition where a step increase in the wavevector of the orbital order reflection was observed. Further investigation using single crystal high energy x-ray diffraction revealed this to be a structural transition, evident through a change in the lattice parameters [11]. This revealed that the orbital order reflection remains at the commensurate position, and the transition is purely structural, rather than a commensurate to incommensurate transition.

Figure 2 shows the relative intensities of the $(0,\frac{1}{2},0)$ and $(\frac{1}{2},0,0)$ reflections. This demonstrates the simultaneous presence of both orbital configurations between 280 K and 325 K. Although this might suggest the presence of a second order transition, as the crystal is twinned with many small domains, it is quite possible that within each domain the transition is of first order. The inset of Figure 1 shows a schematic of the orbital stripes in the ab plane in both phases. Despite the reduction of intensity previously seen in the orbital signal in bilayer manganites[12], we

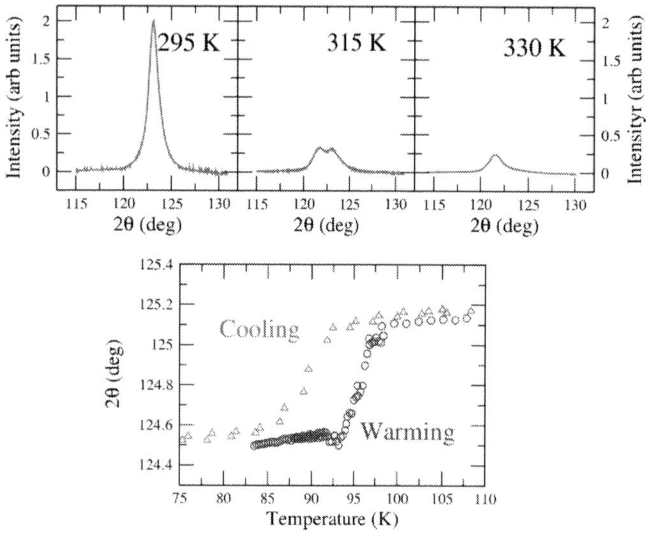

Figure 1. *θ-2θ* scans showing the diffraction signals in the different orbital phases. All scans were measured at 652 eV. The inset shows the change in position of the reflection through the low temperature transition (T_S). The change in position of the reflection between 295 K and 100 K is due to thermal contraction, and reflects a gradual change of lattice parameter, rather than the abrupt change observed at 90 K.

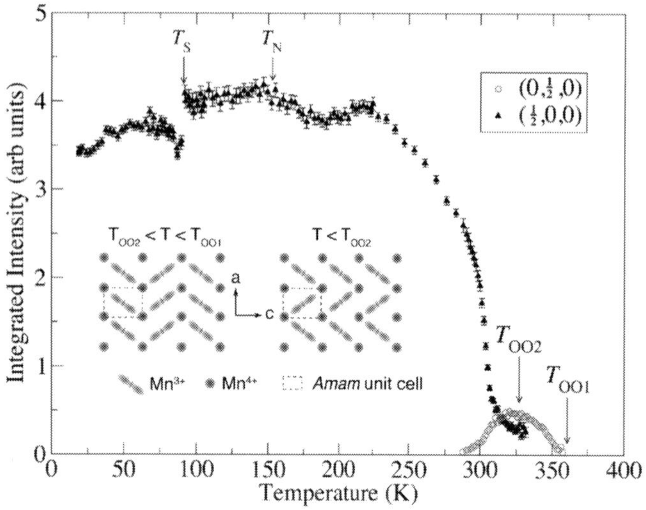

Figure 2. Temperature dependence of the orbital order reflections measured while cooling. The $(0,\frac{1}{2},0)$ reflection appears at T_{OO1}, with the $(\frac{1}{2},0,0)$ reflection forming at T_{OO2}. A further transition is observed at T_S. The orbital schematic insets show the orbital order configuration above and below T_{OO2}. The data was measured at 652 eV at the Mn L_3 edge.

observe no change of intensity at T_N. In addition to the change in position of the of the orbital signal, there is a slight decrease in intensity at T_S.

Although the position of the $(0,\frac{1}{2},0)$ and $(\frac{1}{2},0,0)$ reflections are sufficient to separate the two orbital configurations, it is possible to measure the anisotropy and therefore the direction of the orbital stripes through azimuthal measurements. The sample is rotated around the scattering vector, and the intensity of the reflection is measured while the Bragg condition is maintained. Figure 3 shows the variation of the intensity of these reflections. The expected intensity of the reflection has been calculated using the orbital model shown in the inset of Figure 2, and shows good agreement with the data.

4. Conclusion

Resonant soft x-ray diffraction has been used to observe the orbital transition in $Pr(Sr_{0.1}Ca_{0.9})_2Mn_2O_7$. Using a highly twinned sample the observation of the $(0,\frac{1}{2},0)$ and

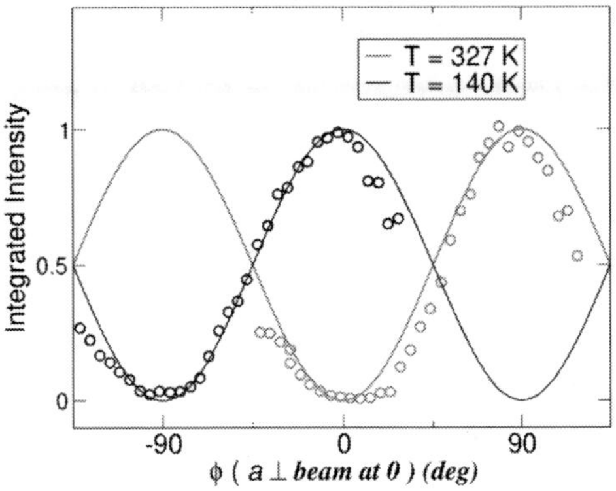

Figure 3. Anisotropy of the ($\frac{1}{2}$,0,0) and (0,$\frac{1}{2}$,0) orbital order reflections, measured by rotating the sample around the scattering vector. The solid line shows the calculated azimuthal dependence. Fitting errors of the peak intensity are within the symbol size.

($\frac{1}{2}$,0,0) has been possible in a single sample. A strong resonant signal was seen, particularly in the low temperature phase, with each orbital reflection showing an azimuthal dependence in agreement with the orbital model. This conclusively demonstrates that orbital order exists in $Pr(Sr_{0.1}Ca_{0.9})_2Mn_2O_7$ and the direction of the orbital chains can be switched through temperature control.

Acknowledgments
The authors would like to thank experimental assistance and stimulating discussion from M von. Zimmermann (BW5, HASYLAB), L. Bouchenoire (XMaS, ESRF), M.D. Roper (Daresbury Laboratory) and A Potenza (I06, Diamond). The authors acknowledge the European Synchrotron Radiation Facility for the provision of synchrotron radiation at beamlines ID08 and BM28, and Diamond Light Source Ltd. for the provision of synchrotron radiation at beamline I06. TAWB, SRB, RDJ, and PDH would like to thank EPSRC and STFC for financial support.

References
[1] Chang C, Hu Z, Wu H, Burnus T, Hollmann N, Benomar M, Lorenz T, Tanaka A, Lin H J, Hsieh H H, Chen C T and Tjeng L H 2009 *Phys. Rev. Lett.* **102** 116401
[2] Paolasini L, Vettier C, de Bergevin F, Yakhou F, Mannix D, Stunault A, Neubeck W, Altarelli M, Fabrizio M, Metcalf P A and Honig J M 1999 *Phys. Rev. Lett.* **82** 4179
[3] Murakami Y, Kawada H, Kawata H, Tanaka M, Arima T, Moritomo Y and Tokura Y 1998 *Phys. Rev. Lett.* **80** 1932
[4] Tokura Y 2003 *Phys. Today* **56** 50
[5] Tokunaga Y, Lottermoser T, Lee Y, Kumai R, Uchida M, Arima T and Tokura Y 2006 *Nature Materials* **5** 937
[6] Tokunaga Y, Kumai R, Takeshita N, Kaneko Y, He J P, Arima T and Tokura Y 2008 *Phys. Rev.* B **78** 155105
[7] Elfimov I S, Anisimov V I and Sawatzky G A 1999 *Phys. Rev. Lett.* **82** 4264
[8] Castleton C W M and Altarelli M 2000 *Phys. Rev.* B **62** 1033
[9] Stojić N, Binggeli N and Altarelli M 2005 *Phys. Rev.* B **72** 104108
[10] Wilkins S B, Spencer P D, Beale T A W, Hatton P D, v Zimmermann M, Brown S D, Prabhakaran D and Boothroyd A T 2003 *Phys. Rev.* B **67** 205110
[11] Beale T A W, Bland S R, Johnson R D, Hatton P D, Cezar J C, Dhesi S S, v Zimmermann M, Prabhakaran D and Boothroyd A T 2009 *Phys. Rev.* B **79** 054433
[12] Wilkins S B, Stojić N, Beale T A W, Binggeli N, Hatton P D, Bencok P, Stanescu S, Mitchell J F, Abbamonte P and Altarelli M 2006 *J. Phys.: Condens. Matter* **18** L323

Polarized Neutrons and Synchrotron X-rays for Magnetism Conference 2009 IOP Publishing
Journal of Physics: Conference Series **211** (2010) 012008 doi:10.1088/1742-6596/211/1/012008

Dipole-quadrupole spherical tensors in resonant x-ray diffraction

Sergio Di Matteo

Equipe de Physique des Surfaces et Interfaces, Institut de Physique de Rennes UMR UR1-CNRS 6251, Université de Rennes 1, F-35042 Rennes Cedex, France

E-mail: `sergio.dimatteo@univ-rennes1.fr`

Abstract. Parity-breaking phenomena in condensed-matter physics, either associated or not to time-reversal-breaking effects, are nowadays considered as an important part of fundamental and applied research, especially in connection to multiferroic materials. Resonant x-ray diffraction is probably the most suited experimental technique to study multiferroic order parameters. In the present note, after some introduction on multipolar order parameters, I provide the polarization and wave-vector dependence of the spherical tensors involved in parity-breaking effects in resonant x-ray diffraction.

1. Introduction

Parity and time-reversal breaking phenomena in condensed matter have recently received a big deal of attention, especially in connection to the possible technological applications of multiferroic materials, eg, in the field of spintronics [1]. However, such materials are very attractive even from a fundamental point of view, because of the interplay between magnetic and electric degrees of freedom, whose microscopic modelization is still far from clear.

In parallel, the research field in resonant x-ray diffraction (RXD) has achieved tremendous progress in the last years. Nowadays RXD is rapidly becoming the crucial technique for investigating the subtleties of microscopic parity and time-reversal symmetry-breaking phenomena in such systems [2].

The possibility to single out magnetic and non-magnetic scattering components through the Bragg diffraction, and to study the polarization dependence of the photon diffracted beam as a function of scattering angles (azimuthal scans) has provided a great amount of experimental data, that in turn stimulated the theoretical interpretation. In particular, the interpretation of the RXD signal in terms of electromagnetic multipoles is, at present, one of the most fashionable ways to classify the experiments. In the light of this interpretation, RXD has been applied to the investigations of many complex materials, such as manganites, actinides, multiferroics and other transition metal oxides, in which different degrees of freedom coexist (orbital, charge, magnetic, dipolar and octupolar, etc.), in such a way as to disentangle the different contributions [2].

The next two sections of the paper are devoted, respectively, to a brief summary of the recent advances in the use of RXD to detect multipolar order parameters (Section 2), and to the detailed expression of the polarization and wave-vector dependence of the RXD signal in the dipole-quadrupole channel (Section 3). Such a dependence had been already implemented in 2003 by me and Yves Joly in Joly's program fdmnes [3] and used several times since then [4, 5, 6].

2. Multipolar expansion in crystals and resonant x-ray diffraction.

The aim of this section is to briefly introduce how it is possible to detect electric and magnetic multipoles by means of resonant x-ray spectroscopies and how to write the one-to-one correspondance of each multipole with the measurable quantities at K edge RXD.

In general the multipole expansion of electric and magnetic fields generated by fixed charges and permanent currents is a widely used tool to characterize the electromagnetic state of a physical system[7]. For example, for a charge distribution $\rho(\vec{x})$ in a given external potential field $\Phi(\vec{x})$, whose electrostatic energy is $W_E = \int d^3x \rho(\vec{x})\Phi(\vec{x})$, one can exploit the well-known expansion: $W_E = q\Phi(0) - \vec{d}\cdot\vec{E} - \frac{1}{6}\sum_{ij} Q_{ij}\frac{\partial E_i}{\partial x_j}(0) + ...$ Here (0) is some properly chosen origin, and $\vec{E} = -\vec{\nabla}\Phi$. Such an expansion shows how the external field couples with the various multipoles of the charge distribution: the potential with the charge q, the electric field with the dipole $\vec{d}$, the gradient of the field with the quadrupole $Q_{ij} \equiv 3x_i x_j - r^2\delta_{ij}$, and so on. A similar expansion holds also for the vector potential. In this case there is no monopole term, of course, and the magnetic energy is: $W_M = \int d^3x\ \vec{J}(\vec{x})\cdot\vec{A}(\vec{x})$. However, due to the vector character of the current density $\vec{J}$ and the potential $\vec{A}$, it is often useful to decompose them, before exploiting the expansion, in a longitudinal (rotor-free) and a transversal (divergenceless) part: $\vec{J} = \vec{J}_{\parallel} + J_{\perp}$. As $\vec{\nabla}\cdot\vec{J} = \vec{\nabla}\cdot\vec{J}_{\parallel} \equiv -\partial_t\rho$, then the longitudinal $\vec{J}_{\parallel}$ is related to the time-derivative of the charge multipoles. On the other side, the solenoidal field $\vec{J}_{\perp}$ is characterized by two independent families of moments, and can be written, in the Helmholtz-Debye representation, as: $\vec{J}_{\perp} = \vec{l}\psi(\vec{x}) + \vec{\nabla}\times(\vec{l}\chi(\vec{x}))$, where $\vec{l}$ is the orbital angular momentum and $\psi(\vec{x})$ and $\chi(\vec{x})$ are a scalar and pseudoscalar function, respectively. The first vector describes the toroidal currents (flowing along the parallels on a sphere), and the second the poloidal currents (flowing along the meridians on a torus). Toroidal currents are widely used, eg, in the physics of electron plasma. Following Ref. [8], we define the magnetic multipoles of the two families in terms of the orbital angular momentum; when also spin quantities play a role, it is sufficient to consider the substitution $\vec{l} \to \vec{l} + g_s\vec{s}$, where g_s is the spin giromagnetic ratio. Then we have:

a) Magnetic moment: $\vec{m} = \frac{e}{2m}\vec{l} = \mu_B\vec{l}$;
b) Magnetic quadrupole: $m_{ij} = \mu_B(x_i l_j + l_i x_j)$;
c) Magnetic (polar) toroidal moment: $\vec{t} = \oint(\vec{x}\times\vec{m})d^3x$;
d) Magnetic toroidal quadrupole: $t_{ij} = x_i t_j + t_i x_j$.

Notice that the magnetic toroidal moment is known in high-energy physics as anapole[9]. A simple geometrical image of $\vec{t}$ is a closed circle of elementary magnets $\vec{m}$ joined to each other. By analogy with this image, one can define a similar quantity[10], by replacing $\vec{m}$ with the electric dipole $\vec{d}$. Then we get:

e) Axial toroidal moment: $\vec{g} = \oint(\vec{x}\times\vec{d})d^3x$;
f) Axial toroidal quadrupole: $g_{ij} = x_i g_j + g_i x_j$.

These moments are scalarly coupled with the corresponding fields, as detailed, eg, in Ref. [8].

How can these order parameters be detected by means of RXD ?

In order to answer this question, we ought to write down the equations used to describe RXD. Resonant x-ray diffraction (or resonant x-ray scattering as it is often termed) is a technique that takes advantage of the sensitivity of usual x-ray diffraction to ordered structure and of the sensitivity of electron resonances to local density of states. It is a core resonant spectroscopies that depends on the virtual processes that allow promoting a core electron to some empty energy levels. All these processes can be described by the transition matrix elements of matter-radiation interaction:

$$M_{ng}^{i(o)}(j) = \langle \psi_n | \hat{O}^{i(o)} | \psi_g(j) \rangle \tag{1}$$

where, in the x-ray regime, the operator $\hat{O}$ is written through the multipolar expansion of the photon field up to electric dipole (E1) and quadrupole (E2) terms [11]:

$$\hat{O}^{i(o)} = \vec{\epsilon}^{i(o)} \cdot \vec{r}\,(1 - \frac{1}{2} i \vec{k}^{i(o)} \cdot \vec{r}) \tag{2}$$

In Eq. (1), $\psi_g(j)$ is the core ground state centered around the jth atom and ψ_n the probed photo-excited state, whereas in Eq. (2), $\vec{r}$ is the electron position measured from the absorbing ion, $\vec{\epsilon}^{i(o)}$ is the polarization of the incoming (outgoing) photon and $\vec{k}^{i(o)}$ its corresponding wave vector. In RXD the global process of photon absorption, virtual photoelectron excitation and photon re-emission, is coherent throughout the crystal, thus giving rise to the usual Bragg diffraction condition:

$$F = \sum_j e^{i\vec{Q}\cdot\vec{R}_j} (f_{0j} + f'_j + i f''_j) \tag{3}$$

Here $\vec{R}_j$ stands for the position of the scattering ion j, $\vec{Q}$ is the diffraction vector and f_0 is the usual Thomson factor. The resonant part, $f'_j + i f''_j \equiv f_j$, is the anomalous atomic scattering factor, given by the expression [12]:

$$f_j(\omega) = \frac{m_e}{\hbar^2} \frac{1}{\hbar\omega} \sum_n \frac{(E_n - E_g)^3 M_{ng}^{o*}(j) M_{ng}^{i}(j)}{\hbar\omega - (E_n - E_g) - i\frac{\Gamma_n}{2}} \tag{4}$$

where $\hbar\omega$ is the photon energy, m_e the electron mass, E_g the ground state energy, and E_n and Γ_n are the energy and inverse lifetime of the excited states. The sum is extended over all the excited states of the system.

The matrix element in Eq. (1) depends only on the electronic part of the operator $\hat{O}$, so that the radiation parameters $\hat{\epsilon}$ and $\vec{k}$ can be factorized out. What one gets is therefore:

$$\begin{aligned} f_j(\omega) &= \sum_{\alpha\beta} \epsilon_\alpha^{o*} \epsilon_\beta^i F_{\alpha\beta}^{DD}(j;\omega) \\ &\quad - \frac{i}{2} \sum_{\alpha\beta\gamma} \epsilon_\alpha^{o*} \epsilon_\beta^i \left(k_\gamma^i F_{\alpha\beta\gamma}^{DQ}(j;\omega) - k_\gamma^o F_{\alpha\beta\gamma}^{QD}(j;\omega) \right) \\ &\quad + \frac{1}{4} \sum_{\alpha\beta\gamma\delta} \epsilon_\alpha^{o*} \epsilon_\beta^i k_\gamma k_\delta F_{\alpha\beta\gamma\delta}^{QQ}(j;\omega) \end{aligned} \tag{5}$$

where we have explicitly written the tensor dependence of the scattering amplitude. The terms $F_{\alpha\beta}^{DD}$, $F_{\alpha\beta\gamma}^{DQ(QD)}$ and $F_{\alpha\beta\gamma\delta}^{QQ}$ represent all what is left of Eq. (4) once the polarization and wave vector components are extracted from the transition matrix elements in the dipole-dipole, dipole-quadrupole and quadrupole-quadrupole channels, respectively, whereas $\alpha\beta\gamma\delta$ represent cartesian components (x, y, z). The interest for such a separation can be easily explained as follows: today polarization detectors in the x-ray range allow a reliable polarization analisys of the diffracted signal. At the same time, the incoming x-rays can also be completely polarized. This allows analysing different cartesian components of the signal by varying polarization and wave-vector directions of incoming and outgoing beams, thereby exploring the intrinsic anisotropies of the charge distribution, what in the earlier literature on the subject [13, 14] was called anisotropic tensor of susceptibilities.

However, it is possible to provide a much deeper physical insight if we re-write Eq. (6) in terms of the scalar product of irreducible tensors [15, 16, 4], instead of using cartesian tensors. Formally one can then re-write Eq. (6) as $f_j(\omega) = \sum_{p,q}(-)^q T_q^{(p)} F_{-q}^{(p)}(j;\omega)$. Here $T_q^{(p)}$ depends only on the incident and scattered polarization and wave vectors, ie, it is the spherical counterpart of the polarization and wave-vector terms in Eq. (6), while $F_q^{(p)}(j;\omega)$ represents the properties of the system under study, ie, it is the spherical counterpart of the F^{DD}, $F^{DQ(QD)}$ and F^{QQ} terms. The advantage obtained with this reformulation is that the rank p of these irreducible tensors represents the order of the multipoles in the electromagnetic field expansion (see, e.g., Refs. [4, 15, 17]). For example, for each p, $F_q^{(p)}(j;\omega)$ is related to a specific term of the multipolar expansion of the system, as shown in Ref. [4]. The allowed p are, in the dipole-dipole channel $p = 0, 1, 2$, in the dipole-quadrupole channel $p = 1, 2, 3$ and in the quadrupole-quadrupole channel $p = 0, 1, 2, 3, 4$. As usual, q is an integer between $-p$ and p. For each p, $F_q^{(p)}(j;\omega)$ is related to a specific term of the multipolar expansion of the system, according to the following table:

tensor	$\hat{T}$	$\hat{P}$	multipole
$F^{(0)}(\mathrm{E1}-\mathrm{E1})$	+	+	electric charge
$F^{(1)}(\mathrm{E1}-\mathrm{E1})$	−	+	magnetic dipole
$F^{(2)}(\mathrm{E1}-\mathrm{E1})$	+	+	electric quadrupole
$F^{(1+)}(\mathrm{E1}-\mathrm{E2})$	+	−	electric dipole
$F^{(2+)}(\mathrm{E1}-\mathrm{E2})$	+	−	axialtoroidalquadrupole
$F^{(3+)}(\mathrm{E1}-\mathrm{E2})$	+	−	electric octupole
$F^{(1-)}(\mathrm{E1}-\mathrm{E2})$	−	−	polar toroidal dipole
$F^{(2-)}(\mathrm{E1}-\mathrm{E2})$	−	−	magnetic quadrupole
$F^{(3-)}(\mathrm{E1}-\mathrm{E2})$	−	−	polar toroidal octupole
$F^{(3)}(\mathrm{E2}-\mathrm{E2})$	−	+	magnetic octupole
$F^{(4)}(\mathrm{E2}-\mathrm{E2})$	+	+	electric hexadecapole

For $p = 0, 1, 2$, E1–E1 and E2–E2 tensors represent the same physical quantities, though referred to states with different angular momentum. The identification illustrated in Table I is dictated by the unique properties under time-reversal ($\hat{T}$) and parity ($\hat{P}$) reflections of $F^{(p)}$ tensors on one side and multipole terms on the other. In fact, for any given tensor rank $p = 1, 2, 3, 4$, there is just one electromagnetic multipole of the same rank (1→dipole, 2→quadrupole, 3→octupole, 4→hexadecapole) with the same $\hat{T}$ and $\hat{P}$ properties. Notice that $\hat{P}$-odd E1–E2 tensors have both $\hat{T}$-odd (-) and $\hat{T}$-even (+) terms for any p ([17]), while $\hat{P}$-even tensors (both E1–E1 and E2–E2) are $\hat{T}$-odd if of odd rank and $\hat{T}$-even if of even rank.

As each $F^{(p)}$ is scalarly related to a corresponding $T^{(p)}$, they have the same $\hat{T}$ and $\hat{P}$ properties. For example, in the E1–E1 or E1–E2 channels, $T^{(0)} \equiv \hat{\epsilon}^{o*} \cdot \hat{\epsilon}^i$ is a scalar and $T^{(1)} \equiv \hat{\epsilon}^{o*} \times \hat{\epsilon}^i$ is a $\hat{T}$-odd and $\hat{P}$-even vector. Analogously, in the E1–E2 channel, the previous quantities combine with the two $\hat{P}$-odd vectors $\vec{k}^i + \vec{k}^o$, which is $\hat{T}$-odd, and $\vec{k}^i - \vec{k}^o$, which is $\hat{T}$-even.

A comment here is required: up to now, we have deliberately avoided the question of whether the local density of the tensors probed by these spectroscopy has anything to do with the properties of the ground state, which we are interested in. At first sight, the presence of the core hole in the final state would prone to give a negative answer to this question, and yet, the wide success of sum rules for magnetic circular dichroism[18] seems to prove the contrary. Moreover, in Ref. [17] in the framework of multiple scattering theory, it was demonstrated that these spectroscopies are sensitive to the expectation values of some tensor operators in the scattering states at the energy of the measurement, so that, by integrating over a specific edge, sum rules can be recovered[19]. However, a definitive answer to this problem appears far from clear, as the introduction of the so-called 'breathing' parameters for 5d radial wave functions in the case of rare-earths $L_{2,3}$-edges [20] stands to indicate (see in particular Refs. [21, 22]). In the following we assume that we may refer to the values of the multipoles in the ground-state, but all these 'warnings' should always be present to the reader.

All the quantities described in Table I are detectable by means of RXD, where the local transition amplitudes are added with a phase factor that can compensate the possible vanishing effect due to the global symmetry. RXD allows us to gain information about the following set of electromagnetic multipoles (the first five are parity-even, E1–E1 or E2–E2, the latters parity-odd, E1–E2):

1) Charge, expressed by $F^{(0)}$. This is usually the strongest contribution. This term allows us to prove (or disprove) the so-called charge-ordering through forbidden diffraction-peaks analysis (see, e.g., Ref. [23]). In fact, any modification of the atomic charge implies a core level shift, and the measure of the energy shift of the threshold allows us to obtain the total valence charge with a good resolution.

2) Magnetic moments (ie, $F^{(1)}(\text{E1}-\text{E1})$) have been treated in a widespread context, due to the big success of the "Carra-Thole-van der Laan" sum-rules[18]. In RXD, this term has been detected several times, for example in V_2O_3 [24] and $KCuF_3$ [25].

3) Electric quadrupole ($F^{(2)}(\text{E1}-\text{E1})$) is the biggest non-magnetic anisotropic term, responsible, e.g., for structural linear dichroism, as well as, in RXD signals, for the anomalous Jahn-Teller and "orbital ordering" peaks in systems like manganites and $KCuF_3$ [25]. It is often called Templeton scattering, from the name of its discoverer [13].

4) Magnetic octupole ($F^{(3)}(\text{E2}-\text{E2})$) was invoked in Ref. [26] to explain the anomalous RXD signal in NpO_2 and was proved to give a contribution in the anomalous magnetic signal in V_2O_3 [24].

5) Electric hexadecapole ($F^{(4)}(\text{E2}-\text{E2})$) was detected in α-haematite by Finkelstein and collaborators, as explained in Ref. [15], and detected in Ref. [24].

6) Electric dipole ($F^{(1+)}(\text{E1}-\text{E2})$) is observable by RXD in antiferroelectric materials, like $ErGe_3$ or $VOMoO_4$, as theoretically predicted in Ref. [27], or experimentally observed in K_2CrO_4 [28].

7) Non-magnetic (axial) toroidal quadrupole was found with XNCD in α-$LiIO_3$,[29] and in V_2O_3 by means of RXD[16].

8) Electric octupole ($F^{(3+)}(\text{E1}-\text{E2})$) has been predicted in Ref. [30] for K_2CrO_4 on the basis of symmetry arguments and detected in Ref. [28].

9-10-11) Finally, magnetic toroidal moment ($F^{(1-)}(\text{E1}-\text{E2})$) and octupole ($F^{(3-)}(\text{E1}-\text{E2})$), together with magnetic quadrupole ($F^{(2-)}(\text{E1}-\text{E2})$) have been found in the anomalous magnetic signal of V_2O_3[24, 31], and, more recently, in the magnetoferroelectric material $GaFeO_3$ [32, 5].

3. Classification of dipole-quadrupole terms: polarization and wave-vector dependence.

This section is devoted to writing down the explicit expression for the polarization and wave-vector dependence of E1-E2 terms in RXD, in order to allow disentangling the different contributions by azimuthal scan in RXD. This is quite simple when the symmetry allows only a few multipoles (as shown, eg, in Ref. [16] for corundum systems at the $(00.3)_h$ reflection where only the electric hexadecapole and axial toroidal quadrupole are present). It might however become very compicate for low-symmetry systems, due to the intrinsic experimental uncertainty in the relative intensity of an experimental azimuthal scan (ie, one should be sure that the sample rotation is ideal in such a way as to probe always the same volume in the sample).

Journal of Physics: Conference Series **211** (2010) 012008 doi:10.1088/1742-6596/211/1/012008

Theoretically, however, the procedure is well defined, once we have the explicit dependence on geometrical conditions (polarization and wave-vector) of each single multipolar term, as given below.

If we write:

$$F^{DQ}_{\alpha\beta\gamma}(j;\omega) = \frac{m_e}{\hbar^2}\frac{1}{\hbar\omega}\sum_n \frac{(E_n - E_g)^3 I_{\alpha\beta\gamma}(j,n)}{\hbar\omega - (E_n - E_g) - i\frac{\Gamma_n}{2}} \tag{6}$$

and

$$F^{QD}_{\alpha\beta\gamma}(j;\omega) = \frac{m_e}{\hbar^2}\frac{1}{\hbar\omega}\sum_n \frac{(E_n - E_g)^3 I^*_{\alpha\beta\gamma}(j,n)}{\hbar\omega - (E_n - E_g) - i\frac{\Gamma_n}{2}} \tag{7}$$

where $I_{\alpha\beta\gamma}(j,n) = \langle\psi_g(j)|r_\alpha|\psi_n\rangle\langle\psi_n|r_\beta r_\gamma|\psi_g(j)\rangle$, then the dipole-quadrupole part of the scattering amplitude in Eq. 6 can be rewritten after some algebraic manipulations as:

$$\begin{aligned} f_j^{DQ}(\omega) &= \frac{i}{2}\sum_n Den(n,\omega,\Gamma)\sum_{\alpha\beta\gamma}\Big((\epsilon^{o*}_\alpha\epsilon^i_\beta k^i_\gamma - \epsilon^{o*}_\beta\epsilon^i_\alpha k^o_\gamma)\mathcal{R}eI_{\alpha\beta\gamma}(j,n) \\ &+ i\,(\epsilon^{o*}_\alpha\epsilon^i_\beta k^i_\gamma + \epsilon^{o*}_\beta\epsilon^i_\alpha k^o_\gamma)\mathcal{I}mI_{\alpha\beta\gamma}(j,n)\Big) \end{aligned} \tag{8}$$

where $Den(n,\omega,\Gamma) = \frac{m_e}{\hbar^2}\frac{1}{\hbar\omega}\frac{(E_n-E_g)^3}{\hbar\omega-(E_n-E_g)-i\frac{\Gamma_n}{2}}$. Finally, Eq. 9 can be rewritten in spherical coordinates, in a form that can be directly related to the multipoles of Table I:

$$f_j^{DQ}(\omega) = \frac{i}{2}\sum_{l=1,2,3}\sum_{m\geq 0}\frac{(-)^m}{2}\left(T^{(l+)}_m F^{(l+)}_m - T^{(l-)}_m F^{(l-)}_m\right) \tag{9}$$

Here the terms $F^{l\pm}$, $l = 1,2,3$, correspond to the physical multipoles of Tables I and the terms $T^{l\pm}$ are the corresponding polarization and wave-vector dependence. The knowledge of these latter terms is therefore useful experimentally in order to single out each single term of the multipolar development.

Consider any cartesian reference frame xyz. In actual experiments, it is usually useful to choose such a frame in such a way that the z direction corresponds to an easy axis of the material, in order to simplify the $F^{(l\pm)}$ tensor, or in such a way that some polarization and/or wave vectors components are zero, in order to simplify the $T^{(l\pm)}$ term. In any of these reference frames, we have the following components for the geometrical x-ray parameters of the experiment:

a) The electric dipole is associated to the following components:

$$\begin{aligned} n_z \sim\ & \frac{3}{2\sqrt{15}}\Big((\epsilon^{o*}_x\epsilon^i_x + \epsilon^{o*}_y\epsilon^i_y)(k^i_z - k^o_z) + \epsilon^{o*}_x\epsilon^i_z k^i_x + \epsilon^{o*}_y\epsilon^i_z k^i_y - \epsilon^{o*}_z\epsilon^i_x k^o_x - \epsilon^{o*}_z\epsilon^i_y k^o_y\Big) \\ & + \frac{1}{\sqrt{15}}\Big(2\epsilon^{o*}_z\epsilon^i_z(k^i_z - k^o_z) + \epsilon^{o*}_x\epsilon^i_z k^o_x + \epsilon^{o*}_y\epsilon^i_z k^o_y - \epsilon^{o*}_z\epsilon^i_x k^i_x - \epsilon^{o*}_z\epsilon^i_y k^i_y\Big) \end{aligned} \tag{10}$$

$$\begin{aligned} n_x \sim\ & \frac{1}{2\sqrt{30}}\Big((4\epsilon^{o*}_x\epsilon^i_x + 3\epsilon^{o*}_y\epsilon^i_y + 3\epsilon^{o*}_z\epsilon^i_z)(k^i_x - k^o_x) - 2(\epsilon^{o*}_x\epsilon^i_y k^i_y + \epsilon^{o*}_x\epsilon^i_z k^i_z - \epsilon^{o*}_y\epsilon^i_x k^o_y - \epsilon^{o*}_z\epsilon^i_x k^o_z) \\ & +3(\epsilon^{o*}_y\epsilon^i_x k^i_y + \epsilon^{o*}_z\epsilon^i_x k^i_z - \epsilon^{o*}_x\epsilon^i_y k^o_y - \epsilon^{o*}_x\epsilon^i_z k^o_z)\Big) \end{aligned} \tag{11}$$

$$\begin{aligned} n_y \sim\ & \frac{1}{2\sqrt{30}}\Big((4\epsilon^{o*}_y\epsilon^i_y + 3\epsilon^{o*}_x\epsilon^i_x + 3\epsilon^{o*}_z\epsilon^i_z)(k^i_y - k^o_y) - 2(\epsilon^{o*}_y\epsilon^i_x k^i_x + \epsilon^{o*}_y\epsilon^i_z k^i_z - \epsilon^{o*}_x\epsilon^i_y k^o_x - \epsilon^{o*}_z\epsilon^i_y k^o_z) \\ & +3(\epsilon^{o*}_x\epsilon^i_y k^i_x + \epsilon^{o*}_z\epsilon^i_y k^i_z - \epsilon^{o*}_y\epsilon^i_x k^o_x - \epsilon^{o*}_y\epsilon^i_z k^o_z)\Big) \end{aligned} \tag{12}$$

b) The polar (magnetic) toroidal moment is associated to the following components:

$$\begin{aligned}\Omega_z \sim & \frac{3}{2\sqrt{15}}\Big((\epsilon_x^{o*}\epsilon_x^i + \epsilon_y^{o*}\epsilon_y^i)(k_z^i + k_z^o) + \epsilon_x^{o*}\epsilon_z^i k_x^i + \epsilon_y^{o*}\epsilon_z^i k_y^i + \epsilon_z^{o*}\epsilon_x^i k_x^o + \epsilon_z^{o*}\epsilon_y^i k_y^o\Big) \\ & + \frac{1}{\sqrt{15}}\Big(2\epsilon_z^{o*}\epsilon_z^i)(k_z^i + k_z^o) - \epsilon_x^{o*}\epsilon_z^i k_x^o - \epsilon_y^{o*}\epsilon_z^i k_y^o - \epsilon_z^{o*}\epsilon_x^i k_x^i - \epsilon_z^{o*}\epsilon_y^i k_y^i\Big) \end{aligned} \tag{13}$$

$$\begin{aligned}\Omega_x \sim & \frac{1}{2\sqrt{30}}\Big((4\epsilon_x^{o*}\epsilon_x^i + 3\epsilon_y^{o*}\epsilon_y^i + 3\epsilon_z^{o*}\epsilon_z^i)(k_x^i + k_x^o) - 2(\epsilon_x^{o*}\epsilon_y^i k_y^i + \epsilon_x^{o*}\epsilon_z^i k_z^i + \epsilon_y^{o*}\epsilon_x^i k_y^o + \epsilon_z^{o*}\epsilon_x^i k_z^o) \\ & +3(\epsilon_y^{o*}\epsilon_x^i k_y^i + \epsilon_z^{o*}\epsilon_x^i k_z^i + \epsilon_x^{o*}\epsilon_y^i k_y^o + \epsilon_x^{o*}\epsilon_z^i k_z^o)\Big) \end{aligned} \tag{14}$$

$$\begin{aligned}\Omega_y \sim & \frac{1}{2\sqrt{30}}\Big((4\epsilon_y^{o*}\epsilon_y^i + 3\epsilon_x^{o*}\epsilon_x^i + 3\epsilon_z^{o*}\epsilon_z^i)(k_y^i + k_y^o) - 2(\epsilon_y^{o*}\epsilon_x^i k_x^i + \epsilon_y^{o*}\epsilon_z^i k_z^i + \epsilon_x^{o*}\epsilon_y^i k_x^o + \epsilon_z^{o*}\epsilon_y^i k_z^o) \\ & +3(\epsilon_x^{o*}\epsilon_y^i k_x^i + \epsilon_z^{o*}\epsilon_y^i k_z^i + \epsilon_y^{o*}\epsilon_x^i k_x^o + \epsilon_y^{o*}\epsilon_z^i k_z^o)\Big) \end{aligned} \tag{15}$$

c) The diagonal component (l=2,m=0) of the axial toroidal quadrupole is associated to the following components of the x-ray tensor:

$$L_z\Omega_z \sim \frac{1}{2}\Big((\epsilon_x^{o*}\epsilon_y^i + \epsilon_y^{o*}\epsilon_x^i)(k_z^i + k_z^o) + \epsilon_x^{o*}\epsilon_z^i k_y^i - \epsilon_y^{o*}\epsilon_z^i k_x^i - \epsilon_z^{o*}\epsilon_x^i k_y^o + \epsilon_z^{o*}\epsilon_y^i k_x^o\Big) \tag{16}$$

d) The diagonal component (l=2,m=0) of the magnetic quadrupole is associated to the following components of the x-ray tensor:

$$n_z\Omega_z \sim \frac{1}{2}\Big((\epsilon_x^{o*}\epsilon_y^i + \epsilon_y^{o*}\epsilon_x^i)(k_z^i - k_z^o) + \epsilon_x^{o*}\epsilon_z^i k_y^i - \epsilon_y^{o*}\epsilon_z^i k_x^i + \epsilon_z^{o*}\epsilon_x^i k_y^o - \epsilon_z^{o*}\epsilon_y^i k_x^o\Big) \tag{17}$$

e) The diagonal component (l=3,m=0) of the electric octupole is associated to the following components of the x-ray tensor:

$$\begin{aligned} n_z(3L_z^2 - L^2) \sim & \frac{1}{\sqrt{10}}\Big((2\epsilon_z^{o*}\epsilon_z^i - \epsilon_x^{o*}\epsilon_x^i - \epsilon_y^{o*}\epsilon_y^i)(k_z^i - k_z^o) \\ & -(\epsilon_x^{o*}\epsilon_z^i + \epsilon_z^{o*}\epsilon_x^i)(k_x^i - k_x^o) - (\epsilon_y^{o*}\epsilon_z^i + \epsilon_z^{o*}\epsilon_y^i)(k_y^i - k_y^o)\Big) \end{aligned} \tag{18}$$

f) The diagonal component (l=3,m=0) of the polar toroidal octupole is associated to the following components of the x-ray tensor:

$$\begin{aligned} \Omega_z(3L_z^2 - L^2) \sim & \frac{1}{\sqrt{10}}\Big((2\epsilon_z^{o*}\epsilon_z^i - \epsilon_x^{o*}\epsilon_x^i - \epsilon_y^{o*}\epsilon_y^i)(k_z^i + k_z^o) \\ & -(\epsilon_x^{o*}\epsilon_z^i + \epsilon_z^{o*}\epsilon_x^i)(k_x^i + k_x^o) - (\epsilon_y^{o*}\epsilon_z^i + \epsilon_z^{o*}\epsilon_y^i)(k_y^i + k_y^o)\Big) \end{aligned} \tag{19}$$

It should be noted that, in general, as shown in Ref. [5], magnetic and non-magnetic components for a given (l,m) are transformed one into the other by the change $\vec{k}^o \to -\vec{k}^o$.

I would like to acknowledge the very stimulating discussions with Yves Joly and C.R. Natoli between 2002 and 2003 that led to the calculations presented here in Section 3.

References

[1] Béa H, Gajek M, Bibes M and Barthélémy A 2008 *J. Phys.: Condens. Mat.* **20** 434221
[2] Santini P, Carretta S, Amoretti G, Caciuffo R, Magnani N and Lander G H 2009 *Rev. Mod. Phys.* **81** 807
[3] Joly Y 2001 *Phys. Rev.* B **63** 125120. The program can be freely downloaded at the web address http:/www-cristallo.grenoble.cnrs.fr/simulation.
[4] Di Matteo S, Joly Y and Natoli C R 2005 *Phys. Rev.* B **72** 144406
[5] Di Matteo S and Joly Y 2006 *Phys. Rev.* B **74** 044625
[6] Mannix D, McMorrow D F, Ewings R A, Boothroyd A T, Prabhakaran D, Joly Y, Janousova B, Mazzoli C, Paolasini L and Wilkins S B 2007 *Phys. Rev.* B **76** 184420
[7] Jackson J D 1975 *Classical Electrodynamics* (New York: Wiley)
[8] Dubovik V M and Tugushev V V 1990 *Phys. Rep.* **187** 145
[9] Zeldovich Ia B 1957 *Zh. Eksp. Teor. Fiz.* **33** 1531 (1958 *Sov. Phys. JETP* **6** 1184)
[10] A complete discussion about polar and axial toroidal moments in solid state physics can be found in Ref. [8].
[11] The magnetic dipole term can be safely neglected in the x-ray range.
[12] Blume M 1994 *Resonant Anomalous X-ray scattering* ed G Materlik, J Sparks and K Fisher (Amsterdam: Elsevier) p 495
[13] Templeton D H and Templeton L K 1980 *Acta Cryst.* **A36** 237; 1982 *Acta Cryst.* **A38** 62
[14] Dmitrienko V E 1983 *Acta Cryst.* **A39** 29
[15] Carra P and Thole B T 1994 *Rev. Mod. Phys.* **66** 1509
[16] Di Matteo S, Joly Y, Bombardi A, Paolasini L, de Bergevin F and Natoli C R 2003 *Phys. Rev. Lett.* **91** 257402
[17] Di Matteo S and Natoli C R 2002 *J. Synchrotron Radiat.* **9** 9
[18] Thole B T, Carra P, Sette F and van der Laan G 1992 *Phys. Rev. Lett.* **68** 1943
[19] For RXD, the presence of the real anomalous term f' adds technical complications related to the inversion of Kramers-Kronig relations.
[20] van Veenendaal M, Goedkoop J B and Thole B T 1997 *Phys. Rev. Lett.* **78** 1162
[21] Kim J W, Lee Y, Wermeille D, Sieve B, Tan L, Bud'ko S L, Law S, Canfield P C, Harmon B N and Goldman A I 2005 *Phys. Rev.* B **72** 064403
[22] Giorgetti Ch, Dartyge E, Baudelet F and Galera R M 2004 *Phys. Rev.* B **70** 035105
[23] Joly Y, Grenier S and Lorenzo J E 2003 *Phys. Rev.* B **68** 104412
[24] Paolasini L, Di Matteo S, Vettier C, de Bergevin F, Sollier A, Neubeck W, Yakhou F, Metcalf P A and Honig J M 2001 *Jour. Elec. Spectr. Rel. Phen.* **120** 1; Paolasini L, Vettier C, de Bergevin F, Yakhou F, Mannix D, Stunault A, Neubeck W, Altarelli M, Fabrizio M, Metcalf P A and Honig J M 1999 *Phys. Rev. Lett.* **82** 4719
[25] Caciuffo R, Paolasini L, Sollier A, Ghigna P, Pavarini E, van den Brink J and Altarelli M 2002 *Phys. Rev.* B **65** 174425
[26] Santini P and Amoretti G 2000 *Phys. Rev. Lett.* **85** 2188
[27] Di Matteo S 2009 *J. Phys.: Conf. Series* **190** 012008
[28] Mazzoli C, Wilkins S B, Di Matteo S, Detlefs B, Detlefs C, Scagnoli V, Paolasini L and Ghigna P 2007 *Phys. Rev.* B **76** 195118
[29] Natoli C R, Brouder Ch, Sainctavit Ph, Goulon J, Goulon-Ginet Ch and Rogalev A. 1998 *Eur. Phys. J.* B **4** 1
[30] Marri I and Carra P 2004 *Phys. Rev.* B **69** 113101
[31] Joly Y, Di Matteo S and Natoli C R 2004 *Phys. Rev.* B **69** 224401
[32] Arima T, Jung J-H, Matsubara M, Kubota M, He J-P, Kaneko Y and Tokura Y 2005 *J. Phys. Soc. Jpn.* **74** 1419

Journal of Physics: Conference Series **211** (2010) 012009 doi:10.1088/1742-6596/211/1/012009

Resonant x-ray scattering and full polarisation analysis of forbidden half-integer reflections in magnetite

S R Bland[1], B Detlefs[2], S B Wilkins[3] T A W Beale[1], C Mazzoli[2], Y Joly[4], P D Hatton[1], J E Lorenzo[4], S D Brown[2] and V A M Brabers[5]

[1] Department of Physics, Durham University, Durham, DH1 3LE, UK
[2] European Synchrotron Radiation Facility, 6 rue Jules Horowitz, BP220, 38043 Grenoble, France
[3] Brookhaven National Laboratory, Condensed Matter Physics & Materials Science Department, Upton, NY, 11973-5000, USA
[4] Laboratoire de Cristallographie, CNRS, BP166, F-38042 Grenoble Cedex, France
[5] Department of Physics, Eindhoven University of Technology, NL-5600, MB Eindhoven, The Netherlands

E-mail: p.d.hatton@durham.ac.uk

Abstract. Magnetite is one of the oldest known magnetic materials, but questions still surround both its crystal and electronic structures at low temperature. The most debated of these low temperature properties regard the presence, or lack of, of charge and orbital order. Using resonant x-ray diffraction at the iron K-edge to probe the long range order present on the iron sites, we have studied $(0\ 0\ \frac{2n+1}{2})_C$ type reflections. By using the technique of full linear polarisation, we have shown that the key features of the reflections can be described merely using the simplified *Pmca* structure, without invoking orbital order.

1. Introduction

Magnetite, Fe_3O_4, is the oldest known magnetic material, but despite its long history, a number of mysteries still surround it. The high temperature system is well understood; magnetite is ferrimagnetic with an inverse spinel structure, with Fe ions sitting at tetrahedral (A) and octahedral (B) co-ordinated sites. Upon cooling below the Verwey transition, $\sim$120 K, the system undergoes a first-order phase transition, accompanied by a change in structure[1], resistivity[2], specific heat[3] and magnetisation[4] properties. Verwey[5, 6] proposed that this transition is also accompanied by a transformation to a charge ordered state. Specifically he proposed an alternate layering along the c-axis of Fe^{2+} and Fe^{3+} ions on the octahedral iron sites. It now appears that this simple explanation is not correct, although it has not stopped magnetite often being described as the classic charge ordered system. It is in this low temperature state where the remaining mysteries lie. Perhaps the most surprising problem is that the low temperature crystal structure is still not fully understood. There have been indications the structure is in fact triclinic[7], however the best refinements to date have been obtained using monoclinic $P2/c$[8] and Cc constraints[9]. However, the further simplified orthorhombic $Pmca$ low temperature structure provides a useful framework, reducing the problem to 4 inequivalent octahedral B sites and 2 inequivalent tetrahedral A sites.

The next most pressing problem regards the charge ordered structure of magnetite. While it has become clear that Verwey's initial model is not correct[10, 11, 12, 13] debate still continues as to the correct arrangement and magnitude of the charge disproportionation (see *e.g.* [14, 15, 16, 11]). An extra level of complexity is then added, when the possibility of orbital order is included[12, 13, 17]. Recently both hard[18] and soft[19] resonant x-ray diffraction experiments have claimed to observe orbital ordering in magnetite. In order to investigate these claims of orbital order we have used the technique of full polarisation analysis[20, 21, 22] to study the origins of these reflections. The most common way to assess such reports is to study the azimuthal dependencies of the reflections, so allowing any anisotropy to become evident. However, this can prove problematic in high quality crystals of magnetite, as the sample is frequently rotated into positions where multiple scattering dominates over the resonant diffraction signal. The solution is to position the sample at an azimuthal position relatively free of multiple scattering, and instead rotate the incident polarisation of the linearly polarised light. This is then followed by post scattering polarisation analysis via the use of the more commonplace analyser crystal.

2. Experimental Results and Discussion

The sample was a synthetic magnetite crystal, produced using the floating zone method. The high quality of the sample was confirmed using heat capacity measurements. These measurements showed the sample to be of high quality, with a purity above $\delta < 0.0007$ (where $Fe_{3(1-\delta)}O_4$), and a Verwey transition of 120.4 $\pm$ 0.3 K [23, 24]. In this paper we have studied the low temperature phase, but index the reflections using the high temperature cubic lattice. Thus $(0\ 0\ 1)_C \equiv (0\ 0\ 2)_O$, where C and O refer to *cubic* and *orthorhombic* respectively.

After mounting the sample on a 0.3 T magnet, in order to fix the monoclinic c-axis, the sample was cooled to 20 K. A scan was conducted at the Fe K-edge along the $[0\ 0\ L]$ direction in reciprocal space in the unrotated σ-σ channel, locating the superlattice reflections. Figure 1, shows reflections at integer and half-integer positions along the c-axis. By contrast in the high temperature structure only the $(0\ 0\ 4)_C$ Bragg reflection and the $(0\ 0\ 2)_C$ ATS[25, 26] (Anisotropic Tensor of Susceptibility) reflection are seen[27, 28]. ATS reflections are forbidden off-resonance due to compound symmetry operations, but at the absorption edge the anisotropic terms in the scattering factors become significant and they can be detected. The $(0\ 0\ \frac{1}{2})_C$, $(0\ 0\ 1)_C$, $(0\ 0\ \frac{9}{2})_C$ and $(0\ 0\ 5)_C$ reflections were all found to resonate close to the iron K-edge;

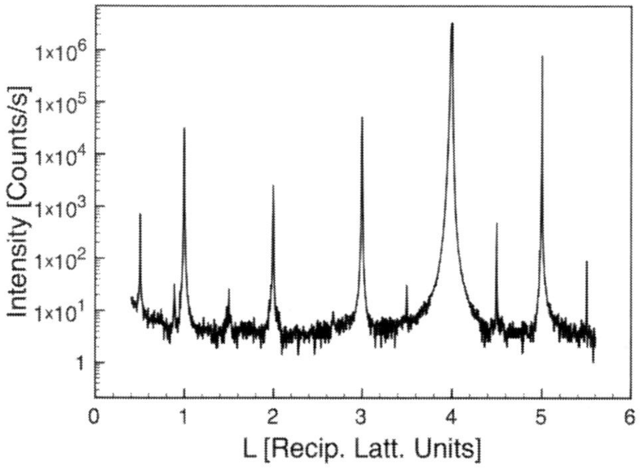

Figure 1. Scan along the L-axis in reciprocal space, at 20 K, with an incident energy of 7.120 keV.

the spectrum of the $(0\ 0\ \frac{9}{2})_C$ reflection is shown in Figure 2. Both half-integer and integer type resonant reflections were found to show the same width, which was of the same order of magnitude as the allowed Bragg reflections.

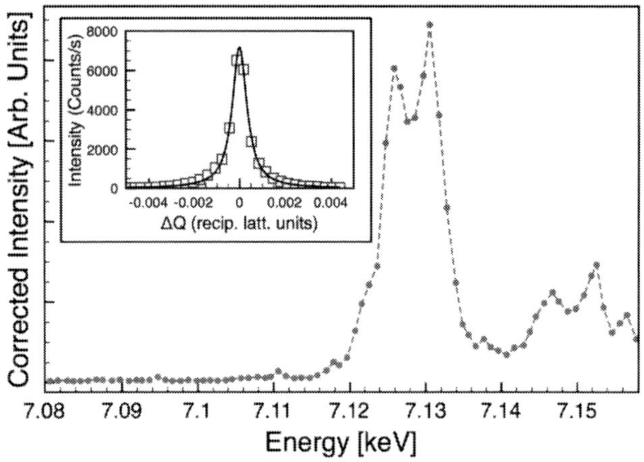

Figure 2. Energy scan of the $(0\ 0\ \frac{9}{2})_C$ reflection at 20 K, measured with fixed wavevector. The scan has been corrected for multiple scattering and absorption effects. Inset: corresponding scan along the L-axis, of the $(0\ 0\ \frac{9}{2})_C$ reflection.

In order to explore the origin of the half-integer type reflections, we then subjected the $(0\ 0\ \frac{9}{2})_C$ reflection to full linear polarisation analysis, the results of which can be seen in Figure 3. Presented in the form of the self-normalising Stokes' Parameters (see *e.g.* [29]), the results demonstrate that the polarisation of the incident light is almost entirely rotated upon scattering. The model shown in Figure 3 represents the expected behaviour of an E1-E1 scattering process for an ATS reflection in the *Pmca* structure, which corresponds to a full rotation of the incident light. In this case the scattering tensor, for all contributing iron sites, is given by

$$\hat{S}_{Pmca} = \begin{pmatrix} 0 & 0 & 0 \\ 0 & 0 & Q_{yz} \\ 0 & Q_{yz} & 0 \end{pmatrix}, \tag{1}$$

in the basis of the *Pmca* structure. We note that while comparing to the Stokes' parameters it is not necessary to consider the absolute value of Q_{yz}. There is clearly a strong agreement with the experimental data, although it is obvious that the model does not fully explain the

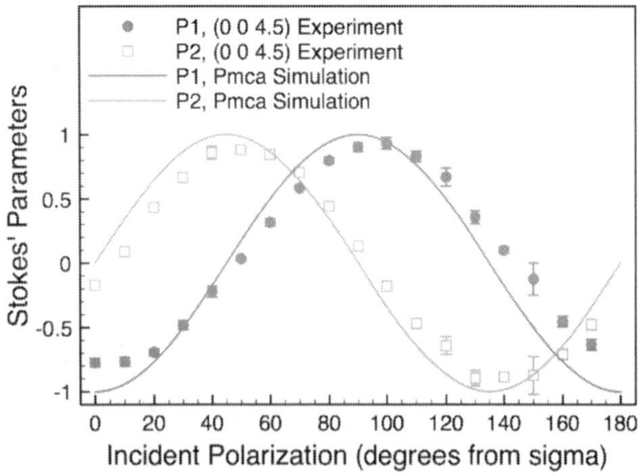

Figure 3. Polarisation dependence of the $(0\ 0\ \frac{9}{2})_C$ reflection at 20 K, tuned to 7.120 keV. The simulations show the expected result for ATS scattering using the *Pmca* structure.

results, as there is a small disagreement with the experiment. There is a slight offset between the simulation and the model, and the experimental results clearly show that, at least initially, the incident light is not fully rotated; indeed, the signal was easily detectable in the unrotated channel in Figures 1 and 2. However, to simulate the ATS reflection in the true low temperature space group goes beyond the scope of this paper, and requires the full crystallographic structure. This remains unknown at present.

Figure 4 shows the temperature dependence of a forbidden half-integer $(0\ 0\ \frac{9}{2})_C$ reflection, as well as the $(0\ 0\ 4)_C$ Bragg reflection, which is allowed in all phases. The transition is clearly first order, and does not appear to show any sign of the continuous transition as measured in other samples [18]. Our temperature dependence of the Bragg reflection also shows a clear first order transition at the same temperature, highlighting the structural change at the transition. Our results in this regard are in agreement with those of García *et al.* [30], who also observed sharp transitions for all reflections. Combined with our full polarisation analysis, it is thus clear that the origin of the $(0\ 0\ \frac{9}{2})_C$ reflection is structural rather than orbital in our high quality sample.

3. Conclusions

The appearance of the $(0\ 0\ \frac{9}{2})_C$ reflection below the Verwey transition can be explained as a consequence of a structural transition to the lower symmetry space group. The appearance of the ATS reflection indicates that there is local anisotropy associated with the iron sites, but this is not a direct indication of orbital order, merely of the structural anisotropy. It is often tempting to attribute ATS reflections (in materials besides magnetite) to orbital order, as they appear at an absorption edge, display the expected anisotropy and appear at the appropriate temperature. Indeed, if the system is orbitally ordered, these reflections are indeed tied to the corresponding electron distribution and associated structural Jahn-Teller distortion. However, the *presence* of an ATS reflection is merely an indication that the scattering factors are neither spherically symmetric or forbidden, whether that be for orbital and/or structural reasons.

In order to definitively explain the results it is necessary to go beyond the *Pmca* approximation, and consider the true low temperature charge ordered structure, and consider the effects of the different scattering factors associated with each of the inequivalent iron sites. It is emphasised that our results do not suggest that magnetite is not orbitally ordered. Rather that the claims of orbital order based on the observation of the $(0\ 0\ \frac{2n+1}{2})_C$ type reflections are not conclusive of orbital order. These reflections arise due to the structural anisotropy of the low-symmetry crystal structure. Further definitive crystallographic studies are needed to

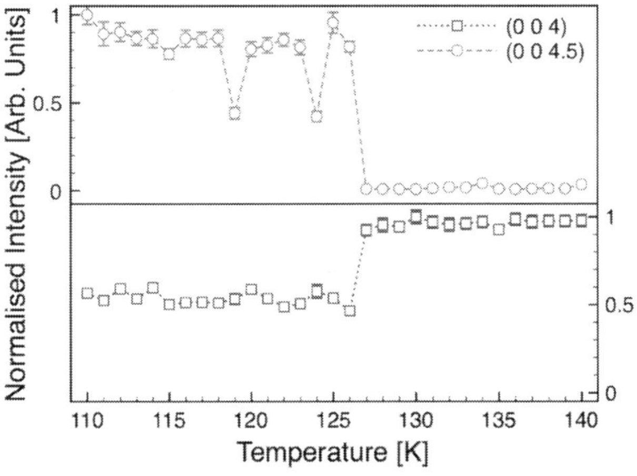

Figure 4. Temperature dependence of the forbidden $(0\ 0\ \frac{9}{2})_C$ and $(0\ 0\ 4)_C$ Bragg reflections. The temperature shown is slightly offset, due to the placement of the temperature sensor. The transition lies at 123.5 K, as revealed by specific heat measurements.

confirm the structure of the low temperature system before conclusive claims of orbital order can be confirmed.

Acknowledgments

The authors wish to thank the ESRF and ID 20 for the beamtime and experimental support. SRB and PDH would like to thank EPSRC for funding. The work at Brookhaven National Laboratory is supported by the Office of Science, U.S. Department of Energy, under contract no. DE-AC02-98CH10886. TAWB and PDH would like to thank STFC for financial support. Part of this work was performed on the EPSRC-funded XMaS beamline at the ESRF, directed by M.J. Cooper and C.A. Lucas. We are grateful to the beam line team of S.D. Brown, P. Normile, O. Bikondoa, L. Bouchenoire and P. Thompson for their invaluable assistance, and to S. Beaufoy and J. Kervin for additional support.

References

[1] Li H 1932 *Phys. Rev.* **40** 1002
[2] Okamura T 1932 *Sci. Rep. Tohoku Imp. Univ.* **21** 231
[3] Millar R W 1929 *J. Am. Chem. Soc.* **51** 215
[4] Weiss P and Forrer R 1929 *Ann. Phys.* **12** 279
[5] Verwey E J W 1939 *Nature* **144** 327
[6] Verwey E J W, Haayman P W and Romeijn F C 1947 *J. Chem. Phys.* **15** 181
[7] Medrano C, Schlenker M, Baruchel J, Espeso J and Miyamoto Y 1999 *Phys. Rev.* B **59** 1185
[8] Wright J P, Attfield J P and Radaelli P G 2002 *Phys. Rev.* B **66** 214422
[9] Joly Y, Lorenzo J E, Nazarenko E, Hodeau J L, Mannix D and Marin C 2008 *Phys. Rev.* B **78** 134110
[10] Leonov I, Yaresko A N, Antonov V N and Anisimov V I 2006 *Phys. Rev.* B **74** 165117
[11] Nazarenko E, Lorenzo J E, Joly Y, Hodeau J L, Mannix D and Marin C 2006 *Phys. Rev. Lett.* **97** 056403
[12] Jeng H T, Guo G Y and Huang D J 2006 *Phys. Rev.* B **74** 195115
[13] Leonov I, Yaresko A N, Antonov V N, Korotin M A and Anisimov V I 2004 *Phys. Rev. Lett.* **93** 146404
[14] Zuo J M, Spence J C H and Petuskey W 1990 *Phys. Rev.* B **42** 8451
[15] Wright J P, Attfield J P and Radaelli P G 2001 *Phys. Rev. Lett.* **87** 266401
[16] Subías G, García J, Blasco J, Proietti M G, Renevier H and Sánchez M C 2004 *Phys. Rev. Lett.* **93** 156408
[17] Uzu H and Tanaka A 2008 *J. Phys. Soc. Jap.* **77** 074711
[18] Lorenzo J E, Mazzoli C, Jaouen N, Detlefs C, Mannix D, Grenier S, Joly Y and Marin C 2008 *Phys. Rev. Lett.* **101** 226401
[19] Schlappa J, Schussler-Langeheine C, Chang C F, Ott H, Tanaka A, Hu Z, Haverkort M W, Schierle E, Weschke E, Kaindl G and Tjeng L H 2008 *Phys. Rev. Lett.* **100**
[20] Detlefs B, Wilkins S B, Caciuffo R, Paixão J A, Kaneko K, Honda F, Metoki N, Bernhoeft N, Rebizant J and Lander G H 2008 *Phys. Rev.* B **77** 024425
[21] Mazzoli C, Wilkins S B, DiMatteo S, Detlefs B, Detlefs C, Scagnoli V, Paolasini L and Ghigna P 2007 *Phys. Rev.* B **76** 195118
[22] Johnson R D, Bland S R, Mazzoli C, Beale T A W, Du C H, Detlefs C, Wilkins S B and Hatton P D 2008 *Phys. Rev.* B **78** 104407
[23] Shepherd J P, Koenitzer J W, Aragón R, Spalek J and Honig J M 1991 *Phys. Rev.* B **43** 8461
[24] Honig J M 1995 *J. Alloys Compd.* **229** 24–39
[25] Dmitrienko V E 1983 *Acta Cryst. A* **39** 29–35
[26] Templeton D H and Templeton L K 1994 *Phys. Rev.* B **49** 14850
[27] García J, Subías G, Proietti M G, Renevier H, Joly Y, Hodeau J L, Blasco J, Sánchez M C and Bérar J F 2000 *Phys. Rev. Lett.* **85** 578
[28] Kanazawa M, Hagaiwara K, Kokubun J and Ishida K 2002 *J. Phys. Soc. Jap.* **71** 1765–1770
[29] Hecht E 2002 *Optics* 4th ed (Addison-Wesley)
[30] García J, Subías G, Herrero-Martín J, Blasco J, Cuartero V, Sánchez M C, Mazzoli C and Yakhou F 2009 *Phys. Rev. Lett.* **102**

Resonant magnetic x-ray diffraction study on the successive metamagnetic transitions of TbB_4 up to 30 T

T Inami[1], K Ohwada[1], Y H Matsuda[2], Z W Ouyang[3], H Nojiri[3], T Matsumura[4], D Okuyama[5,6] and Y Murakami[5]

[1] Synchrotron Radiation Research Unit, Japan Atomic Energy Agency, Sayo 679-5148, Japan
[2] Institute for Solid State Physics, University of Tokyo, Kashiwa 277-8581, Japan
[3] Institute for Materials Research, Tohoku University, 2-1-1 Katahira, Sendai 980-8577, Japan
[4] AdSM, Hiroshima University, 1-3-1 Kagamiyama, Higashi-Hiroshima 739-8530, Japan
[5] Department of Physics, Tohoku University, Aoba, Sendai 980-8578, Japan

E-mail: inami@spring8.or.jp

Abstract. We report the results of resonant magnetic x-ray diffraction experiments in pulsed high magnetic fields on TbB_4, which exhibits successive metamagnetic transitions between 16 and 28 T. The observed field, polarization, and scattering angle dependence of the diffracted intensity indicates that magnetic moments perpendicular to the magnetic field exist in the high-field plateau phases. This result is clearly inconsistent with normal fractional magnetization plateau phases that have magnetic moments only parallel or antiparallel to the field. In order to resolve this discrepancy, we propose magnetic structures consisting of Ising spins and XY spins. A possible mechanism that produces these novel magnetic structures is also discussed.

1. Introduction

Recently, the magnetic properties of rare-earth tetraborides RB_4, which crystallize in a tetragonal structure (space group: $P4/mbm$), have been intensively investigated. In RB_4, the rare-earth ions form a network that is equivalent to a frustrated Shastry-Sutherland (SS) lattice, and thus several pieces of work have been carried out from the viewpoint of frustration, particularly that between rare-earth quadrupole moments [1, 2].

One of the most fascinating experimental results obtained in the study of the RB_4 family is the multistep magnetization curve observed for TbB_4 [3]. When a magnetic field parallel to the c-axis was applied, TbB_4 exhibited 9-fold successive metamagnetic transitions between 16 T and 28 T. The magnetization increased stepwise with increasing magnetic field, and magnetization plateaus were observed between the transition fields, as shown in Fig.1(b). Although similar multiple metamagnetic transitions are sometimes observed in the easy-axis magnetization of Ising magnets such as CeSb and $PrCo_2Si_2$ [4, 5], a qualitative difference exists between these magnets and TbB_4. The magnetic structures of the Ising magnets comprise a complicated sequence of up and down spins, and, basically, the metamagnetic transitions correspond to a reversal of the spins in some sections in the crystal. A neutron powder diffraction experiment at

[6] Present Address: Cross-Correlated Materials Research Group, RIKEN, Wako, Saitama 351-0198, Japan

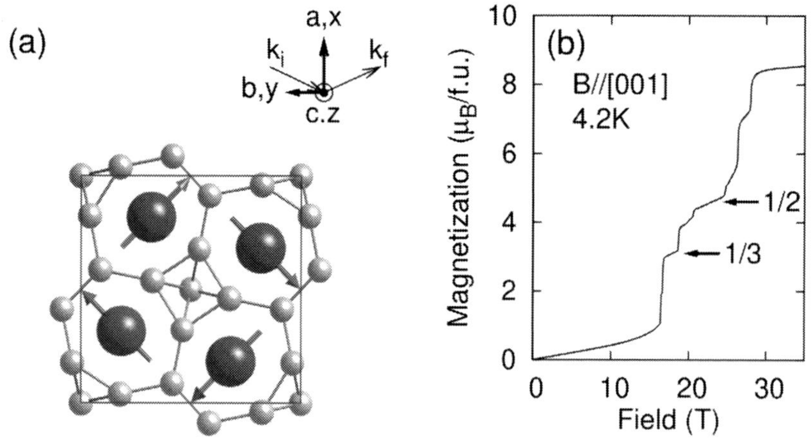

Figure 1. (a) Crystal and magnetic structure of TbB_4 between T_{N1} and T_{N2}. Small and large spheres are B and Tb atoms, respectively. Sample coordination (a, b, c, x, y and z) and incident and scattered x-rays (wave vectors k_i and k_f) are also shown. (b) Magnetization process of TbB_4 taken from Ref. [3].

0 T, however, revealed that the magnetic moments of TbB_4 are confined to the basal ab-plane [6]. Hence, the magnetic moments should simply cant toward the c-axis with increasing magnetic field ($\parallel c$), and a simple linear curve is expected for the c-axis magnetization of TbB_4.

Magnetic structures in the high-field phases provide valuable information on the mechanism that stabilizes a number of magnetization plateaus in TbB_4. In the last five years, x-ray diffraction experiments under pulsed high magnetic field have been intensively developed at several places. Field-induced structural changes have been observed for many samples in magnetic fields over 30 T [7, 8, 9]. In order to examine the magnetic structures of TbB_4 under high magnetic fields, we utilized this technique with the aid of resonance enhancement to increase the small magnetic x-ray scattering cross section. In this paper, we report the recent results of resonant magnetic x-ray diffraction experiments on TbB_4 under pulsed high magnetic fields up to 30 T [10].

Here we briefly summarize the magnetic properties of TbB_4. The Tb ion has the large total angular momentum of J=6, and thus the Tb spin is approximated by a classical spin. Two transition points T_{N1}=44 K and T_{N2}=24 K were reported in susceptibility and resistivity measurements [11]. The magnetic structure between T_{N1} and T_{N2} is a non-collinear coplanar structure [6], as shown in Fig. 1(a). The magnetic moments are of XY-type and lie in the ab-plane. Additional in-plane anisotropy orients the moments parallel to the [110] and [1$\bar{1}$0] directions. The nearest- and next-nearest-neighbor interactions are antiferromagnetic and form a frustrated SS lattice. Below T_{N2}, the magnetic moments rotate by 23° about the c-axis alternately and the magnetic structure becomes an orthorhombic non-collinear structure.

2. Experimental

The resonant x-ray diffraction (RXD) experiments were carried out at BL22XU in SPring-8. The x-ray energy was tuned to the Tb L_{III} absorption edge ($\approx$7.514 keV). Magnetic fields up to 30 T were generated using a small pulsed magnet (20 mm in outer diameter and 24 mm in length). The magnet was a split-pair magnet, and the pulse duration was about 0.6 ms.

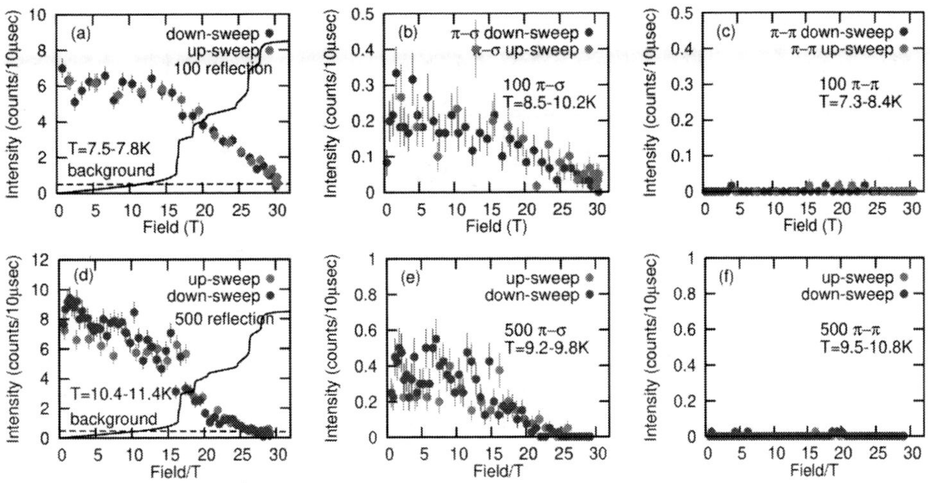

Figure 2. Field dependence of the peak intensity of magnetic reflections below T_{N2}. (a-c) (100) reflection; (a) without polarization analysis, (b) $\pi-\sigma'$ channel and (c) $\pi-\pi'$ channel. (d-f) (500) reflection; (d) without polarization analysis, (e) $\pi-\sigma'$ channel and (f) $\pi-\pi'$ channel. The magnetization curve is also shown in (a) and (d).

The magnet was attached to the first stage of a conventional closed-cycle refrigerator, while the sample was attached to the cold head (second stage) of the cryostat. The cryostat was mounted on a conventional diffractometer with a horizontal scattering plane, and we set the c-axis normal to the scattering plane. Single crystals of TbB_4 were grown by the floating zone method. The sample was a thick plate with dimensions of 1 mm×1 mm×0.5 mm with a polished (100) surface. Magnetic fields were applied parallel to the c-axis. The peak intensity of the (100) and (500) magnetic reflections was monitored as a function of magnetic field. We also separated diffracted x-rays into π-polarization (parallel to the scattering plane) and σ-polarization (normal to the scattering plane) using a PG (006) crystal analyzer and measured their intensities as a function of magnetic field. The polarization of incident x-rays was only π-polarization. A few tens of field scans were necessary to obtain satisfactory statistics. Details of the experimental setup are also found in Refs. [7] and [10].

3. Results and Discussion

The peak intensity of the (100) and (500) reflections below T_{N2} exhibited a huge enhancement at the Tb L_{III} main edge, indicating that the resonance is ascribed to the electric dipole (E1) transition. The diffraction intensity appeared below T_{N1} and increased with decreasing temperature, indicating a magnetic origin of the diffraction intensity [10].

The field dependence of the peak intensity of the (100) and (500) reflections below T_{N2} is shown in Fig. 2. The photon energy was tuned to the peak energy of the energy spectra. The intensity of the (100) reflection without polarization analysis (Fig. 2(a)) is constant up to 16 T and gradually decreases above 16 T as the field is increased. Consistency of the up-sweep and down-sweep data suggests that induction heating of the metallic TbB_4 is negligible. The data of the (500) reflection (Fig. 2(d)) also show a similar tendency, although an abrupt jump is observed at 16 T. This is probably due to the high sensitivity of such large-2θ reflections to the small structural change expected at 16 T.

An important finding in these data is that a considerable amount of intensity remains in the

Table 1. Scattering amplitude of magnetic and ATS scattering components for the $\pi\pi'$ and $\pi\sigma'$ channels.

	m_x	m_y	m_z	O_{xy}	O_{yz}	O_{zx}	O_{22}	O_{20}
$\pi\pi'$	0	0	$i\sin 2\theta$	0	0	0	-1	$\cos 2\theta/\sqrt{3}$
$\pi\sigma'$	$i\sin\theta$	$-i\cos\theta$	0	0	$-\sin\theta$	$\cos\theta$	0	0

magnetization plateau phases above 16 T. Furthermore, polarization analysis clearly indicated that the diffraction intensity of the (100) reflection is dominated by the $\pi-\sigma'$ channel, as shown in Figs. 2(b) and 2(c). For the (500) reflection, although the data above 25 T are not available due to low intensity, we can reach the same conclusion for the diffraction intensity below 25 T.

The resonant x-ray scattering amplitude of rare-earth ions at the E1 transition is given by [12]

$$f_{\mathrm{res}} \propto \alpha_0(\omega)\vec{\epsilon}_{\mathrm{f}}\cdot\vec{\epsilon}_{\mathrm{i}} - i\alpha_1(\omega)(\vec{\epsilon}_{\mathrm{f}}\times\vec{\epsilon}_{\mathrm{i}})\cdot\vec{m} + \alpha_2(\omega)\,\vec{\epsilon}_{\mathrm{f}}\,O\,\vec{\epsilon}_{\mathrm{i}}, \tag{1}$$

where $\vec{m}$ is the magnetic moment and $\vec{\epsilon}_{\mathrm{i}}$ and $\vec{\epsilon}_{\mathrm{f}}$ are the polarization vectors of the incident and scattered x-rays, respectively. The symmetric second-rank tensor O describes the anisotropy of the Tb $5d$ orbital caused by an anisotropic crystal environment or a quadrupole order and is represented by a linear combination of the five elements O_{xy}, O_{yz}, O_{zx}, O_{22} and O_{20}. The first term in Eq. (1) is the conventional anomalous scattering, which does not contribute to the forbidden reflection in this case, and the second term is the magnetic scattering. The last term corresponds to the anisotropy of the tensor of susceptibility (ATS) scattering.

The scattering amplitudes of magnetic and ATS components are listed in table 1 for the geometry of the current experiments (c-axis $\perp$ scattering plane), where θ is the Bragg angle. The Bragg angles of the (100) and (500) reflections are 6.6° and 35.2°, respectively. The absence of the $\pi-\pi'$ process for both high and low scattering-angle reflections signifies that m_z, O_{22} and O_{20} do not contribute to the (h00) reflections (h:odd integer). In addition, a RXD experiment on GdB_4, which has a magnetic structure identical to that of TbB_4 at 0 T, revealed that resonant intensity at (h00) reflections (h:odd integer) arises only from the magnetic moment m_y ($\propto\cos\theta$) for this geometry[13, 14]. The similarity of the field dependence of the (100) and (500) reflections suggests that the θ dependence of the diffraction intensities in the high-field phases and in the low-field phase is the same. We can thus conclude that m_y and/or O_{zx} give rise to the resonant intensity in the high-field phases. Quite recently, a neutron diffraction experiment on TbB_4 under pulsed high magnetic fields succeeded and reported similar field dependence of the magnetic diffraction intensity of the (100) reflection [15]. Hence the pure O_{zx} order is excluded, and we can conclude that the major order parameter in the high-field phases is m_y, that is, the magnetic moment perpendicular to the applied magnetic field.

This result is obviously inconsistent with ordinary magnetization plateau phases, in which the magnetic moments are parallel or antiparallel to the magnetic field. Usually, the coexistence of antiferromagnetic perpendicular components and ferromagnetic parallel components is indicative of a canted antiferromagnetic (CAF) structure. However, the direction of the moments in CAF structures is parallel to the vector sum of the internal and external fields. Hence, a slight increase of the external field always results in a small increase of the parallel component, and magnetization plateaus cannot be observed in CAF structures. Therefore, the perpendicular components and parallel components of the magnetic moments must spatially separate at different sites. We propose magnetic structures that consist of very hard XY-type magnetic moments and Ising-like magnetic moments for the plateau phases in TbB_4. An example of the magnetic structure in the one-third plateau phase is depicted in Fig. 3(b).

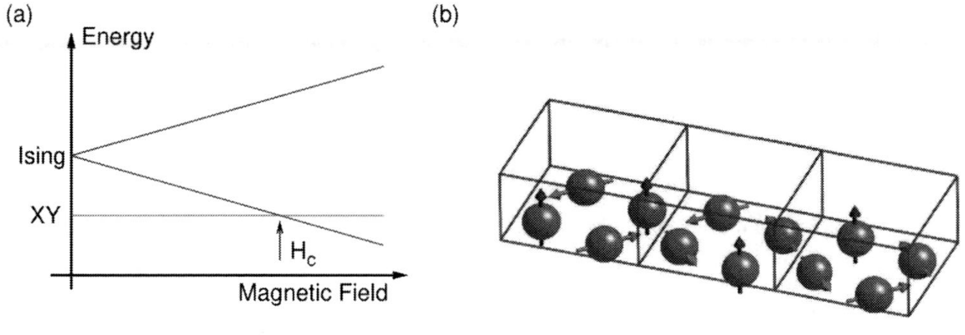

Figure 3. (a) Schematic energy diagram for level crossing. (b) Example of the magnetic structure of the one-third plateau phase. Blue and red arrows, respectively, denote a spin parallel and normal to the c-axis (and magnetic field).

A possible interpretation for these novel magnetic structures is energy level crossing between low-lying multiplets. In Tb ions, the $J = 6$ ground-state is split into several multiplets owing to crystalline electric fields (CEF). If the expected value of J_z in upper levels is larger than that in the lowest levels, level crossing happens at some magnetic field $H_c(\| z)$ (Fig. 3(a)). Such level crossing and accompanying metamagnetic transitions are observed in a number of rare-earth compounds [16, 17]. Recently, Tanaka pointed out that the fourth-order Stevens operator equivalent O_4^0 in the CEF Hamiltonian plays an important role in the magnetic properties of $R\mathrm{B}_4$ compounds [18]. He interpreted major features of the susceptibility and the magnetization of $R\mathrm{B}_4$ (R=Tb, Dy, Ho, Er) by introducing O_4^0 to the spin Hamiltonian. The effect of the usual single-ion type anisotropy $O_2^0 = 3J_z^2 - J(J+1)$ is established. Replacing J_z by $J\cos\Theta$, where Θ is the polar angle, we find that the $B_2^0O_2^0$ has a minimum at $\Theta = 0$ and favors Ising spins when the coefficient B_2^0 is negative. If the B_2^0 is positive, the lowest energy is obtained at $\Theta = \pi/2$ and XY spins are selected. On the other hand, the effect of $O_4^0 = 35J_z^4 - 30J(J+1)J_z^2 + 25J_z^2 - 6J(J+1) + 3J^2(J+1)^2$ is in marked contrast to that of O_2^0. The $B_4^0O_4^0$ has two minima at $\Theta = 0$ and $\pi/2$ when the coefficient B_4^0 is negative, while a single minimum exists at an intermediate angle when B_4^0 is positive. Hence, the negative B_4^0 prefers both XY and Ising spins, and spins with intermediate angles have high energy. In TbB_4, it is most likely that XY-type ground states and Ising-type first excited states are selected by negative B_4^0. It is considered that the level scheme shown in Fig. 3(a) is highly probable.

Normally, level crossing occurs simultaneously at all sites. However, if there are interactions between the excited states, the interactions can split the single metamagnetic transition into successive metamagnetic transitions by stabilizing several particular arrangements of excited Ising states. RKKY-type long-range interactions are expected in metallic TbB_4 and may cause the successive transitions.

4. Summary

We have investigated the high-field magnetization plateau phases of TbB_4 using RXD. The magnetic intensity observed at the (100) and (500) reflections in the plateau phases indicated that large magnetic moments normal to the magnetic field coexist with forced ferromagnetic moments parallel to the magnetic field. We proposed magnetic structures that consist of XY spins and Ising spins for the plateau phases, and also inferred that the origin of the novel magnetic structures and the successive metamagnetic transitions of TbB_4 is energy level crossing between the low-lying multiplets of the Tb ions.

Acknowledgments
This work was supported by Grant-in-Aid for Scientific Research on Priority Areas "High Field Spin Science in 100T" (No.451) from the Ministry of Education, Culture, Sports, Science and Technology (MEXT).

References
[1] Watanuki R, Sato G, Suzuki K, Ishihara M, Yanagisawa T, Nemoto Y and Goto T 2005 *J. Phys. Soc. Jpn.* **74** 2169
[2] Okuyama D, Matsumura T, Nakao H and Murakami Y 2005 *J. Phys. Soc. Jpn.* **74** 2434
[3] Yoshii S, Yamamoto T, Hagiwara M, Michimura S, Shigekawa A, Iga F, Takabatake T and Kindo K 2008 *Phys. Rev. Lett.* **101** 087202
[4] Rossat-Mignod J, Burlet P, Villain J, Bartholin H, Tcheng-Si W, Florence D and Vogt O 1977 *Phys. Rev.* B **16** 440–461
[5] Shigeoka T, Fujii H, Yonenobu K, Sugiyama K and Date M 1989 *J. Phys. Soc. Jpn.* **58** 394
[6] Matsumura T, Okuyama D and Murakami Y 2007 *J. Phys. Soc. Jpn.* **76** 015001
[7] Matsuda Y H, Inami T, Ohwada K, Murata Y, Murakami Y, Nojiri H, Murakami Y, Ohta H, Zhang W and Yoshimura K 2006 *J. Phys. Soc. Jpn.* **75** 024710
[8] Terada N, Narumi Y, Katsumata K, Yamamoto T, Staub U, Kindo K, Hagiwara M, Tanaka Y, Kikkawa A, Toyokawa H, Fukui T, Kanmuri R, Ishikawa T and Kitamura H 2006 *Phys. Rev.* B **74** 180404
[9] Detlefs C, Duc F, Kazeĭ Z A, Vanacken J, Frings P, Bras W, Lorenzo J E, Canfield P C and Rikken G L 2008 *Phys. Rev. Lett.* **100** 056405
[10] Inami T, Ohwada K, Matsuda Y H, Ouyang Z W, Nojiri H, Matsumura T, Okuyama D and Murakami Y 2009 *J. Phys. Soc. Jpn.* **78** 033707
[11] Fisk Z, Maple M B, Johnston D C and Woolf L D 1981 *Solid State Commun.* **39** 1189
[12] Nagao T and Igarashi J 2005 *Phys. Rev.* B **72** 174421
[13] Fernández-Rodríguez J, Blanco J A, Brown P J, Katsumata K, Kikkawa A, Iga F and Michimura S 2005 *Phys. Rev.* B **72** 052407
[14] Ji S, Song C, Koo J, Lee K B, Park Y J, Kim J Y, Park J H, Shin H J, Rhyee J S, Oh B H and Cho B K 2003 *Phys. Rev. Lett.* **91** 257205
[15] Yoshii S, Ohyama K, Kurosawa K, Nojiri H, Matsuda M, Frings P, Duc F, Vignolle B, Rikken G L, Regnault L P, Michimura S and Iga F 2009 *Phys. Rev. Lett.* **103** 077203
[16] Kazeĭ Z A, Snegirev V V, Broto J M, Rakoto H and Kozeeva L P 2006 *JETP Letters* **84** 441
[17] Kirste A, Puhlmann N, Stolpe I, Mueller H U, von Ortenberg M, Tatsenko O M, Platonov V, Kazeĭ Z A, Kolmakova N P and Sidorenko A A 2001 *Physica* B **294-295** 132
[18] Tanaka A 2009 private communication

Journal of Physics: Conference Series **211** (2010) 012011 doi:10.1088/1742-6596/211/1/012011

Neutron diffraction and X-ray magnetic circular dichroism studies of the Co_2SiO_4 magnetic properties

Andrew Sazonov[1,2], **Vladimir Hutanu**[1,2], **Martin Meven**[1], **Gernot Heger**[2], **Arsen Gukasov**[3] **and Andrei Rogalev**[4]

[1] Forschungsneutronenquelle Heinz Maier-Leibnitz (FRM II), Technische Universität München (TUM), D-85747 Garching, Germany
[2] Institut für Kristallographie, Rheinisch-Westfälische Technische Hochschule (RWTH) Aachen, D-52056 Aachen, Germany
[3] Laboratoire Léon Brillouin (LLB), CEA/Saclay, F-91191 Gif-sur-Yvette, France
[4] European Synchrotron Radiation Facility (ESRF), F-38043 Grenoble Cedex, France

E-mail: andrew.sazonov@frm2.tum.de

Abstract. Synthetic Co_2SiO_4 crystallizes in the olivine structure (space group *Pnma*) with two crystallographically non-equivalent Co positions and shows antiferromagnetic ordering below 50 K. We have studied the magnetic properties of Co_2SiO_4 by means of both non-polarized and polarized neutron diffraction as well as X-ray magnetic circular dichroism. Measurements with non-polarized neutrons were made at 2.5 K whereas polarized neutron diffraction experiments were done at 70 K and 150 K in an external magnetic field of 7 T parallel to the *b* axis. Symmetry analysis of the Co_2SiO_4 magnetic structure proves that it corresponds to the magnetic (Shubnikov) group *Pnma* which allows the antiferromagnetic configuration (G_x, C_y, A_z) for the 4*a* site with inversion symmetry $\bar{1}$ (Co1 position) and $(0, C_y, 0)$ for the 4*c* site with mirror symmetry *m* (Co2 position). The polarized neutron study of the magnetization induced by an applied field indicates a delocalization of the magnetic moment from Co towards neighbouring O due to the superexchange coupling. This is also in agreement with X-ray magnetic circular dichroism measurements performed at the Co K-edge in the temperature range 7 K–150 K in an external magnetic field of 6 T applied close to the *b* axis of Co_2SiO_4. An important orbital contribution to the total magnetic moment of Co^{2+} in Co_2SiO_4 predicted by non-polarized neutron diffraction was independently proven by means of the X-ray magnetic circular dichroism and polarized neutron diffraction studies which indicate an orbital to spin magnetic moment ratio $\mu_L/\mu_S \approx 0.3$.

1. Introduction

The general chemical formula for olivine, which is a major phase of the earth's upper mantle, is M_2SiO_4 (M – divalent transition metal). The olivine-type silicates have an orthorhombic crystal structure with the space group *Pnma* (No. 62 according to [1]), in which four formula units are contained in the unit cell ($Z = 4$).

In the crystal structure, the Si atoms are coordinated by four O atoms to form isolated SiO_4 tetrahedra connected via the cations M. These cations are surrounded by six O atoms and occupy two distinct octahedral sites; $M1$ is located on a centre of symmetry (Wyckoff position 4*a*) and $M2$ on a mirror plane (Wyckoff position 4*c*).

The magnetic properties of olivines have been studied a number of times over the past 40 years. The magnetic structure of olivines is found to be quite complex and strongly depends on the type of the cation M.

Below about 50 K, the less studied Co-olivine was found to be magnetically ordered with the magnetic cell equal to the crystallographic cell. It was initially decided that the moments of Co^{2+} on both $M1$ an $M2$ sites are parallel to the b axis and antiferromagnetically coupled [2, 3]. Lottermoser *et al* [4] and Lottermoser and Fuess [5] found a slight deviation of the magnetic moment on $M1$ from the b axis for the whole antiferromagnetic region.

In the present paper the magnetic properties of Co_2SiO_4 were studied in more detail by means of both non-polarized and polarized neutron diffraction as well as X-ray magnetic circular dichroism (XMCD) measurements.

2. Experiment

Crystal growth was performed by the floating-zone method using a mirror furnace (see [6] for more detailed information). For the measurements, small Co_2SiO_4 single crystals were cut from the large as-grown crystal.

To obtain high accuracy in determination of both the nuclear and magnetic structure parameters, a Co_2SiO_4 single crystal was studied on the four-circle diffractometer HEiDi [7] at the hot neutron source of the FRM II reactor, TUM, Germany. Measurements using a short wavelength of 0.552(1) Å up to $\sin\theta/\lambda < 1.1$ Å^{-1} in the temperature range from 300 K down to 2.5 K were performed. Detailed experimental and refinement data can be found elsewhere [6].

The magnetization density distribution for Co_2SiO_4 was obtained from polarized neutron single-crystal diffraction. These measurements were performed at the ORPHÉE reactor, LLB, France. Polarized neutron flipping ratios were measured on the lifting-counter diffractometer 5C1 using neutrons with wavelength $\lambda = 0.845$ Å. Data were collected in an external field of 7 T above the ordering temperature ($T_N \approx 50$ K) at 70 K and 150 K. For both temperatures, one set of flipping ratios was measured with the field applied parallel to the crystallographic b axis of Co_2SiO_4.

The X-ray magnetic circular dichroism (XMCD) measurements were performed at beamline ID12 at the ESRF, France. Co K-edge absorption spectra were obtained from total fluorescence yield measurements using circularly polarized X-rays in the temperature range 7 K–150 K. The XMCD signal was obtained from the difference in absorption for parallel and antiparallel alignments of the helicity of the X-ray beam with respect to an external magnetic field of 6 T applied in a direction close to the b axis of Co_2SiO_4. To avoid experimental artefacts, the external magnetic field direction and the circular polarization of X-rays were alternately flipped.

3. Results and Discussion

3.1. Non-polarized neutron diffraction

The neutron diffraction experiments on the single crystal were performed in order to verify the model of the Co_2SiO_4 magnetic structure suggested in Refs. [4, 5]. A total of 3226 Bragg reflection intensities were measured at 2.5 K. After averaging 1424 independent reflections remained. After applying the criterion $I > 2\sigma(I)$, 1070 observed reflections were used for the nuclear and magnetic structure refinement. Structural parameters of Co_2SiO_4 were obtained from Rietveld refinements using the program FULLPROF [8]. The nuclear and magnetic structures were refined simultaneously.

All observed superstructure reflections were indexed with a magnetic propagation vector $\mathbf{k} = (0, 0, 0)$. The chemical space group of Co_2SiO_4 remains $Pnma$ down to 2.5 K according to our synchrotron and neutron diffraction data [6]. In the framework of the representation analysis, which was developed by Bertaut [9], it was shown that there are only eight possible one-dimensional irreducible representations for $Pnma$ associated with the propagation vector $\mathbf{k} = (0, 0, 0)$. The only irreducible representation consistent with the observed magnetic reflections of

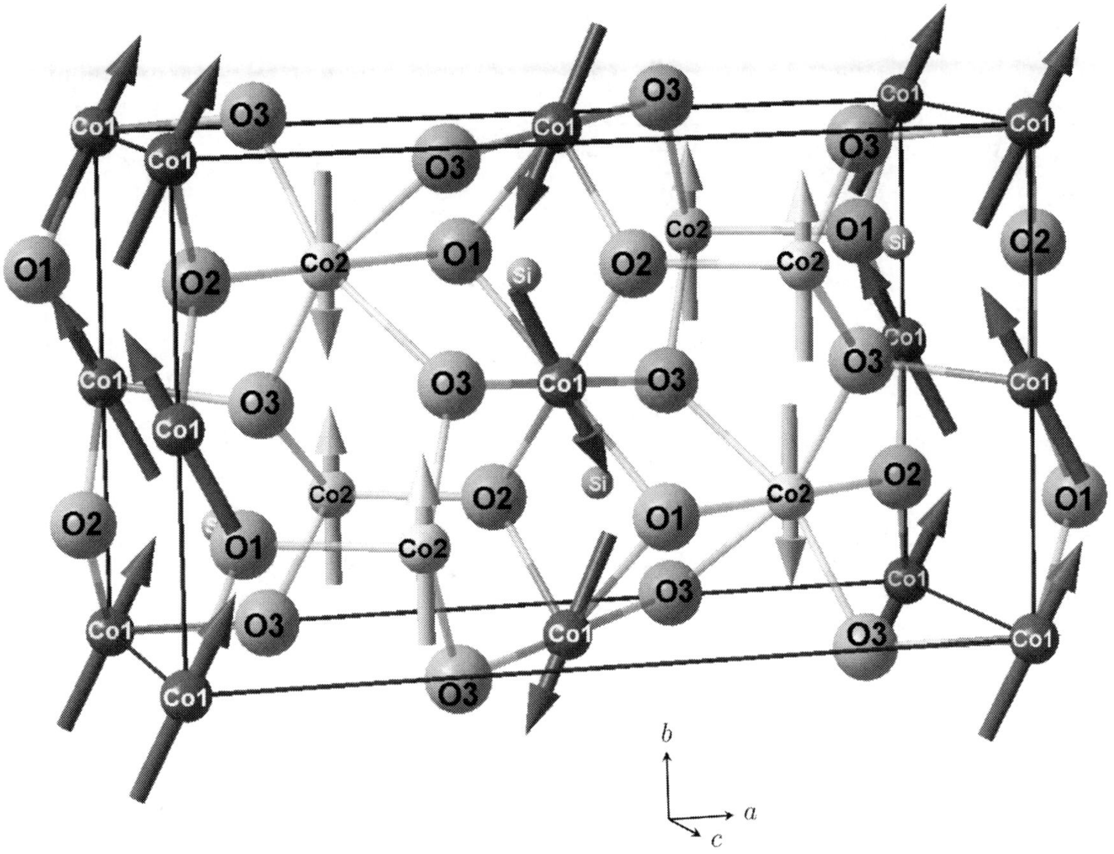

Figure 1. Graphical representation of the crystal and magnetic structures of Co_2SiO_4.

Co_2SiO_4 corresponds to the magnetic (Shubnikov) group *Pnma*. It allows the antiferromagnetic configuration C_y for both 4*a* and 4*c* sites as well as additional antiferromagnetic G_x and A_z components in the case of 4*a*.

The magnetic moments of Co2 have a collinear arrangement along *b*, whereas the Co1 moments are canted with respect to all crystallographic axes (see figure 1). This confirms the magnetic model which was obtained by fitting the neutron diffraction data of Co_2SiO_4 collected at 4.2 K [4, 5]. However, in those works the magnetic moments on both Co sites were found to be similar ($3.90 \pm 0.10\,\mu_B$ and $3.84 \pm 0.08\,\mu_B$ for Co1 and Co2, respectively) in contrast to our results ($3.87 \pm 0.03\,\mu_B$ and $3.35 \pm 0.02\,\mu_B$ for Co1 and Co2, respectively, with $R = 0.033$ and $wR = 0.044$). It is worth noting that the spin-only magnetic moment of Co^{2+} in the high-spin state ($t_{2g}^5 e_g^2$, $S = 3/2$) is $3\mu_B$.

3.2. Polarized neutron diffraction

In order to get further information on the magnetic interactions in Co_2SiO_4, experiments with polarized neutrons were carried out. The classical polarized neutron diffraction technique [10] is used to study the magnetization distribution around magnetic atoms in ferromagnetic and paramagnetic materials. It is much more difficult to make such measurements in antiferromagnetic systems [11]. One way to study the magnetization density distribution in antiferromagnetic compounds is to perform polarized neutron flipping ratio measurements in

Table 1. Induced ferromagnetic moments of Co_2SiO_4 from polarized neutron flipping ratio measurements at 70 K.

Atom	Magnetic moment (μ_B)
Co1: μ_S	0.76 ± 0.02
μ_L	0.24 ± 0.09
Co2: μ_S	0.15 ± 0.02
μ_L	0.04 ± 0.01
O1	0.09 ± 0.02
O2	0.01 ± 0.02
O3	0.04 ± 0.01

$R = 0.045$, $wR = 0.023$

special conditions, namely in the paramagnetic state (above T_N) and in an applied external magnetic field. This gives an information about the magnetization induced by the field. The flipping ratios R measured by polarized neutron diffraction are, for each Bragg reflection, the ratio between the intensity diffracted with spin up (I^+) and the intensity diffracted with spin down (I^-):

$$R = \frac{I^+}{I^-} = \frac{(F_N + \mathbf{F}_M^{\perp})^2}{(F_N - \mathbf{F}_M^{\perp})^2},$$

where F_N is the nuclear structure factor and $\mathbf{F}_M^{\perp}$ is the projection of the magnetic structure factor $\mathbf{F}_M$ into the scattering plane.

For the data collection at 70 K, a total of 207 Bragg reflection flipping ratios for Co_2SiO_4 were measured up to $\sin\theta/\lambda \approx 0.62$ Å^{-1} ($\lambda = 0.845$ Å). These yielded 113 independent reflections. A data collection at 150 K was also performed up to $\sin\theta/\lambda \approx 0.62$ Å^{-1} and a total of 218 Bragg reflection flipping ratios were measured yielding 114 independent reflections. Subsequently, we calculated the magnetic structure factor, assuming that the nuclear one is the structure factor known from our accurate non-polarized single-crystal neutron diffraction measurements [6]. The experimental flipping ratios were used directly in the refinement using the MAGLSQ program from CCSL [12]. The results of the refinement for the measurement at 70 K in a magnetic field of 7 T parallel to the b axis are summarized in table 1.

The Co^{2+} ions in the high-spin state have a relatively strong spin-orbit coupling that makes the orbital angular momentum contribute to the total magnetic moment. Therefore, we tried to refine the orbital part of the Co magnetic moment in addition to the spin one. Thus, the refined average ratio μ_L/μ_S is found to be about 0.3. This ratio can be used now to estimate the spin and orbital components of the Co magnetic moment using the non-polarized neutron diffraction data. The refinement also revealed a non-negligible amount of magnetic moment on the three different oxygen positions. We interpret this effect as a delocalization of magnetic moment from the cations towards neighbouring oxygen ions on superexchange pathways. Polarized neutron diffraction data also suggest that the exchange interactions between the Co1 ions via O are weaker compared to those between Co2. That is, the most important exchange interactions are antiferromagnetic and act through the paths Co2–O–Co2.

3.3. X-ray magnetic circular dichroism

In order to evidence the existence of the orbital contribution to the total magnetic moment of Co^{2+} in Co_2SiO_4, the XMCD experiments were performed at the Co K-edge. The experimental XMCD spectra are shown in figure 2 in comparison with isotropic X-ray absorption near edge

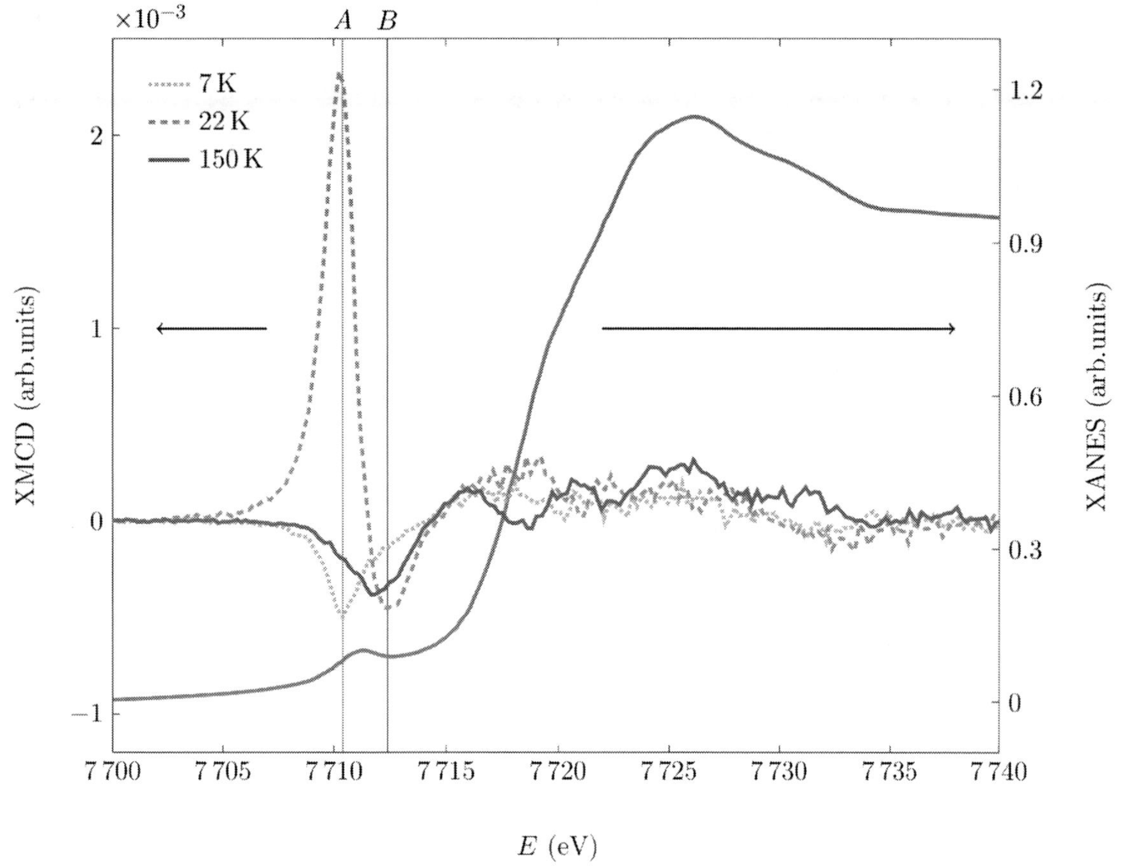

Figure 2. XMCD (left Y-scale) and XANES (right Y-scale) spectra of Co_2SiO_4 measured at the Co K-edge under a magnetic field of ±6 T along the beam (close to the b axis). Only three selected temperatures are shown for simplicity.

structure (XANES) spectra. The isotropic XANES spectra were obtained as an average of the X-ray absorption spectra recorded with left- and right-circular polarization. Characteristic XANES structures which were observed at all measured temperatures give rise to a rather weak pre-edge maximum at 7711 eV and a shoulder at 7720 eV whereas the main absorption edge peaks at 7726 eV.

The observed pre-edge structure is a typical feature of XANES K-edge spectra of $3d$ transition metals with rather localized $3d$-states, and are usually associated with admixtures of $1s \rightarrow 3d$ transitions. While being formally electric-quadrupole transitions which are usually very weak compared to the dipole transition $1s \rightarrow 4p$, a strong electric-dipole character can mix in due to indirect hybridization between $3d$ and $4p$ orbitals on adjacent sites through the ligand states. Appearance of the pre-edge peaks in Co_2SiO_4 can be explained by taking into account a hybridization between the $3d$ states of neighbouring Co ions via the $2p$ state of O. This is in agreement with our polarized neutron diffraction data.

We now consider the XMCD signals shown in figure 2. The prominent part of the Co XMCD consists of a double peak structure with a splitting of about 2 eV, centred around the pre-edge feature in the X-ray absorption spectrum. The pre-edge peaks at various temperatures have almost the same energies of about 7110 eV (peak A) and 7112 eV (peak B). On the other hand, a dramatic change in sign and intensity of these peaks is found with increase of temperature (see figure 2). At the moment, there is no clear understanding of such strong change of the XMCD spectra and additional experiments are required to clarify this phenomenon.

The effective operators responsible for XMCD at the K-edge in both electric dipole and in electric quadrupole transitions have a pure orbital nature [13, 14]. Therefore, the rather intense XMCD signals ($\sim 0.2\%$ with respect to the edge jump) at the pre-edge indicate a strong orbital contribution to the total Co magnetic moment in Co_2SiO_4. However, its quantitative determination from the K-edge XMCD can only be obtained by a combination of experiment and theoretical calculations. As a matter of comparison, the ratio XMCD/XANES of about 0.2% was roughly estimated as the moment of the order of $1\,\mu_B$ per Co^{2+} in the Co-doped TiO_2 film from the Co K-edge measurements [15].

4. Conclusion

Non-polarized neutron diffraction measurements show that the ordered magnetic structure of orthorhombic Co_2SiO_4 corresponds to the magnetic (Shubnikov) group $Pnma$ which allows the antiferromagnetic configuration (G_x,C_y,A_z) for the $4a$ site with inversion symmetry $\bar{1}$ (Co1 position) and (0,C_y,0) for the $4c$ site with mirror symmetry m (Co2 position). The magnetic moments were found to be $3.87 \pm 0.03\,\mu_B$ and $3.35 \pm 0.02\,\mu_B$ for Co1 and Co2, respectively. Both Co1 and Co2 moments exceed the spin-only value of $3\mu_B$ for Co^{2+} in the high-spin state. Compared to literature data, an excellent agreement between the magnetic moments of Co1 is seen, whereas the values for Co2 are different [5].

An important orbital contribution to the total magnetic moment of Co^{2+} in Co_2SiO_4 was proven by the X-ray magnetic circular dichroism as well as polarized neutron diffraction studies, which indicate an orbital to spin magnetic moment ratio $\mu_L/\mu_S \approx 0.3$. According to this, the excess magnetic moment of Co^{2+} above the theoretical expectation of the spin-only value of $3\mu_B$ is caused by the orbital part. Both XMCD and polarized neutron flipping ratio measurements also suggest a hybridization between the $3d$ states of neighbouring Co ions via $2p$ state of O due to the superexchange interaction.

Acknowledgments

This work was partially supported by the German Federal Ministry for Education and Science (BMBF project 03HE7AAC).

References

[1] Hahn T (ed) 1995 *International tables for crystallography, Vol. A: Space group symmetry* (London: Kluwer Academic Publishers)
[2] Nomura S, Santoro R, Fang J and Newnham R 1964 *J. Phys. Chem. Solids* **25** 901–905
[3] Santoro R P, Newnham R E and Nomura S 1966 *J. Phys. Chem. Solids* **27** 655–666
[4] Lottermoser W, Müller R and Fuess H 1986 *J. Magn. Magn. Mater.* **54-57** 1005–1006
[5] Lottermoser W and Fuess H 1988 *Phys. Status Solidi* A **109** 589–595
[6] Sazonov A, Meven M, Hutanu V, Kaiser V, Heger G, Trots D and Merz M 2008 *Acta Cryst.* B **64** 661–668
[7] Meven M, Hutanu V and Heger G 2007 *Neutron News* **18** 19–21
[8] Rodríguez-Carvajal J 1993 *Physica* B **192** 55–69
[9] Bertaut E F 1968 *Acta Cryst.* A **24** 217–231
[10] Nathans R, Shull C G, Shirane G and Andresen A 1959 *J. Phys. Chem. Solids* **10** 138
[11] Brown P J, Forsyth J B and Tasset F 1999 *Physica* B **267-268** 215–220
[12] Brown P J and Matthewman J C 2008 *The Cambridge crystallography subroutine library: MARK-4.4 users' manual* (http://www.ill.eu/sites/ccsl/html/ccsldoc.html)
[13] Goulon J, Rogalev A, Wilhelm F, Jaouen N, Goulon-Ginet C, Goujon G, Youssef J B and Indenbom M V 2007 *J. Electron Spectr. Rel. Phenom.* **156-158** 38–44
[14] Carra P, König H, Thole B T and Altarelli M 1993 *Physica* B **192** 182–190
[15] Shimizu N, Sasaki S, Hanashima T, Yamawaki K, Udagawa H, Kawamura N, Suzuki M, Maruyama H, Murakami M, Matsumoto Y and Koinuma H 2004 *J. Phys. Soc. Jpn.* **73** 800–803

Journal of Physics: Conference Series **211** (2010) 012012 doi:10.1088/1742-6596/211/1/012012

Principal interactions in the helimagnetic system $Mn_{1-y}Fe_ySi$

V A Dyadkin[1], S V Grigoriev[1], D Lamago[2,3], Th Wolf[3], E V Moskvin[1,4], S V Maleyev[1] and H Eckerlebe[5]

[1]Petersburg Nuclear Physics Institute, Gatchina, St. Petersburg, 188300 Russia
[2]Laboratoire Léon Brillouin, CEA Saclay, 91191 Gif-Sur Yvette, France
[3]Institut für Festkorperphysik, Forschungszentrum Karlsruhe, 76021 Karlsruhe, Germany
[4]Helmholtz-Zentrum Berlin, 14109 Berlin, Germany
[5]GKSS Forschungszentrum, 21502 Geesthacht, Germany

E-mail: `dyadkin@lns.pnpi.spb.ru`

Abstract. The magnetic structure of the noncentrosymmetrical cubic ferromagnets with Dzyaloshinskii-Moriya interaction $Mn_{1-y}Fe_ySi$ with $y = 0$, 0.06, 0.08, and 0.1 has been studied by means of the small angle neutron diffraction and magnetization measurements. These compounds are shown to be ordered into the helix structure below the critical temperature T_C that decreases linearly with the Fe doping and approaches zero at $y \approx 0.13$. We build the $H-T$ (magnetic field – temperature) phase diagrams for each compound and interpret them taking into account the Bak-Jensen hierarchical model of the principal interactions. Following the concentration evolution of these principal interactions (such as the spin wave stiffness A, the Dzyaloshinskii constant D, and the gap Δ in spin wave spectrum) we show that A resembles behavior of T_C approaching zero at $y \approx 0.15$. Thus, it is the isotropic ferromagnetic exchange interaction that plays a dominating role in determination of the critical temperature T_C of the compounds under study. The Dzyaloshinskii constant D has been found to be the same either in the system under study and in the related compounds $Fe_{1-x}Co_xSi$.

The magnetic and transport properties of MnSi, the noncentrosymmetrical cubic magnet with the space group $P2_13$, have been a subject of intensive investigations during the last forty years due to their peculiar and intriguing properties. Its magnetic structure orders below $T_c = 29$ K in a left-handed spiral along the $\langle 111 \rangle$ directions with a propagation vector $k = 0.36$ nm^{-1} at $T = 4$ K [1, 2, 3]. The spin helicity had been understood as a result of the equilibrium between the isotropic ferromagnetic (IF) exchange interaction and the antisymmetric Dzyaloshinskii-Moriya (DM) exchange interaction caused by lack of a center of symmetry in Mn atomic arrangement [4, 5, 6, 7]. These two interactions are considered isotropic themselves but another weak anisotropic exchange (AE) interaction fixes a direction of the helix along one of the cube diagonals [5].

The magnetic structure under an applied magnetic field for MnSi [8, 9] and the similar system $Fe_{1-x}Co_xSi$ [10, 11] was recently studied and interpreted within the theory [12] which takes into account the interactions mentioned above. The neutron diffraction experiments enable to provide an important parameters of the magnetic system: the wavevector of the structure $\mathbf{k}$, the critical field of the transformation to the ferromagnetic phase H_{C2} and appearence of the A-phase in the $H-T$ phase diagram of this compound. (A-phase is the phenomenon of the 90°-rotation of the helix wave vector **k** from the direction parallel to the field axis to the direction perpendicular to it in a small pocket of the $H-T$ phase diagram near T_C in the narrow range of the fields $g\mu_B H_{\rm gap} \sim \sqrt{2}\Delta$. This so-called **k**-flop phase will be minutely considered below in the text).

The Author of Ref. [12] developed the microscopic theory for the cubic magnets without center of symmetry under magnetic field. The ground state energy and the spin wave spectrum were evaluated showing, in accord with Ref. [5], that the system finds its equilibrium in the helical ordering with the helix wave vector

$$k = SD/A. \tag{1}$$

Here D is the strength of DM interaction, A is the spin wave stiffness at a distance much shorter than the helix period (it is a quantitative characteristic of the IF interaction), and S is the average spin per a unit cell. The structure was also shown to be intrinsically unstable with respect to the small magnetic field applied perpendicular to **k** unless it is stabilized via a small gap in the spin wave spectrum Δ which is caused by the DM interaction. Luckily for these systems, the spin wave interaction in presence of the DM interaction guarantees enhancement of a small but positive gap of order of $\Delta^2 \sim (D^2/2A)^2$. The existence of this gap can also provide an explanation for the appearance of A-phase [8]. Vise versa, the existence of the A-phase can be considered as an indirect proof for existence of the spin wave gap predicted by the theory [12].

The theory [12] is able to relate the spin wave stiffness A and the spin wave gap Δ to the critical fields H_{C2} and $H_{\rm gap}$, respectively, through the following expressions

$$g\mu_B H_{C2} = Ak^2 \tag{2}$$

$$\Delta \simeq g\mu_B H_{\rm gap}/\sqrt{2}. \tag{3}$$

Combining these expression with the definition of k (Eq. (1)) one is able to estimate the principal interactions of these systems from the parameters of the magnetic structure measured in the experiment (k, $H_{\rm gap}$, and H_{C2}).

The present paper is aimed to follow the change of the principal interactions in the $Mn_{1-y}Fe_ySi$ compounds. The single crystals with the concentrations y = 0, 0.06, 0.08, and 0.1 were chosen for the study. These samples were the discs with a thickness of 3 mm and a diameter of 30 mm. The single crystals were grown using the Bridgmann method at the Forschungszentrum Karlsruhe in Germany. The crystal mosaic is about 0.5 degree. The structure and the homogeneity of the crystals were controlled by the x-ray Laue diffraction at Helmzoltz-Zentrum Berlin. All these compounds are found to have the same structure of the space group $P2_13$.

The polarized SANS experiments were carried out at the SANS-2 scattering facility of the FRG-1 research reactor in Geesthacht (Germany). A polarized neutron beam with an initial polarization of $P_0 = 0.93$, the neutron wavelength $\lambda = 0.58$ nm ($\Delta\lambda/\lambda = 0.1$), and a divergence of 2.5 mrad was used. The scattered neutrons were detected by a position sensitive detector with 256×256 pixels. The detector-sample distance was set such that the q range was covered from $6 \cdot 10^{-2}$ to 1 nm^{-1} with a step of 0.01 nm^{-1}. We used the magnetic field which was ranged from 0 to 800 mT and directed along [111] axis. The scattering intensity was measured in the temperature range $8 \leqslant T \leqslant 60$ K with an accuracy better than 0.05 K. The magnetization was measured with the Quantum Design physical properties measurement system at HZB.

The SANS map for the sample with y = 0.06 at T = 8 K (that is well below T_C) for the neutron polarization $\mathbf{P_0}$ along ("up") the guide field is shown in Fig. 1. This map demonstrates a typical magnetic Bragg scattering from the helical spin structure. The peaks in a range of small-angle scattering are visible, proved the fact that the Bragg condition is fulfilled:

Journal of Physics: Conference Series **211** (2010) 012012 doi:10.1088/1742-6596/211/1/012012

$2d\sin(\Theta_B/2) = \lambda$ where $d \gg \lambda$, with d as the period of the helix and Θ_B as the scattering angle. The helix wave vector **k** of the doped samples has an ill-defined direction since the Bragg peaks are smeared around either $\langle 111\rangle$ or $\langle 100\rangle$ axes. This ill-defined orientation of the wave vector **k** is caused by a disorder induced by Fe doping into the pure MnSi compound. The doping can be supposed to induce randomness in the local anisotropy and, therefore, unpinns partially the directions of spirals. The helix wavevector of the pure MnSi is clearly oriented along the $\langle 111\rangle$ axes only with the magnetic mosaic of order of 2–3 degrees [9].

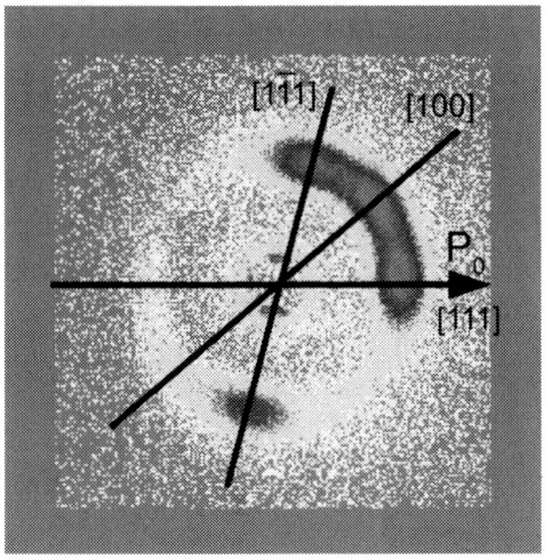

Figure 1. SANS map of $Mn_{0.94}Fe_{0.06}Si$ at $T = 8$ K.

The transformations of the magnetic system under applied magnetic field are studied and have found to be very typical for all compounds. Figure 2 shows magnetization and susceptibility curves of the sample $Mn_{0.94}Fe_{0.06}Si$ taken at $T = 2$ K. The magnetization changes the slope from γ_1 to γ_2 in the range of small field at H_{C1}. Further on, the magnetization demonstrates the linear dependence in the range from H_{C1} to H_{C2}. It saturates above the critical field H_{C2} indicating the field-induced phase transition from the conical to the ferromagnetic state. The susceptibility shows clearly the features at these two critical fields H_{C1} and H_{C2}.

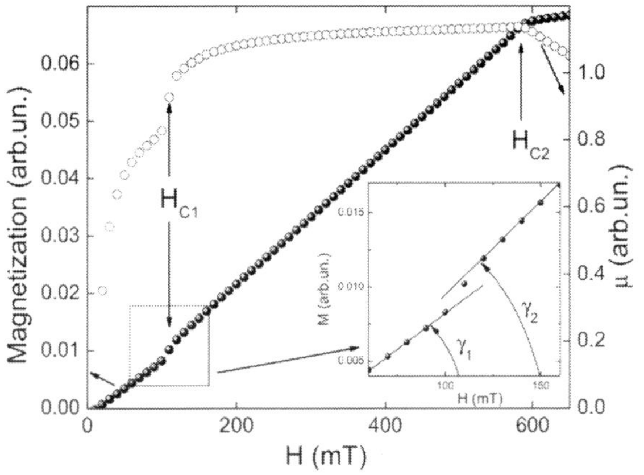

Figure 2. Magnetic field dependence of the magnetic susceptibility μ (∘) and the magnetization (•) of $Mn_{0.94}Fe_{0.06}Si$ at $T = 2$ K. Inset shows the crossover at $H = H_{C1}$ in more details.

The small angle diffraction measurements show that in absence of the magnetic field the magnetic structure consists of spiral domains with ill-defined orientation of the spiral wavevector **k**. An external magnetic field provides the process of reorientation of spiral domain along the field axis. This process starts from the threshold field H_{C1} defined as a field which dominates over the local anisotropy and forms in the sample the single domain state with conical spirals oriented along the field. In the scattering picture, it is seen as an accomulation of the different Bragg spots to the single one with **k** collinear to the magnetic field axis. Further increase of the magnetic field destroys the spiral order and forms the ferromagnetic state at $H = H_{C2}$, where the Bragg spot vanishes. The system remains a stable conical structure in the range $H_{C1} < H < H_{C2}$. The critical fields H_{C1} and H_{C2} are plotted as a function of temperature in the $H - T$ phase diagram (Fig.3) for the representative compound with $y = 0.06$.

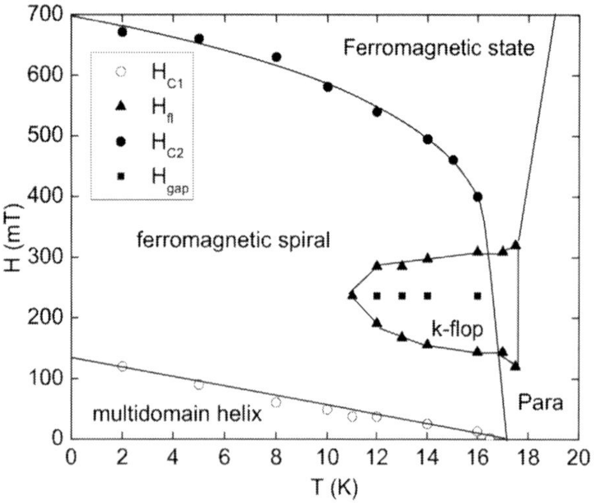

Figure 3. $H - T$ phase diagram of $Mn_{0.94}Fe_{0.06}Si$.

The important feature of the $H-T$ phase diagram (Fig. 3) is a **k**-flop of the helix wave vector that occurs in a certain field range close to T_C. The **k**-flop is seen in the diffraction experiment as a 90° jump of the wavevector from $\mathbf{k} \parallel \mathbf{H}$ to $\mathbf{k} \perp \mathbf{H}$. Going deeper into the k-flop phase, the intensity of the Bragg reflection at $\mathbf{q} = \mathbf{k} \parallel \mathbf{H}$ decreases, while a new Bragg spot appears at $\mathbf{k} \perp \mathbf{H}$. The field H_{fl} in Fig. 3 (triangles) shows the boundary of the new **k**-flop phase. The field value associated with the minimum of the intensity for $\mathbf{k} \parallel \mathbf{H}$ is denoted as H_{gap}.

The $H - T$ phase diagram shown in Fig. 3 is typical for the cubic helimagnets including the ternary compounds under study $Mn_{1-y}Fe_ySi$ and the compounds $Fe_{1-x}Co_xSi$ studied in [10, 11]. The similarity of the $H - T$ phase diagrams of different compounds suggests that the magnetic system of these compounds are governed by the same set of the interactions which interplay determines the values of the critical fields H_{C2} and H_{gap} as well as the value of the helix wavevector k (see Eqs. (1 – 3)). The concentration dependences of these three parameters are shown in Fig. 4. Here H_{C2} is determined as a value taken at the lowest measured temperature. As seen from Fig. 4 values of H_{C2} and H_{gap} demonstrate rather small alteration. The value of the helix wavevector $|\mathbf{k}| = k = 2\pi/d$ changes significantly showing a linear increase from 0.35 to 0.63 nm^{-1} with doping y, i.e. the helix length shortens from 185 to 100 Å.

Taking into account the theory [12] and using Eqs. (1 - 3) we have calculated the changes of the major driving interactions of the magnetic system with doping y: the spin wave stiffness A, the Dzyaloshinskii constant SD, and the spin wave gap Δ. This theory was successfully applied to the pure MnSi [9] and to the ternary compounds $Fe_{1-x}Co_xSi$ [10].

The calculated energies of the principal interactions A/a^2, SD/a, and Δ are shown in Fig. 5 as a function of y (here a is the lattice parameter). The exchange energy A/a^2 decreases linearly from 2.4 to 1 meV. The extrapolation (red doted line in Fig. 5) gives the critical concentration $y \approx 0.15$ where the exchange interaction becomes zero. The energy of DM interaction behaves similarly and has a moderated decrease from 0.38 to 0.25 meV. Opposite to these two, the spin wave gap Δ linearly increases from 12 to 32 μeV (see the inset in Fig. 5).

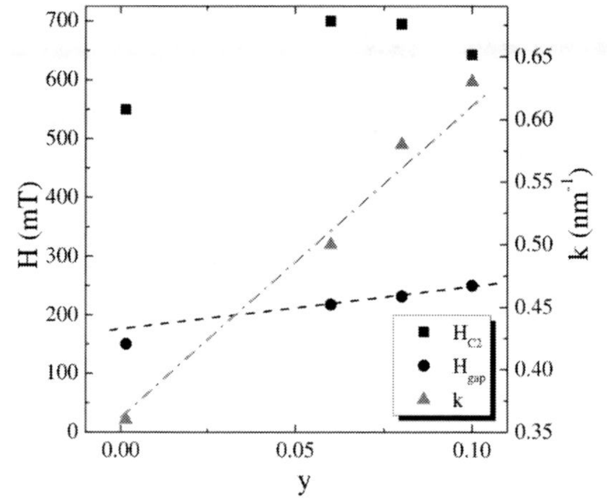

Figure 4. Concentration dependences of the following principal parameters of $Mn_{1-y}Fe_ySi$: (■) the second critical field H_{C2}; (•) the **k**-flop field $H_{\rm gap}$; (▲) the value of the helix wave vector $|\mathbf{k}|$.

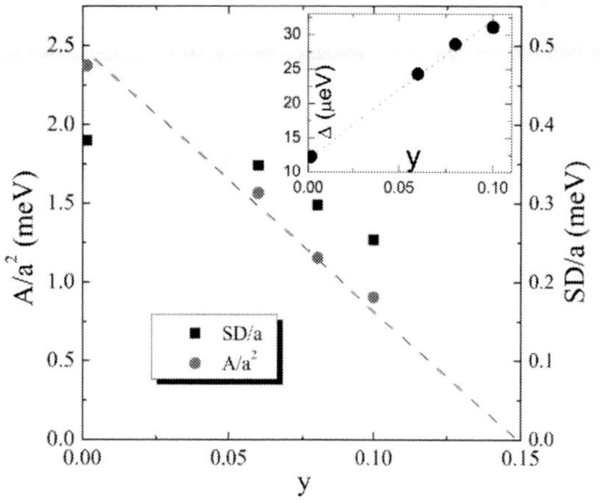

Figure 5. Concentration dependences of the following principal interactions of $Mn_{1-y}Fe_ySi$: (•) the spin wave stiffness energy A/a^2; (■) the Dzyaloshinskii energy. Inset shows the concentration dependence of the spin wave gap Δ.

Remarkable though that this set of the principal interactions is able to explain the concentration dependence of the critical temperature T_C plotted in Fig. 6. The critical temperature T_C has been obtained from the neutron scattering measurements as the temperature, where the magnetic Bragg reflection transforms from Gaussian to Lorenzian. The critical temperature of this transformation was determined $T_C = 28.80(5)$, $16.50(5)$, $10.55(5)$, and $6.80(5)$ K for the samples with $y = 0.0, 0.06, 0.08$, and 0.10, respectively. T_C is well seen to decrease linearly with the Fe concentration y in agreement with the early studies (see Ref. [13] and references therein). Comparing concentration dependences of T_C (Fig. 6) and A (Fig. 5) one can see that the critical temperature resembles the behavior of the spin wave stiffness. This fact leads to the conclusion that the isotropic ferromagnetic exchange interaction governes the transition temperature in $Mn_{1-y}Fe_ySi$.

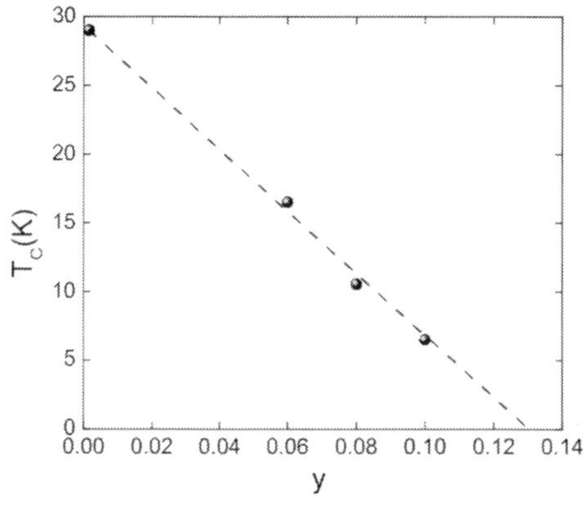

Figure 6. Critical temperature T_C as a function of the Fe concentration y. Solid line is the linear extrapolation of the T_C that approaches $T = 0$ at $y \approx 0.13$.

In our estimations using Eqs. (1 – 3) the value of the DM interaction is coupled to the average spin per unit cell SD/a. In fact the average spin changes significantly with the concentration

y and, therefore, it is interesting to obtain the true numbers for the Dzyaloshinskii constant $D\mu_B/a$. The values of the average spin can be found in Ref. [13] and after division to the value of $\langle S \rangle$ one obtains the numbers of $D\mu_B/a$ for different concentrations of y. For completeness, we calculated thus obtained values of $D\mu_B/a$ for the other compound $Fe_{1-x}Co_xSi$ described in Refs. [10] and [11]. The combined data are shown in Fig. 7. The obtained numbers for the Dzyaloshinskii constant $D\mu_B/a$ fluctuate around value of 1.15 ± 0.1 meV. Thus, one can conclude that the Dzyaloshinskii constant D does not depend on the concentration of these compounds (either $Mn_{1-y}Fe_ySi$ or x for $Fe_{1-x}Co_xSi$) and is the intrinsic number inherent to the crystallographic structure with the symmetry $P2_13$ and the lattice constant of order of 0.45 nm.

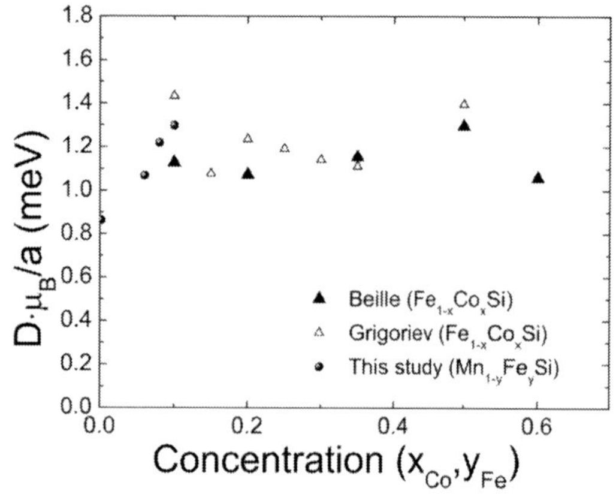

Figure 7. Dependence of the DM constant $D\mu_B/a$ on the Fe concentration y for $Mn_{1-y}Fe_ySi$ compounds and Co concentration x for $Fe_{1-x}Co_xSi$ compounds.

In this paper we have considered the concentration evolution of the principal interactions in the compound $Mn_{1-y}Fe_ySi$. These interactions have been calculated taking into account the theory [12] and combined data of the neutron diffraction and magnetization measurements. Firstly, we have established the Dzyaloshinskii constant D is independent on the concentration of the dope in both cases $Mn_{1-y}Fe_ySi$ and $Fe_{1-x}Co_xSi$. Secondly, we have obtained that the transition temperature is governed by the energy of isotropic ferromagnetic exchange interaction. The tendency $T_C \rightarrow 0$ suggests that the Fe doping can lead the system $Mn_{1-y}Fe_ySi$ to the quantum phase transition at $y \approx 0.13$. Indeed, the increase of the non-thermal parameter, the concentration y, results in T_C approaching 0, which is, by definition, the Quantum Critical Point (QCP). This concept can help to model the behavior of these systems close to the QPT using the Fe concentration as a non-thermal parameter. We believe that following the evolution of the system on the way to the QPT, using the changes of the principal interactions with doping y estimated in this work, and combining all these data with a those of the critical behavior with y up to 0.15 can shed a new light on the quantum phase transition in MnSi.

The PNPI-team acknowledges GKSS and HZB for their hospitality. The work is partly supported by the RFBR (project No 07-02-01318). V.A.D. acknowledges PNSXM2009 Orginizing Committee and the government of Saint-Petersburg for the partial support.

References

[1] Ishikawa Y, Tajima K, Bloch D and Roth M 1976 *Soild State Commun.* **19** 525
[2] Ishikawa Y, Shirane G, Tarvin J A and Kohgi M 1977 *Phys. Rev.* B **16** 4956
[3] Shirane G, Cowley R, Majkrzak C, Sokoloff J B, Pagonis B, Perry C H and Ishikawa Y 1983 *Phys. Rev.* B **28** 6251
[4] Dzyaloshinskii I E 1964 *Zh. Eksp. Teor. Fiz.* **46** 1420, *Sov. Phys. JETP* **19** 960
[5] Båk P and Jensen M H 1980 *J. Phys.* C **13** L881
[6] Nakamishi D, Janase A, Hasejawa A and Kitaoka M 1980 *Solid State Commun.* **35** 995
[7] Kataoka M and Nakanishi O 1981 *J. Phys. Soc. Jpn.* **50** 3888

[8] Grigoriev S V, Maleyev S V, Okorokov A I, Chetverikov Yu O and Eckerlebe H 2006 *Phys. Rev.* B **73** 224440

[9] Grigoriev S V, Maleyev S V, Okorokov A I, Chetverikov Yu O, Böni P, Georgii R, Lamago D, Eckerlebe H and Pranzas K 2006 *Phys. Rev.* B **74** 214414

[10] Grigoriev S V, Dyadkin V A, Menzel D, Schoenes J, Chetverikov Yu O, Okorokov A I, Eckerlebe H and Maleyev S V 2007 *Phys. Rev.* B **76** 224424

[11] Grigoriev S V, Maleyev S V, Dyadkin V A, Menzel D, Schoenes J and Eckerlebe H 2007 *Phys. Rev.* B **76** 092407

[12] Maleyev S V 2006 *Phys. Rev.* B **73** 174402

[13] Manyala N, Sidis Y, DiTusa J F, Aeppli G, Young D P and Fisk Z 2000 *Nature* (London) **404** 581

[14] Ishimoto K, Yamaguchi Y, Suzuki J, Arai M, Furusaka M and Endoh Y 1995 *Physica* B **213&214** 381

[15] Chattopadhyay M K, Roy S B, Chaudhary S, Singh K J and Nigam A K 2002 *Phys. Rev.* B **66** 174421

Magnetic correlations in pyrochlore spin ice as probed by polarized neutron scattering

L J Chang[1], Y Su[2], H Schneider[2], R Mittal[2], M R Lees[3], G Balakrishan[3] and Th Brückel[2,4]

[1] Nuclear Science and Technology Development Center, National Tsing Hua University, Hsinchu 300, Taiwan;

[2] Jülich Centre for Neutron Science (JCNS), IFF, Forschungszentrum Jülich, Outstation at FRM-II, Lichtenbergstrasse 1, D-85747 Garching, Germany;

[3] Department of Physics, University of Warwick, Coventry CV4 7AL, UK;

[4]Institut für Festkörperforschung (IFF), Forschungszentrum Jülich, D-52425 Jülich, Germany.

Email: ljchang@mx.nthu.edu.tw

Abstract. Polarized neutron diffuse scattering experiments have been carried out on (1-10) and (111) oriented single crystals of the pyrochlore spin ice [1] compound $Ho_2Ti_2O_7$ in order to investigate the complex spin correlations in this material at low temperature. The incident neutron beam was polarized perpendicular to the horizontal scattering plane. Thus, the corresponding components of the spin correlations in and out of the scattering plane can be probed with spin-flip (SF) and non-spin-flip (NSF) scattering respectively via neutron polarization analysis. A strong anisotropy of the in-plane and out-of-plane spin scattering has been observed in both samples. The temperature dependence of magnetic diffuse scattering has also been measured. Here we concentrate on reporting the experimental results. In addition, a brief discussion of the possible nature of spin correlations in this pyrochlore spin ice will also be given.

1. Introduction

Geometrically frustrated magnets, in which the subtle balance between the energy terms of the exchange, dipolar interactions, single ion anisotropy, and the special local lattice geometries such as triangles or tetrahedrons can drive the systems into novel types of quantum-mechanical magnetic ground states with no classical equivalent, have attracted a great deal of attention from both experimental and theoretical physicists [1-13]. Pyrochlore oxides are regarded as ideal candidates for studying geometrically magnetic frustration in three dimensions because of their unusual structure of corner-sharing tetrahedrons which can lead to remarkable low-temperature magnetic properties [3]. These properties include a collective paramagnetism or spin liquid behavior [2,4-6], spin glass behavior but with very little structural disorder [7-9], and perhaps the most interesting and extensively studied the spin ice behavior [10-13].

$Ho_2Ti_2O_7$, $Ho_2Sn_2O_7$ and $Dy_2Ti_2O_7$ have been identified as spin ice compounds [13-16]. As in the case of water ice, the spin ice has a degeneracy of energetically preferred states which generates a residual finite entropy as the temperature approaches absolute zero [17]. The magnetic ground-states' organizing principles, or "ice rules", require that two spins should point in, and two out of each elementary tetrahedron in the lattice occupied by the rare earth atoms. The "two in, two out" ice rule was generally believed to be the result of a delicate balance between the interaction of the spins and the special geometry of the pyrochlore lattice. Theoretical simulations by Bramwell *et al.* suggested that $Ho_2Ti_2O_7$ can be described using a dipolar spin ice model [1], which indicates that the "ice-rule" has the configuration of two spins in and two spins out at the tetrahedral corners, and is from [111]

[1] To whom any correspondence should be addressed.

anisotropy and ferromagnetic next neighbor coupling. In recent neutron diffuse scattering studies on the spin ice compounds, scattering intensities on the zone boundary have been observed for the spin ice $Dy_2Ti_2O_7$ [18]. A cluster-like correlation model, which was used to explain the zone boundary scattering intensities in the spinel $ZnCr_2O_4$ [19], was also used to understand the scattering intensities on the zone boundary in spin ice. However, Yavors'kii *et al.* showed that an equally satisfactory description is obtained via a microscopic theory in which the interaction terms in the Hamiltonian of the Monte Carlo simulations are fine-tuned [20] when includes the second and third nearest neighbor exchange couplings.

Pyrochlore oxides with the chemical formula $A_2B_2O_7$, where A is a rare-earth ion and B is a transition-metal ion, have a face-centered-cubic structure with space group *Fd-3m* and eight molecules of its general formula in a unit cell. In the spin ice $Ho_2Ti_2O_7$, the Ho^{3+} ions occupy the 16*d* site with a local trigonal symmetry *3m*, and are the only magnetic ions in the specimen. Fig.1 shows the projections of the Ho^{3+} sublattice of $Ho_2Ti_2O_7$ on (a) the (001), (b) the (110), and (c) the (111) planes; the black-lines indicate a unit cell, and the dark-blue balls are Ho^{3+} which are located at the corners of tetrahedrons which are shown in yellow. The projection on the (001) plane displays a chessboard lattice; the (110) plane a Kagomé lattice; and the (111) a hexagonal lattice. These three orientations display the possible anisotropies in a cubic pyrochlore lattice.

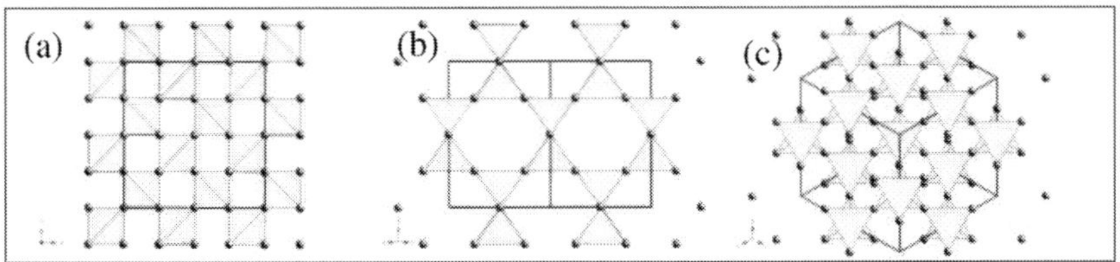

Figure 1. The projections (from left to right) of the Ho^{3+} sublattice of $Ho_2Ti_2O_7$ on (a) the (001) plane, a chessboard lattice; (b) the (110) plane, a Kagomé lattice; and (c) the (111) plane, a hexagonal lattice. The black-lines indicate a unit cell, and the dark-blue balls are Ho^{3+} ions which are located at the corners of tetrahedrons, and the tetrahedrons are shown in yellow.

Neutron diffuse scattering is a powerful technique to probe short-range spin correlations in frustrated magnets, such as the pyrochlore spin ice. By combining this technique with longitudinal neutron polarization analysis, it is possible to separate the contributions of the components of spin correlations in and out of the scattering plane. Here we report the experimental results of neutron polarization analysis on the diffuse scattering for the single crystals $Ho_2Ti_2O_7$ aligned in the (1-10) and the (111) orientations.

2. Experimental

A $Ho_2Ti_2O_7$ single crystal ~0.6 cm in diameter, ~2 cm in length, and amber-red in color was grown in an infrared double mirror image furnace. The details of the crystal growth are given in reference [21]. Pieces of the crystals were cut, oriented by x-ray Laue diffraction, and mounted on an aluminum sample stick so that either the [1-10] or [111] zone-axis was perpendicular to the horizontal scattering plane. Neutron diffuse scattering was carried out at the newly commissioned polarized diffuse scattering spectrometer DNS at FRM-II. A $^3He/^4He$ dilution insert was used for the measurements in the mK range. The high-temperature data (T>1.7 K) were obtained with a standard closed cycle cryostat. A neutron wavelength of 4.74 Å was selected for all the experiments. Magnetic diffuse scattering with z-direction polarization analysis was used to separate the contributions from the in-plane (neutron SF scattering), and out-of-plane i.e. along the [1-10] or [111] zone-axis (neutron NSF scattering) components of the spin correlations. The details for the neutron polarization analysis technique are described in reference [22].

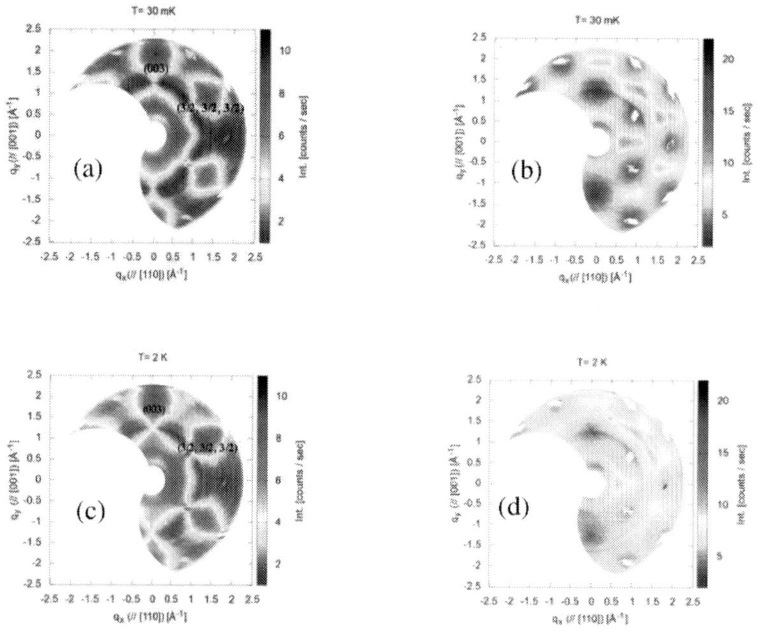

Figure 2. Contour plots of the neutron diffuse scattering at 30 mK for (a) neutron SF scattering, (b) neutron NSF scattering; and at 2 K for (c) neutron SF scattering, (d) neutron NSF scattering in the (1-10) zone plane. The color code scales for SF and NSF are different for the patterns clarified.

3. Results and Discussion

Contour plots of neutron diffuse scattering at 30 mK are shown in Fig. 2(a) SF scattering, 2(b) NSF scattering in the (1-10) zone plane at 30 mK.The residual intensities at nuclear scattering positions for SF channel are due to the imperfect neutron polarization in the experiments. In Fig. 2(a), the results correspond to the magnetic correlations projected on the (1-10) plane, and are very similar to the simulated pattern for the dipolar spin ice model [1] calculated from three dimensional magnetic correlation. There is an unusual diffuse scattering related to long range spin correlation, along transverse Q at pinch points. The NSF scattering shown in Fig. 2(b) reveals the correlations among spin components perpendicular to the (hhl) plane. The zone boundary scattering and\or strong antiferromagnetic diffuse scattering was observed at the positions of h = k = odd and l = odd. The strong intensity at (003) and by comparison the weaker features seen at (3/2, 3/2, 3/2) in Fig. 2(b) are consistent with the relative intensities predicted by the dipolar spin ice model for the region around (003) and (3/2, 3/2, 3/2), and the spread of the broad features along at (3/2, 3/2, 3/2) [1]. Yavors'kii *et al.* indicated that the zone boundary scattering pattern can be interpreted using either an independent small scale hexagonal spin cluster model [20], or an extensive generalized dipolar spin ice model (g-DSM), which has included the second and third nearest neighbor exchange couplings in the Monte Carlo simulations for another spin ice compound $Dy_2Ti_2O_7$. Pinch points were argued as the main differences between these two models. However, it is worth noting that the diffuse scattering patterns of g-DSM around (1/2, 1/2, 1/2) and (3/2, 3/2, 3/2) shown in reference [20] are made up of three separate features. Our results shown in Fig. 2(b) also contain three separate features at these two positions, although the antiferromagnetic correlations for Ho^{3+} are not predicted in this g-DSM model. Fig. 2(c) shows the contour plot for the SF scattering in the (1-10) zone plane at 2 K. This pattern is reminiscent of the nearest neighbor spin ice model simulated by Bramwell *et al* [1]. The main differences between dipolar and nearest neighbor spin ice models are the four intense regions around (000) in the dipolar model, the relative intensities of the region around (003) and (3/2, 3/2, 3/2), and the spread of the broad features along the diagonal [1]. With the exception of the variation of intensity around (000), which we cannot probe due to the limited q range in our experiments, Fig. 2(a) and Fig. 2(c) reflect these differences. The diffuse scattering pattern for the NSF scattering at 2 K is not clear (see Fig. 2(d)). This means that the projected magnetic correlations perpendicular to the (hhl) plane are very weak, or that the fluctuations out of the plane are faster than our measurements at this temperature. Fig. 2(c) and Fig. 2(d) seem to reveal the two-dimensional character of spin correlations in the (hhl) plane at 2 K at the first sight. However, the same magnitude of spin components in plane and out of plane, leading to the SF and NSF intensities respectively, is expected from the cubic symmetry of the compound. The diffuse scattering is clearly more pronounced in the SF scattering may be explained, for example, that in a projection along one edge of the tetrahedral, and within a tetrahedral, the vertical components in the nearest neighbor spin correlations is equal to zero in average for the ice-rule. A further study is necessary for clarifying.

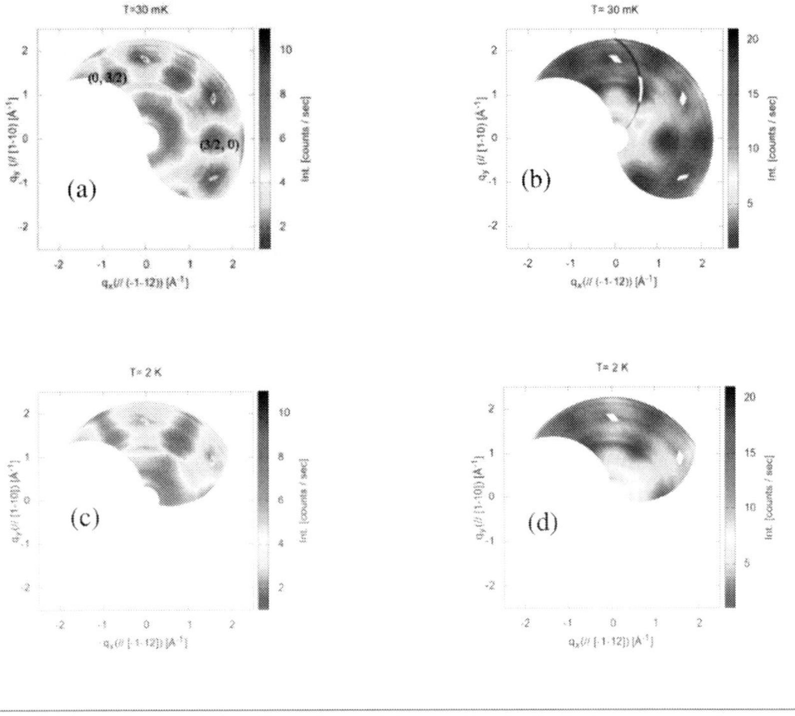

Figure 3. Contour plots of the neutron diffuse scattering at 30 mK for (a) neutron SF scattering, (b) neutron NSF scattering; and at 2 K for (c) neutron SF scattering, (d) neutron NSF scattering in the (111) zone plane. The color code scales for SF and NSF are different for the patterns clarified.

Figure 3 (a) – (d) manifest the SF and NSF scatterings in the (111) zone plane at different temperatures. In Fig. 3(a) and 3(b), a hexagonal and a stellate pattern were displayed around the center

(000) for SF and NSF respectively at 30 mK. Strong diffuse scattering intensities were also observed at (3/2,0), (3/2,3/2), and (0,3/2) in Fig. 3(a) when the diffraction plane is indexed according to a hexagonal lattice symmetry. Fig. 3(c) and 3(d) shows the contour plots of the SF and NSF scattering respectively at 2 K. These patterns are very similar to those shown in Fig. 3(a) and Fig. 3(b) for 30 mK except for the weaker overall intensities. The results are different to what we observed in the (1-10) scatterings, which display different patterns at 30 mK and 2 K in both SF and NSF channels. These results suggest that more detailed models are required to fully understand the spin ice freezing process and to completely explain the nature of the magnetic correlations in this material. Theoretical simulations and calculations are under way to investigate the magnetic correlations in the (111) orientation.

4. Conclusion

We have investigated the complex magnetic correlations in spin ice $Ho_2Ti_2O_7$ crystals aligned in the (1-10) and (111) orientations via polarized neutron diffuse scattering experiments. In the neutron SF scattering, the diffuse patterns suggest that there is a crossover from a regime above 2 K where nearest neighbor exchange interactions are most important to a dipolar-dipolar dominated regime below ~2 K. However, in the NSF scattering, a clear diffuse scattering at the Brillouin zone boundary was observed below 2 K. The NSF scattering patterns are also reminiscent of the zone boundary diffuse scattering pattern proposed in reference [20], in which the second and third nearest neighbor exchange interactions (the g-DSM model) [20] or the independent small scale hexagonal spin cluster model [19] need be taken into consideration. Further neutron diffuse scattering studies for different crystal orientations and non-magnetic doped specimens are processing to better understand the magnetic correlations appearing in the NSF scattering channel.

After submission of this paper the work of Fennel *et al.* appeared for scattering near pinch points which are indicative of separated dipoles and resemble the behavior of magnetic monopoles [23].

Acknowledgements

This work was supported by the National Science Council, Taiwan under the project number NSC 96-2739-M-213-001.

References

[1] Bramwell S T, Harris M J, den Hertog B.C, Gingras M J P, Gardner J S, McMorrow D F, Wildes A R, Cornelius A L, Champion L D M, Melko R G and Fennell T 2001 *Phys. Rev. Lett.* **87** 047205

[2] Gingras M J P, den Hertog B C, Faucher M, Gardner J S, Dunsinger S R, Chang L J, Gaulin B D, Raju N P and Greedan J E 2000 *Phys. Rev.* B **62** 6496

[3] for example, Diep H T (ed) 2004 *Frustrated Spin System* (Singapore: World Scientific); Greedan J E 2006 *J. Alloys Compd.* **408-412** 444

[4] Enjalran M and Gingras M J P 2004 *Phys. Rev.* B **70** 174426

[5] Gardner J S, Dunsiger S R, Gaulin B D, Gingras M J P, Greedan J E, Kiefl R F, Lumsden M D, MacFarlane W A, Raju N P, Sonier J E, Swainson I and Tun Z 1999 *Phys. Rev. Lett.* **82** 1012

[6] Molavian H R, Gingras M J P and Canals B 2007 *Phys. Rev. Lett.* **98** 157204

[7] Gardner J S, Gaulin B D, Lee S-H, Broholm C, Raju N P and Greedan J E 1999 *Phys. Rev. Lett.* **83** 211

[8] Gaulin B D, Reimers J N, Mason T E, Greedan J E and Tun Z 1992 *Phys. Rev. Lett.* **69** 3244

[9] Greedan J E, Reimers J N, Stager C V and Penny S L 1991 *Phys. Rev.* B **43** 5682

[10] Harris M J, Bramwell S T, McMorrow D F, Zeiske T and Godfrey K W 1997 *Phys. Rev. Lett.* **79** 2554

[11] Kadowaki H, Ishii Y, Matsuhira K and Hinatsu Y 2002 *Phys. Rev.* B **65** 144421

[12] Matsuhira K, Hinatsu Y, Tenya K and Sakakibara T 2000 *J. Phys.: Condens. Matter* **12** L649

[13] Ramirez A P, Hayashi A, Cava R J, Siddharthan R and Shastry B S 1999 *Nature* **399** 333

[14] Bramwell S T and Gingras M J P 2001 *Science* **294** 1495
[15] den Hertog B C and Gingras M J P 2000 *Phys. Rev. Lett.* **84** 3430
[16] Bramwell S T and Harris M J 1998 *J. Phys.: Condens. Matter* **10** L215
[17] Lau G C, Freitas R S, Ueland B G, Muegge B D, Duncan E L, Schiffer P and Cava R J 2006 *Nature Physics* **2** 249
[18] Fennell T, Petrenko O A, Fak B, Bramwell S T, Enjalran M, Yavors'kii T, Gingras M J P, Melko R G and Balakrishnan G 2004 *Phys. Rev.* B **70** 134408
[19] Lee S-H, Broholm C, Ratcliff W, Gasparovic G, Huang Q, Kim T H and Cheong S-W 2002 *Nature* **418** 856
[20] Yavors'kii T, Fennell T, Gingras M J P and Bramwell S T 2008 *Phys. Rev. Lett.* **101** 037204
[21] Balakrishnan G, Petrenko O A, Lees M R and Paul D M 1998 *J. Phys.: Condens. Matter* **10** L723
[22] for example, Brückel Th and Schweika W (ed) 2002 *Polarized Neutron Scattering* (Jülich: Forschungszentrum Jülich); Moon R M, Riste T and Koehler W C 1969 *Phys. Rev.* **181** 920
[23] Fennell T, Deen P P, Wildes A R, Schmalzl K, Prabhakaran D, Boothroyd A T, Aldus R J, McMorrow D F and Bramwell S T 2009 *Science* **326** 415

Polarized Neutrons and Synchrotron X-rays for Magnetism Conference 2009 IOP Publishing
Journal of Physics: Conference Series **211** (2010) 012014 doi:10.1088/1742-6596/211/1/012014

Polarized neutron studies on exchange-bias micro-stripe pattern

Katharina Theis-Broehl[1], Christine Hamann[2], Jeffrey McCord[2], Boris P. Toperverg[3] and Hartmut Zabel[3]

[1]University of Applied Sciences Bremerhaven, An der Karlstadt 8, D-27568 Bremerhaven, Germany

[2]Institute for Metallic Materials, IFW Dresden, Postfach 270116, D-01171 Dresden, Germany

[3]Institute for Experimental and Solid State Physics, Ruhr-University, D-44780 Bochum, Germany

E-mail: ktheis-broehl@hs-bremerhaven.de

Abstract. In the present study we examined the results of competing interactions in laterally exchange biased modulated thin films. For studying the magnetic interactions between neighboring magnetic elements we designed a model system. Considering that the model system should be of technological interest on the one hand, and should allow the application of several different and efficient experimental techniques for complementary analysis on the other hand, we have chosen films with in-plane stripe-like magnetic domains. The evolution of the magnetic domain structure along the hysteresis loop was analyzed with a combination of experimental techniques: magneto-optical Kerr effect, Kerr microscopy, polarized neutron reflectometry, and off-specular scattering of polarized neutrons with polarization analysis.

Due to the competition between interfacial exchange bias at the ferromagnetic-antiferromagnetic interface and intra-layer exchange interactions between neighboring ferromagnetic stripes the alignment is not perfectly antiparallel, but is periodically canted with respect to the stripe axis. Thus the net magnetization of the ferromagnetic film turns almost perpendicular to the stripes. At the same time the projection of the magnetization vector onto the stripe axis has a periodically alternating sign. The experimental observations are explained and quantitatively described within the frame of a phenomenological model. The model defines conditions which can be used for tailoring nano- and micro-patterned EB systems with different types of magnetic order.

1. Introduction

Patterned magnetic materials are extremely important for achieving maximum package densities in magnetic recording media. We used non-topographic magnetic patterning as an alternative approach to topographic patterning resulting in a local variation of only magnetic properties of thin films, which otherwise remain structurally homogeneous and maintain flat surfaces. With such systems one can avoid, e.g. the problem of very small feature sizes, where the long-term thermal stability of the magnetic elements is lost due to superparamagnetic fluctuations[1]. For non-topographic magnetic patterning the exchange bias (EB) effect between ferromagnetic (F) and antiferromagnetic (AF) bilayers can be used [2]. In those films the magnetization of the ferromagnet locally is pinned differently to the antiferromagnet at the interface implying that the complete film will adopt this lateral varying magnetization.

We applied two routes for locally changing the EB. First, we prepared a CoFe/MnIr film by using Ion Beam Induced Magnetic Patterning (IBMP) [3][4] for changing the EB between F and AF layers in one part of the film, while keeping the EB properties of the untreated regions unchanged. Second, we used a slightly changed film structure consisting of an extended soft ferromagnetic NiFe layer locally exchange biased to a periodic antiferromagnetic MnIr stripe array and studied several films with varied ferromagnetic film thickness and stripe width.

For a detailed study of the magnetic interactions between neighboring magnetic elements in a continuous film we designed model films which allowed the efficient use of different experimental techniques for complementary analysis. Therefore, we did not approach very small length scales but designed the magnetic elements to be in the micrometer range and used methods as the magneto-optical Kerr effect (MOKE), Kerr microscopy (KM) and polarized neutron reflectometry (PNR).

From the study of the first sample we found that while the EB in neighboring striped lateral areas was set antiparallel to each other the magnetization of each stripe appears to be tilted with respect to the EB frozen-in field so that the stripe magnetizations are aligned not completely antiparallel to each other. The reason is the additional ferromagnetic exchange interaction between the striped regions which favors a parallel orientation between the magnetization in the film volume[5]. With the second set of samples we were able to investigate the overall remagnetization process in dependence of different EB and F exchange contributions.

2. Experimental Details

2.1. Sample preparation

The first sample studied is an EB $Co_{70}Fe_{30}$(28 nm)/$Mn_{83}Ir_{17}$(15 nm) F/AF bilayer with buffer layers Cu(28 nm)/SiO_2(50 nm) on Si(111) and a TaO-Ta(9 nm) top layer prepared by magnetron sputtering. The initial EB direction was set by field cooling in a magnetic in-plane field of 1 kOe after an annealing step for 1 h at 275 °C which is above the blocking temperature of the antiferromagnetic material. Subsequently the film was covered by a photoresist, which was patterned into equally spaced stripes of width $d = 2.5$ μm and a periodicity of $\Lambda = 5$ μm. Then the sample was bombarded with He^+ ions with a fluency of 1×10^{15} ions/cm^2 at 10 keV. The resist thickness of 0.5 μm is sufficient to protect the covered regions from the He^+ exposure. During bombardment a magnetic field of 1 kOe aligned opposite to the initial EB direction was applied. This resulted in a stripe-like pattern with alternate sign of the unidirectional anisotropy, and hence the sign of the EB in neighboring stripes[6].

The second set of samples are polycrystalline full $Mn_{77}Ir_{23}(t_{AF})/Ni_{81}Fe_{19}(t_F)$ films deposited by magnetron sputtering on oxidized Si-wafers with a buffer layer of 4 nm Ta and covered by a TaO-Ta(3 nm) top layer. In order to obtain a 2-dimensional lateral modulation of the unidirectional exchange anisotropy, each layer stack was subjected to a subsequent lithography patterning and ion etching to remove the Ta cover layer in stripe-like parts of the sample. Therefore, only the uncovered parts of the sample were subsequently exposed to air for undergoing an oxidation process of MnIr resulting in a superimposed stripe array of MnIr and MnIrO (for more details see Ref. [7]). MnIr is antiferromagnetic while MnIrO is non-magnetic. At a temperature of 250°C, which is above the blocking temperature of the system, an external field of 10 kOe was applied to set the unidirectional anisotropy direction of the F-AF stripes to be along the stripe axis. The resulting EB patterned film of alternating stripes with nominal TaO $-$ $Ta/Mn_{77}Ir_{23}(t_{AF})/Ni_{81}Fe_{19}(t_F)$ and $MnIrO_x/Ni_{81}Fe_{19}(t_F)$ comprises exchange biased (AF-F) stripes and unbiased ferromagnetic stripes (F) through an underlying extended ferromagnetic layer. The stripe period Λ was chosen to be 4 μm and 12 μm, respectively, to thus vary the amount inter-stripe interfaces. The magnitude of the unidirectional anisotropy was tuned by adjusting the ferromagnetic layer thickness t_F to be either 20 nm or 30 nm while the antiferromagnetic layer thickness $t_{AF} \geq 7$ nm was always above the critical thickness for the existence of EB.

2.2. Magneto-optical Kerr effect

We have analyzed the magnetization reversal of the first sample with a vector-MOKE set up as described in Ref. [8]. For measuring the magnetization reversal process of the second set of samples we employed inductive magnetometry to map the macroscopic properties.

Further insight into the microscopic rearrangement of magnetization during cycling with an external field along the EB axis was achieved by a high resolution magneto-optical Kerr microscope (KM) [9] that is sensitive to directions orthogonal to the field [10]. The KM images provide good input data for constructing a remagnetization model which was then used to refine the physical parameters required for fitting polarized neutron reflectometry (PNR) measurements.

2.3. Polarized neutron scattering

Neutron scattering experiments were carried out with the ADAM reflectometer at the Institut Laue-Langevin, Grenoble, France [11], and with the HADAS reflectometer at the FRJ-2 reactor in Jülich, Germany. The measurements were performed with fixed neutron wavelengths of $\lambda = 0.441$ and 0.452 nm, respectively. To apply the external field H along the stripe axis, an electromagnet with the field direction parallel to the sample surface and parallel to the neutron polarization and perpendicular to the scattering plane was used. The incoming neutron polarization was either set parallel (+) or antiparallel (-) to the external magnetic field. For specular reflectivity measurements, non-spin-flip (NSF) reflectivities R^{++} and R^{--} as well as neutrons with flipped polarization after the reflection, i.e. two spin-flip (SF) intensities R^{+-} and R^{-+}, were analyzed.

The magnetic induction vector $\boldsymbol{B} = \boldsymbol{H} + 4\pi\boldsymbol{M}$ (which is mainly given by the magnetization vector) is assumed to lie in the sample plane (in-plane shape anisotropy). Furthermore, $\boldsymbol{B}$ may have an angle γ with respect to the external magnetic field. We assume that a monochromatic neutron beam incident onto the sample surface under the angle α_i is scattered under the glancing angle α_f, so that for specular reflection $\alpha_i = \alpha_f$.

3. Experimental results

The first sample (CoFe/MnIr) was studied by vector-MOKE measurements with the field applied along the EB axis which is parallel to the IBMP magnetically patterned stripe edges, see Fig. 1(a). A two step magnetization reversal is clearly visible in the longitudinal hysteresis, corresponding to a successive remagnetization of the two regions with alternative directions of the EB field. The transverse hysteresis loop in Fig. 1(a) shows a considerable transverse component of magnetization with plateaus in the field range where we expect an antiparallel alignment of the magnetization in irradiated and non-irradiated parts of the sample.

A detailed KM analysis of this sample was recently published by McCord et al.[10]. With the microscope tuned to longitudinal magnetic contrast it was shown that starting from saturation magnetization the reversal proceeds through an antiparallel alignment of magnetization by head-on domain wall propagation along the stripes. In addition, the transverse magnetization contrast provides a rather detailed insight into the microscopic arrangement at different stages of the remagnetization process. Some representative examples of KM images taken with the Kerr sensitivity set close to the transverse direction, are reproduced in Fig. 1 (b)-(e). The images were taken in the ascending loop branch at fields just below the first reversal at H_{c1} (b), just above H_{c1} (c), in the antiparallel state between H_{c1} and H_{c2} (d), and at the second reversal at H_{c2} (e). Ripple domains can also be observed in Fig. 1 (b), (c) and (e) in stripes which are close to reversal. Note that rippling is typical for Co and Co rich CoFe alloys, [12, 13] and in our sample always occurs around both coercive fields $H_{c,1}$ and $H_{c,2}$.

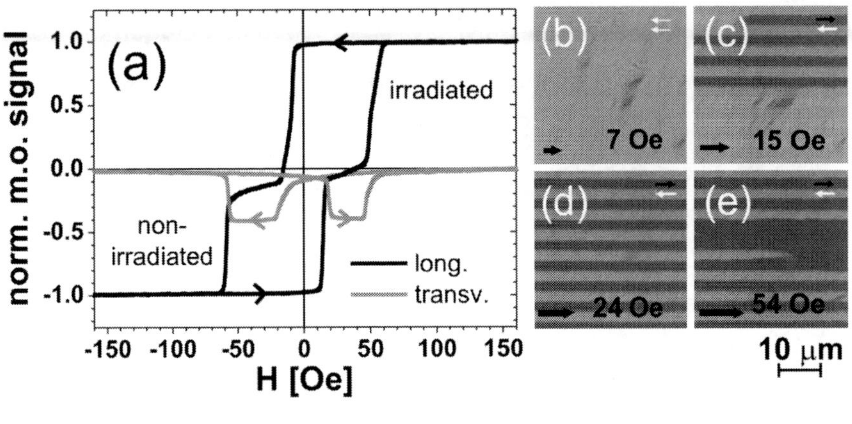

Figure 1. (a) Vector-MOKE hysteresis loops of the CoFe/MnIr sample measured with the field applied along the EB axis and showing the longitudinal and transverse components of the magnetization vector. (b)-(e) Corresponding longitudinal Kerr images taken along the ascending magnetization loop.

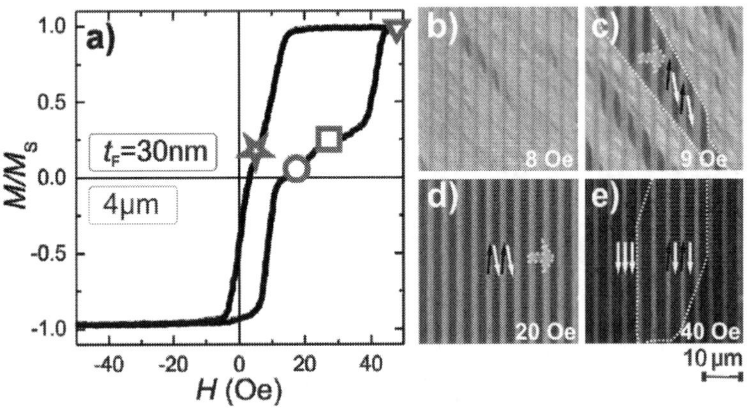

Figure 2. (a) Magnetic hysteresis loop of the MnIr - MnIrO/NiFe (20 nm) film with $\Lambda = 4$ μm from the second set of samples. (b)-(e) Corresponding longitudinal Kerr images along the ascending magnetization loop.

In Fig. 2 the remagnetization process of one of the EB modulated films of the second set of samples with a thickness of $t_F = 20$ nm and a stripe period of $\Lambda = 4$ μm is presented. The inductively measured hysteresis loop and corresponding high-resolution Kerr images are depicted in Figures 2(a) and (b-e), respectively. The film structure also exhibits a two-step switching process, typical for such a hybrid magnetic two-phase system. Coming from negative magnetic fields, first, below 10 Oe the free F stripe magnetization starts to ripple (b), followed by the reversal of those stripes by head-on-domain wall motion (c). On the hysteresis step, a rather continuous change of the magnetization is observed, indicating rotation and relaxation

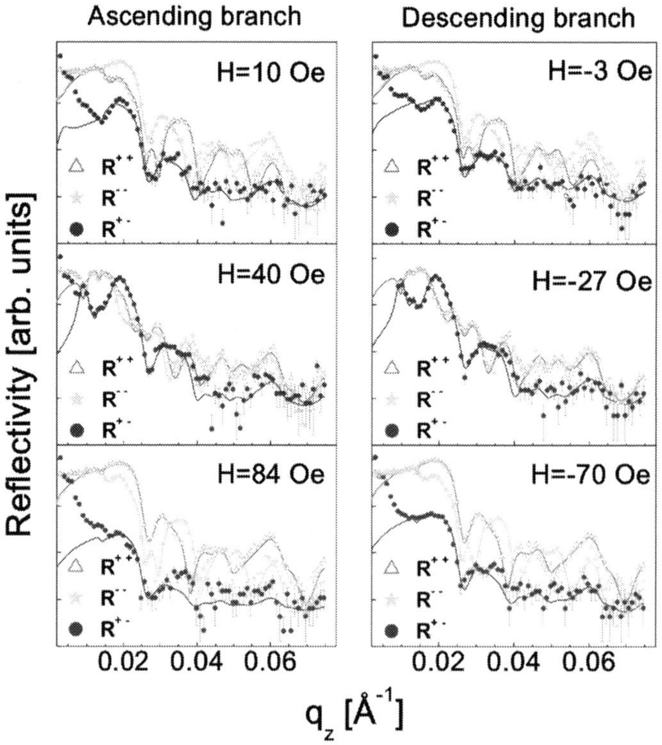

Figure 3. (Color online) Polarized neutron reflectivity measurements performed at the CoFe/MnIr sample at different applied magnetic field. The symbols present measurements of non-spin flip reflectivities $\mathcal{R}^{++}$ and $\mathcal{R}^{--}$ and one spin flip reflectivity $\mathcal{R}^{+-}$. The lines represent fits to the data points. The sign of the field value corresponds to that in MOKE measurements in Fig. 1.

along the positive field direction. During the second step at 40 Oe, the inter-stripe domain walls move into the exchange biased stripes. Accompanied by head-on-domain wall motion (e) the pinned magnetization is finally reversed. During this reversal process, stripes of the same magnetic properties reverse in a correlated manner such that so called quasi-domains[14, 7] (or hyper-domains[15]) occur.

Specular polarized neutron reflectivity was measured at several field values on the ascending and descending branches of the hysteresis loops. In Fig. 3, several representative specular reflectivity curves of the CoFe/MnIr sample are displayed. The NSF reflectivity curves in the first row look quite similar to those in the last one, if the NSF reflectivity $\mathcal{R}^{++}$ in the latter row is substituted by its counterpart $\mathcal{R}^{--}$, and vice versa. This symmetry is just due to the opposite magnetization state of the sample in the upper and lower rows. The SF reflectivity $\mathcal{R}^{+-}$ in the first row, however, is stronger than that measured in saturation, where it is mostly due the limited efficiency of the polarization devices. An appreciable SF signal in the specular channel signifies that the magnetization averaged over the coherence length is tilted by the angle $\bar{\gamma} \neq 0$ with respect to the polarization direction, and hence, to the stripe axis.

For measurements carried out at field value between $H_{c,1}$ and $H_{c,2}$ (middle row of Fig. 3), the NSF reflectivity curves almost merge together. The spin asymmetry $\mathcal{R}^{++} - \mathcal{R}^{--}$ is close to zero, indicating that the mean magnetization within the neutron coherence range has almost no projection onto the direction of the applied field, in agreement with observations from the

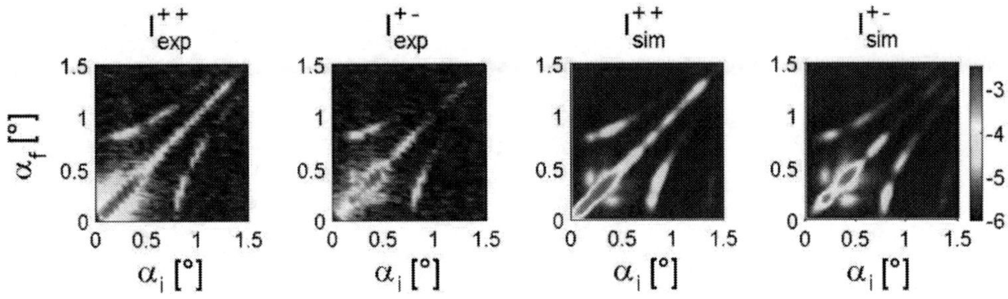

Figure 4. (Color online) Experimental and simulated maps of the polarized neutron scattering intensity on a logarithmic scale measured on the EB $Co_{70}Fe_{30}$(28 nm)/$Mn_{83}Ir_{17}$(15 nm) F/AF bilayer at the descending loop branch at a magnetic field of - 27 Oe. The intensities are plotted as a function of α_i, and α_f.

longitudinal hysteresis loop in Fig. 1 (a) for the plateau region. Such results can be interpreted in favor of an antiparallel alignment of the magnetization vector in neighboring stripes. However, at the same field values we observe a very strong SF signal comparable to the NSF intensities. This is an unambiguous proof for the fact that the mean magnetization over a number of stripes in the coherence length is tilted by a large angle $\overline{\gamma}$ with respect to the polarization direction, and hence, to the stripe axis. Taking into account the degeneracy of NSF channels one concludes that $\overline{\gamma} \approx 90°$. The existence of a large magnetization component perpendicular to the applied field was already noted in the transverse MOKE and KM data.

A quantitative information about the magnetization distribution over the stripes is gained via fitting the specular PNR data to a theoretical model which takes into account that due to variations in the tilt angles $\gamma = \overline{\gamma} + \Delta\gamma$ against the applied field one may observe local periodic variations $\Delta\gamma$ with respect to the mean value $\overline{\gamma}$ within the coherence length. These deviations reduce the mean magnetic optical potential and modify the reflectivity profiles with respect to that in saturation. The ratio between SF and NSF reflectivities is regulated by the mean values of $\langle\cos\overline{\gamma}\rangle$ and $\langle\sin^2\overline{\gamma}\rangle$, where angular brackets denote additional incoherent averaging. Such an averaging of reflected intensities is due to the fact that the coherence area of the neutron beam covers only a small portion of the sample surface and magnetization direction, and hence the tilt angle $\overline{\gamma}$, may vary from one coherence area to the next.

The fitting procedure, basically determining $\langle\cos(\Delta\overline{\gamma})\rangle$, $\langle\cos\overline{\gamma}\rangle$ and $\langle\sin^2\overline{\gamma}\rangle$ was carried out using an originally developed least-squares software package [16], which allows simultaneous evaluation of all four measured reflectivities in one cycle. It is based on the super-iterative algorithm, [17] which generalizes the Parratt algorithm [18].

It should be stressed that from MOKE and specular PNR results alone one cannot draw conclusions about the origin of the microscopic arrangement of magnetic states along the hysteresis loop. However, from fits to the Bragg diffraction and diffuse scattering of polarized neutron intensity maps (one representative example for experimental and simulated maps are shown in Fig. 4) we find a periodic deviation of the magnetization vector in neighboring stripes with an angle in the "antiparallel state" of only 120° (instead of expected 180°) to each other corresponding of $\Delta\gamma = 60°$. In principle, deviation angles $\Delta\gamma$ consist of two parts. The first one is due to periodic alternation of $\Delta\gamma$. It contributes into superstructural off-specular Bragg reflection. The random fluctuations around this periodic alternation is a source of diffuse scattering.

In the maps in Fig. 4 the specular reflection ridge runs along the diagonal, where $\alpha_i = \alpha_f$. Besides little diffuse scattering Bragg diffraction concentrated along curved lines $\cos\alpha_i - \cos\alpha_f \approx$

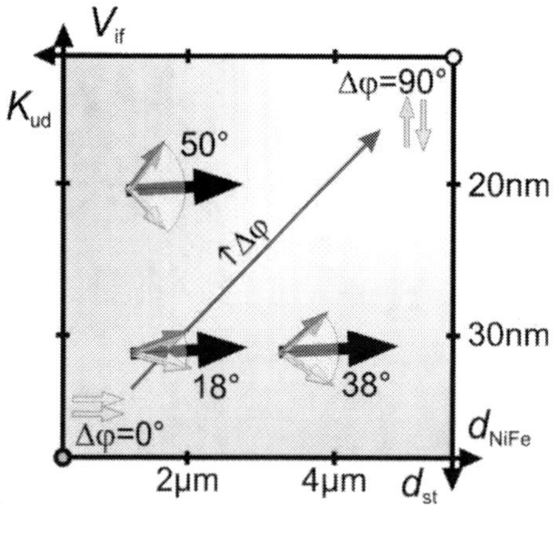

Figure 5. (Color online) Results of the fits to the PNR data of the second set of samples. The graph shows the results with respect to $\Delta\gamma$ as function of ferromagnetic film thickness and EB stripe width d_{st}.

$n(\lambda/\Lambda)$ can be seen at $\alpha_i \neq \alpha_f$, where n denotes the order of diffraction and $\Lambda = 5$ μm is the period. Strong Bragg reflections for $n = \pm 1$ can be recognized, whereas for $n = \pm 2$ they are suppressed. This is caused by the structure factor, resulting from the equal width of adjacent magnetic stripes in a unit cell.

Similar results were also found from the second set of samples. For all three samples in the "antiparallel state" at the hysteresis step, the mean magnetization is canted about 90° away from the stripe axis. The sample with the lower unidirectional anisotropy of the biased stripes, i.e. a F thickness of 30 nm, and smallest pattern period of $\Lambda = 4$ μm shows the smallest splitting of the neighboring stripe magnetizations of $\Delta\gamma = 18°$. With increasing the period to $\Lambda = 12$ μm and thus decreasing the number of stripe interfaces, the perpendicular magnetization component for both stripe types is reduced. For this case, the angle $\Delta\gamma$ is 38°. Keeping the stripe period constant but increasing the unidirectional anisotropy of the AF-F stripes, $\Delta\gamma$ also increases, now to 50°. The results from the three samples are summarized in Fig.5.

4. Discussion

We developed a simple phenomenological model explaining the reason for the intermediate state stable below the coercive field characterized by a canted domain structure in easy axis configuration. It is explained by the interplay between interfacial and bulk exchange coupling. The model also gives an estimate for the range of physical parameters (stripe width and film thickness) where one can expect a pure antiferromagnetic ordering of the stripe magnetization (the model is described in detail in Ref. [5]).

The basic result of the model is that the EB exchange field contribution is proportional to the number of interfacial ions, i.e., in first approximation being proportional to the width of the stripes d times its length l. On the other hand, the bulk F exchange energy across the F domain wall is proportional to the number of sites belonging to the F domain wall running through the F film , i.e., in first approximation being proportional to the film thickness t times stripe length l. Therefore the ratio between bulk and interfacial contributions is roughly proportional to $(t \cdot l)/(d \cdot l) = t/d$. This means that one can suppress the undesirable tilt instability by either increasing the width of the stripes or reducing the F film thickness, while keeping interfacial and bulk F exchange couplings unchanged.

5. Summary

In the first set of samples we have investigated a continuous ferromagnetic CoFe film alternately exchange-biased (with positive and negative EB, respectively) to an antiferromagnetic MnIr layer. In the second set of samples we studied three continuous NiFe layers with variable F film thickness alternately exchange-biased to MnIr layer or free to rotate. We have analyzed the magnetization reversal mechanism of these films in detail, which exhibit rich hysteresis and domain structures due to competing interfacial and intralayer exchange interactions. For a given set of parameters, including interfacial and bulk exchange coupling, stripe width, and film thickness, the intermediate state is stable below the coercive field and is characterized by a canted domain structure which results in a macroscopic magnetic moment directed perpendicular to the stripe direction and the EB axis. A simple phenomenological model suggests a reason for this phenomenon and gives an estimate for the range of physical parameters where one can expect a pure antiferromagnetic ordering of the stripe magnetization. This may provide directions for further advances in the design of magnetic micro- and nano-patterns with tailored properties.

6. Acknowledgments

We also would like to express our thanks to M. Wolff, A. Westphalen, V. Hink, J. Schmalhorst, T.Weis, D. Engel, A. Ehresmann, U. Rücker, R. Kaltofen, I. Mönch and R. Schäfer for collaboration and fruitful discussions. We acknowledge funding by DFG through SFB 491. The PNR measurements were carried out at the ADAM reflectometer of the ILL, which is funded by BMBF Grant No. 03ZA7BOC and at the HADAS reflectometer at FZ Jülich.

References

[1] Moser A, Takano K, Margulies D T, Albrecht M, Sonobe Y, Ikeda Y, Sun S and Fullerton E E 2002 *J. Phys.* D **35** R157
[2] Chappert C, Bernas H, Ferré J, Kottler V, Jamet J-P, Chen Y, Cambril E, Devolder T, Rousseaux F, Mathet V and Launois H 1998 *Science* **280** 1919
[3] Ehresmann A 2004 *Recent Res. Dev. Phys.* **7** 401
[4] Fassbender J and McCord J 2008 *J. Magn. Magn. Mater.* **320** 579
[5] Theis-Bröhl K, Wolff M, Westphalen A, Zabel H, McCord J, Höink V, Schmalhorst J, Reiss G, Weis T, Engel D, Ehresmann A, Rücker U and Toperverg B P 2006 *Phys. Rev.* B **73** 174408
[6] Ehresmann A, Krug I, Kronenberger A, Ehlers A and Engel D 2004 *J. Magn. Magn. Mater.* **280** 369
[7] Hamann C, Mönch I, Kaltofen R, Schäfer R, Gemming T, Schultz L and McCord J 2008 *J. Appl. Phys.* **104** 0021
[8] Schmitte T, Theis-Bröhl K, Leiner V, Zabel H, Kirsch S and Carl A 2002 *J. Phys.: Condens. Matter* **14** 7525
[9] Hubert A and Schäfer R 1998 *Magnetic domains* (Heidelberg: Springer)
[10] McCord J, Schäfer R, Theis-Bröhl K, Zabel H, Schmalhorst J, Höink V, Brückl H, Weiss T, Engel D and Ehresmann A 2005 *J. Appl. Phys.* **97** 10K102
[11] Schreyer A, Siebrecht R, Englisch U, Pietsch U and Zabel H 1998 *Physica* B **248** 349
[12] Theis-Bröhl K, Toperverg B P, Leiner V, Westphalen A, Zabel H, McCord J, Rott K and Brückl H 2005 *Phys. Rev.* B **71** 020403(R)
[13] McCord J, Schäfer R, Mattheis R and Barholz K-U 2003 *J. Appl. Phys.* **93** 5491
[14] McCord J, Kaltofen R, Schmidt O G and Schultz L 2008 *Appl. Phys. Lett.* **92** 0003
[15] Theis-Bröhl K, Westphalen A, Zabel H, Rücker U, McCord J, Höink V, Schmalhorst J, Reiss G, Weis T, Engel D, Ehresmann A and Toperverg B P 2008 *New Journal of Physics* **10** 1367
[16] Deriglazov V et al (to be published)
[17] Toperverg B P 2002 in *Polarized Neutron Scattering, Series Matter and Materials* Vol 12 ed T Brückel nd W Schweika (Jülich: Forschungszentrum) 249
[18] Parratt L G 1954 *Phys. Rev.* **95** 359

Journal of Physics: Conference Series **211** (2010) 012015 doi:10.1088/1742-6596/211/1/012015

Soft x-ray resonant magnetic reflectivity studies for in- and out-of-plane magnetization profile in ultra thin films

J-M Tonnerre[1*], N Jaouen[2], E Bontempi[3], D Carbone[4], D Babonneau[5], M De Santis[1], H C N Tolentino[1], S Grenier[1], S Garaudee[1] and U Staub[6]

[1] Institut Néel, CNRS & Université J. Fourier, F-38042 Grenoble, France
[2] Synchrotron SOLEIL, L'Orme des Merisiers, Saint-Aubin, F-91192 Gif-sur-Yvette, France
[3] Laboratorio di Chimica per le Tecnologie, Universita di Brescia, I-25123 Brescia, Italy
[4] ESRF, F-38043 Grenoble, France
[5] Laboratoire de Physique des Matériaux (PHYMAT), CNRS, Université de Poitiers, SP2MI, Téléport 2, Bvd M. et P; Curie, F-86962 Futuroscope Chasseneuil, France
[6] Paul Scherrer Institut, Swiss Light Source, CH-5232 Villigen, Switzerland

E-mail: jean-marc.tonnerre@grenoble.cnrs.fr

Abstract. The possibility to investigate complex magnetic profiles throughout an ultrathin magnetic film or an interface by soft x-ray resonant magnetic reflectivity is presented. The determination of in- and out-of-plane magnetic profile is shown to be possible with a subnanometer resolution by measuring the reflectivity over a wide angular range. The technique is applied to a granular magnetic multilayer and to a perpendicular exchange bias coupled system.

1. Introduction

The investigation of magnetism in reduced dimension systems attracts a considerable interest. Motivated by the development of new materials in spintronics, a detailed understanding of the organization of magnetic moments at surfaces, interfaces and in heterostructures is mandatory. When magnetic properties in an embedded ultra thin film are changing at the unit cell level, most of the conventional or advanced methods [1] are limited either by the lack of in-depth spatial resolution or by surface limited sensitivity. In this paper, we examine the capability of the Soft X-ray Resonant Magnetic Reflectivity (SXRMR) in specular geometry over a wide angular range to probe the distribution of in- and out-of-plane magnetization components in an ultra thin film. In particular, only few techniques allow one to probe the perpendicular magnetization distribution in thin films. In specific cases, nuclear resonant scattering can be sensitive to the out-of-plane magnetization distribution, and lateral resolution may be achieved with imaging techniques as, for example, spin polarized scanning tunneling microscopy or magnetic force microscopy but none provide in-depth spatial resolution with chemical selectivity.

* To whom any correspondence should be addressed.

2. Soft X-ray resonant magnetic reflectivity

The x-ray resonant magnetic reflectivity exploits the magnetization sensitivity of the atomic scattering factor (ASF) at an absorption edge [2-4], which may be expressed on the basis of the two polarization states chosen either parallel or perpendicular to the scattering plane [5]:

$$f = f_c\begin{pmatrix} 1 & 0 \\ 0 & \cos(2\theta) \end{pmatrix} - i\ f_m\begin{pmatrix} 0 & m_y\cos(\theta)+m_z\sin(\theta) \\ -m_y\cos(\theta)+m_z\sin(\theta) & m_x\sin(2\theta) \end{pmatrix} \quad (1)$$

The first term of (1) is the Thomson and resonant charge scattering with $f_c = f_0 + f'(E) + if''(E)$ where E is the incident photon energy. The second term is the resonant magnetic scattering where $f_m = m'(E) + im''(E)$ [3]. The second terms in brackets shows the dependence on the scattering angle θ of the transverse (m_x), longitudinal (m_y), polar (m_z) components of the magnetic moment. In order to analyze the x-ray reflectivity data, it is usually convenient to consider the change of the refractive index or of the dielectric tensor at an interface and the propagation through a continuous layer. These complex quantities are related to the x-ray atomic scattering factor for the forward scattering (q=0) [4].

X-ray resonant magnetic reflectivity has been mainly used in the hard x-ray range for investigating the magnetism of rare-earths [6], 5*d* transition metal [7] and 5*f* actinides [8] in superlattices. In order to probe the magnetism of 3*d* levels of transition metals, it is necessary to conduct experiments in the soft x-ray range (200 to 2000 eV) in the vicinity of their $L_{2,3}$ edges ($2p \rightarrow 3d$ transition). These transitions exhibit sharp and strong resonances [9]. They provide a high charge and magnetic contrast allowing multi-element heterostructures to be investigated [10]. The wavelength of the photons being one order of magnitude larger than in the hard x-rays range, the reflectivity measurements are possible up to large incident scattering angles (up to 85 degree). We point out that such capability is essential to probe the m_z component.

The SXRMR experiments have been conducted on the SIM beamline of the Swiss Light Source at the Paul Scherrer Institut using circularly polarized light and the RESOXS endstation [11]. In order to probe the in-plane magnetization, the measurements are performed in the longitudinal geometry using an electromagnet for sample magnetization. At each incidence angle θ the reflectivity $I(+)$ and $I(-)$ are obtained by reversing an applied magnetic field (μ_0H up to 0.2T) while keeping the x-ray helicity unchanged. In order to probe the out-of-plane magnetization, the measurements are performed in the polar geometry where the sample is magnetized by a permanent perpendicular magnet μ_0H=0.4T brought to the sample surface followed by data collection in remanence. In this case, $I(+)$ and $I(-)$ are obtained by reversing the x-ray helicity.

We turn next to the quantitative analysis of SXRMR aiming at the determination of the magnetic profile along the growth axis. A crucial issue is the knowledge of the resonant terms in the atomic scattering factor. The charge $f''(E)$ and magnetic $m''(E)$ imaginary parts are obtained from x-ray absorption and XMCD spectra, respectively. Ideally, these parameters have to be obtained from the sample under investigation. Alternatively, they can be derived from a reference system provided the XMCD and SXRMR measurements are collected with a similar energy resolution. $f'(E)$ and $m'(E)$ are calculated by using the Kramers-Kronig relation. Another important issue is the determination of the structural parameters. In the first step of the analysis, the geometrical film structure is derived by fitting the average signal $[I(+) + I(-)]/2$ measured at energies far and near the resonance. The magnetic profile is derived from the refinement of the magnetic asymmetry while keeping the structure parameters constant. The magnetic film is subdivided into slices and the magnetization vector of each slice can be described by an amplitude term and two angles to take into account arbitrary orientations. The magnetic amplitude of each slice is refined by varying a weighting factor w_m applied to $m'(E)$ and $m''(E)$. The magnetic moment amplitude is estimated from the value of w_m and from sum rules applied to the XMCD data [12]. The simulations are performed using a recursive calculation involving the reflection and transmission coefficients at each interface for magneto-optics [13].

3. Applications of soft x-ray resonant magnetic reflectivity

In this section, we discuss two specific applications of soft x-ray resonant magnetic reflectivity to the study of in- and out-of-plane magnetic profile.

3.1. FePt/C granular multilayer

Nano-particles are attracting considerable attention due to their potential in the field of ultrahigh density magnetic recording media [14]. To achieve high storage densities beyond 1 Tbit in^{-2}, assemblies of ferromagnetic monodomain particles with uniform and small size below 10 nm are required. However, as the magnetic bit size is further reduced to the superparamagnetic limit, the magnetization of the particles is easily perturbed by thermal agitation. In order to overcome superparamagnetism, recent studies have been focused on CoPt and FePt nanoparticle arrays owing mainly to the existence of chemically ordered phases with large magnetocrystalline anisotropy constant above 10^7 erg cm^{-3} [15]. Additional requirements for future magnetic recording media are that the nanoparticles have to be magnetically isolated but also chemically and mechanically stable. Therefore, much research has been done on granular thin films consisting of CoPt and FePt nanoparticles embedded in various nonmagnetic matrices [16]. Among the prospective materials, C/FePt granular thin films have attracted great interest because carbon not only impedes the growth of the nanoparticles, but also provides protection against outside degradations and reduces interparticle exchange interactions [17].

In this contribution, we describe the SXRMR technique applied to a C/FePt multilayer where the FePt granular layer is analyzed as a continuous film. The motivation is to probe the possibility of a core/shell magnetic structure. A change of the magnetization at the surface of the particles could be detected as a change along the z-axis at the interface between the FePt and C layers stronger than the modification at the core of the layer. The discussion is limited to the as-prepared sample. The investigation of the change under annealing will be published elsewhere [18].

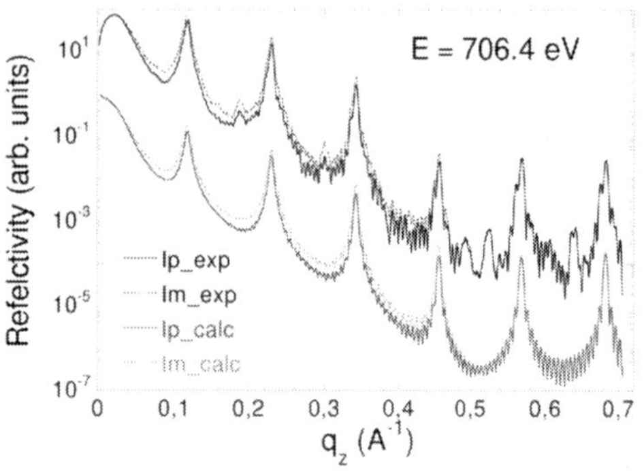

Figure 1. Experimental and corresponding refined (shifted for clarity) q dependent x-ray reflectivity curves (see text).

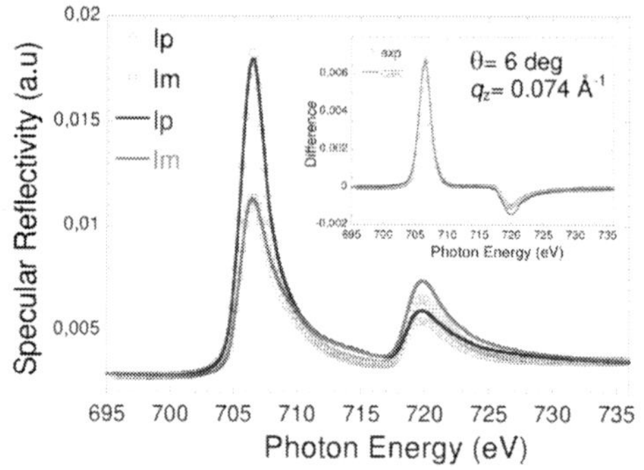

Figure 2. Experimental and refined energy dependent x-ray reflectivity curves. Inset: corresponding difference curves.

The structural properties of the granular multilayer were characterized by several techniques [16,17]. The x-ray reflectivity has been carried out far from Fe resonances (E=8051 eV and 740 eV) as well as at the Fe L_3 absorption edge (E=706.4 eV). The scans were recorded at room temperature with unpolarized light (8051 eV) and circularly polarized light (740 and 706.4 eV). Figure 1 displays the experimental intensities for two opposite directions of an applied magnetic field at 706.4 eV. The refinement of all the reflectivity data leads to the following structure C(57±5Å)/[FePt(12.8±0.2Å)/C(42.4±0.2Å)]$_{\times 20}$/SiO_2(27±3Å)/Si(001) with a period Λ=55.2±0.1Å with and roughnesses $\sigma_{C/vac}$=4.5 Å, $\sigma_{FePt/C}$=2.5 Å, $\sigma_{C/FePt}$=2.9 Å (σ_{SiO2} has been set to 2 Å and has no influence on the fit). The peaks in between the main Bragg peaks (BP) are not related to third harmonic photons. They are observable in the soft and hard x-ray range. Although it is possible to reproduce nicely these peaks by considering a superstructure (Λ'=3Λ) with three different thicknesses of the C layer (±0.5Å), which is highly unlikely, they are ascribed to complex scattering process between particles, not taken into account in the analysis. In the refinement, all the FePt and C layers in the multilayer stack were assumed to be identical except for the very top C layer, which was found to be thicker. It turned out to be very important to reproduce the magnetic asymmetry in the 0-0.1 $Å^{-1}$ range (see hereafter in Fig. 3). The best agreements impose to reduce the effective density of the FePt layers which is a granular layer. Indeed, a layer with the composition $(FePt)_xC_{1-x}$ should be taken into account. However, because of the strong contrast between the FePt and the carbon layers, there is a large uncertainty on the x parameter as determined from x-ray reflectometry. Also, using three slices to describe the FePt layers, it was not possible to evidence a change of the density along the growth axis inside the FePt layers and the concentration is found to be the same within 2%. Finally, the analysis indicates an increase of the density of the C layer in agreement with the presence of isolated Fe and Pt atoms in the C layers [17].

The magnetic reflectivity scans I(+) and I(-) were collected by reversing a 0.2T applied magnetic field both with respect to the incident photon energies at different angular position and with respect to the x-ray incidence angle at 706.4 eV. The analysis of the reflectivity and of the difference across the Fe $L_{2,3}$ edges, measured at 6 deg (Fig. 2), allows the determination of the averaged magnetization of Fe atoms inside the FePt layers [12]. The difference can be adjusted by reducing w_m to 0.73±0.03 which yield a magnetic moment of 1.58±0.06 μB.

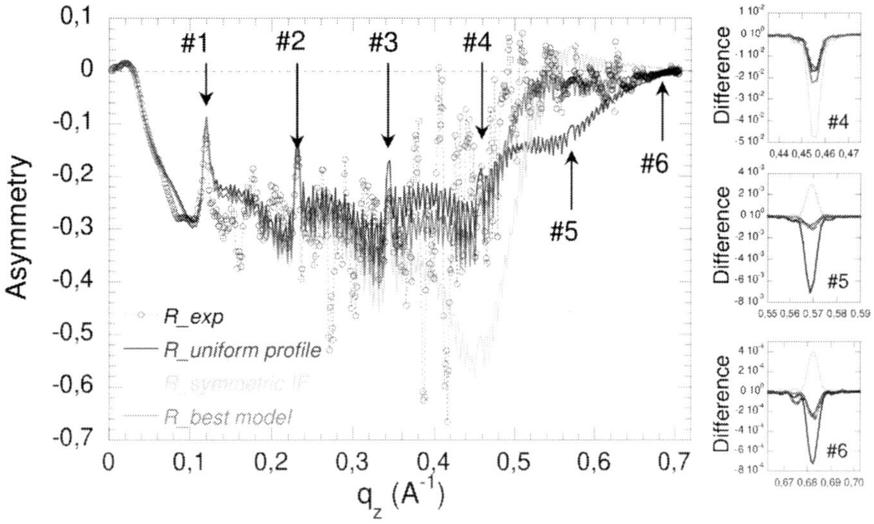

Figure 3. Experimental asymmetry measured at 706.4 eV along with three different models. The difference are also displayed on a linear scale for 4th, 5Th and 6th BP.

The magnetic profile across the FePt layer is determined from the analysis of the angular dependence of the asymmetry [I(+)-I(-)]/[I(+)+I(-)] or of the difference at the Bragg peaks (BP) position. We started from a uniform Fe magnetization profile inside the FePt layers (Fig. 3). The agreement between the experiment and the calculation is valid up to the 4th BP and strongly deviates above. Then, the FePt layers have been divided in three identically thick sublayers. The green curve in Fig. 3 shows also that a symmetric reduction of the magnetization from the interfaces, while keeping the averaged value equal to 0.73, does not improve the agreement. In particular, it is difficult to reproduce the area where the asymmetry converges to zero. The best result is obtained with w_m values from the bottom to the top equal to 0.8/0.82/0.6. It indicates that the profile is mostly constant inside the FePt layers except close to the top surfaces with a 25% reduction of the magnetization.

We point out that XRMR is indeed sensitive to the product concentration time magnetization [6]. Therefore, our result may indicate alternatively that the Fe concentration varies at the top of the layer either due to inhomogeneity or to the shape of the particles (reduction of the diameter during growth). However, this concentration change was not evidenced using x-ray reflectivity refinement. Hence, the change in magnetization is tentatively ascribed to a variation of the disorder inside the particle from the bottom to the top, not detectable by hard x-ray diffraction [17] or by x-ray reflectivity. We encounter here a limitation inherent to the approach, showing that the details of the structure must be known in order to draw firm conclusions about the structure of the magnetization. However, SXRMR turned out to be the only technique to detect such a small change on a subnanometer scale.

3.2. Out-of-plane exchange bias system

The depth dependence of the out-of-plane magnetization across the AF/FM interface of the ultra-thin $[NiO(3ML)/CoO(3ML)]_{\times 3}/PtCo/Pt(111)$ bilayer with perpendicular exchange bias has been investigated by SXRMR [19]. The measurements were carried out in remanence at T = 340 K, right above T_B and below T_N, where the FM layer (1 ML of Co below 1ML of Pt) has a strong perpendicular magnetic anisotropy (PMA) and where the interfacial magnetic coupling with the AF layer constrains the reversal process, as indicated by the increased H_C.

Angle dependent reflected intensities were collected at 776.9 and 778.5 eV (Fig. 4) close to the Co L_3 edge. At both energies, a separation of the curves with right and left polarized light is observed. The magnetic asymmetries are close to zero at small angles and exhibit larger amplitude at high angles (Fig. 5), as expected for an out-of-plane magnetization component. The interference in the angle dependence of R indicates the Co magnetization extends beyond the ultra thin FM Pt-Co layer inside the oxide layer. A further indication of the net magnetization in the AF oxide layers came from measurements at the Ni L_3 edge, which showed a small net magnetic moment on Ni atoms, either due to an extended polarization through the first 3 ML CoO layer or by contact with the PtCo interface due to discontinuity of this layer. Both situations indicate that the Co atoms in the oxide layer, at least close to the interface, are magnetically polarized.

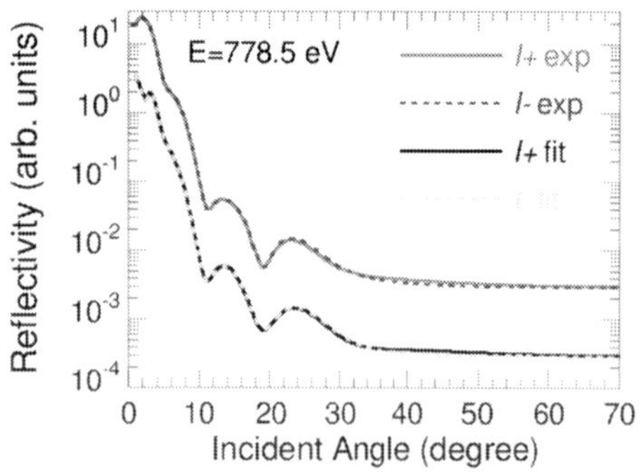

Figure 4. Experimental reflectivity curves and fits at 340 K.

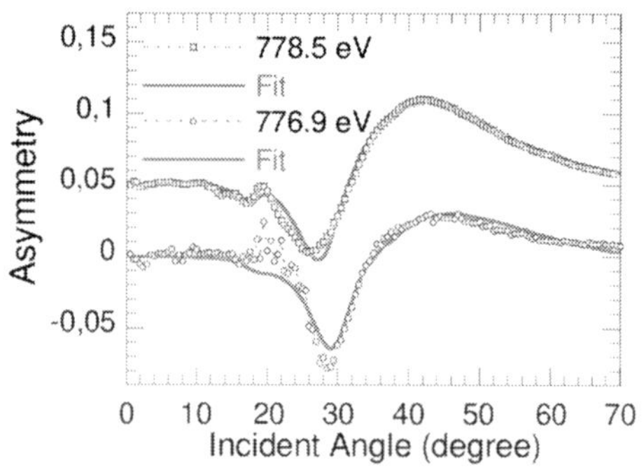

Figure 5. Experimental asymmetries and fits at 340 K.

The structural parameters of the film were derived from the refinement of the average reflectivity [19] and found in good agreement with other structural data from SXRD and AFM [20]. The magnetic profile comes from the analysis of the R at both energies. A model assuming that Co atoms in the oxide may be magnetically ordered up to the first CoO/NiO interface works well. The refinement of the magnetic structure was performed by dividing the first CoO layer in three slices and by adjusting their thicknesses as well as the magnetic moments carried by the Co atoms. The results indicate the out-of-plane magnetization is distributed beyond the 3.6 Å Pt-Co layer and extends over 10 Å in the oxide layer. The coupling to the FM Co spins is parallel in a 3Å thick slice, roughly the first CoO monolayer. Then, it is antiparallel in a 7 Å thick one, with similar amplitude. The last slice that completes the CoO layer is found to be 2 Å thick and has no net magnetization. Considering models with no net magnetization in the oxide layer, completely parallel or antiparallel magnetic slices and more extended magnetization with reduced magnetic amplitude do not fit the interference effect observed experimentally.

This result can be understood as follows: the Co atoms in the first CoO slice, right on top of the Pt-Co layer, have a mixed electronic character in between metallic and oxidized state. Their net induced magnetizations are likely to be dominated by the proximity of the Pt-Co layer and are ferromagnetically coupled to it. For the next CoO slice, the Co atoms are fully oxidized and the net magnetization is antiferromagnetically coupled to the first one. The net magnetization found equal to zero in the third slice means that beyond the second oxide layer the AF material break into domains that are not biased by the field cooling processes.

4. Conclusion

It is well established that combining spectroscopy and scattering brings magnetic spatial selectivity. SXRMR has been measured from ultrathin films over the maximum accessible angular range (or q range) with high signal to noise ratio. In depth in- and out-of-plane magnetic profiles have been obtained with sub-nanometer resolution. The sensitivity to magnetic configurations with an antiparallel component has been shown. Because of its sensitivity to the magnetization distribution, SXRMR is a technique of choice either to determine the magnetic profile inside thin films or to distinguish between different proposed magnetic models. This approach can be used to probe the extension of interfacial magnetic properties, in a “non-magnetic” or spacer layer, on each side of two layers with different magnetic phase. It should be also very useful to probe magnetic layers exhibiting both in- and out-of-plane magnetic components like systems exhibiting spin reorientation transition.

References

[1] Zhu Y (Ed.) 2005 *Modern techniques for characterizing magnetic materials* (Berlin: Springer)
[2] Kao C C *et al* 1990 *Phys. Rev. Lett.* **65** 373
[3] Tonnerre J M *et al* 1995 *Phys. Rev. Lett.* **75** 740
[4] Kortright J B and Kim S K 2000 *Phys. Rev.* B **62** 12216
[5] Hill J P and McMorrow D F 1996 *Acta Crystallogr. Sect.* A **52** 236
[6] Sève L *et al* 1999 *Phys. Rev.* B **60** 9662; Jaouen N *et al* 2002 *Phys. Rev.* B **66** 134420; Kravtsov E *et al* 2009 *Phys. Rev.* B **79** 134438
[7] Jaouen N *et al* 2004 *Phys. Rev.* B **70** 094417
[8] Brown S D *et al* 2008 *Phys. Rev.* B **77** 014427
[9] Stöhr J and Siegmann H C 2006 *Magnetism: From Fundamentals to Nanoscale Dynamics*, Springer Series in Solid-State Sciences, Vol. **152** (Berlin: Springer)
[10] Tonnerre J M *et al* 1998 *J. Appl. Phys.* **83** 6293; Roy S *et al* 2007 *Phys. Rev.* B **75** 014442; Brück S *et al* 2008 *Phys. Rev. Lett.* **101** 126402
[11] Jaouen N *et al* 2004 *J. Synchrotron Radiat.* **11** 353
[12] Déchelette A *et al* 1999 *Phys. Rev.* B **60** 6636
[13] Zak J *et al* 1991 *Phys. Rev.* B **43** 6423; *ibid* 1992 *Phys. Rev.* B **46** 5883
[14] Weiss N *et al* 2005 *Phys. Rev. Lett.* **95** 157204
[15] Liou S H *et al* 1996 *J. Appl. Phys.* **79** 5060; Sun S *et al* 2000 *Science* **287** 1989
[16] see Babonneau D *et al* 2007 *Rev. Adv. Mater. Sci.* **15** 198 and references herein
[17] Babonneau D *et al* 2008 *J. Phys. Condens. Matter.* **20** 035218
[18] Carbone D *et al* unpublished
[19] Tonnerre J M *et al* 2008 *Phys. Rev. Lett.* **100** 157202
[20] Tolentino H C N *et al* 2009 *Brazilian J. of Physics* **39** 150

AC-polarized neutron reflectometry: Application to domain dynamics in thin Fe film

K Zhernenkov[1,3], S Klimko[2], B P Toperverg[1,4] and H Zabel[1]

[1] Ruhr-Universität Bochum, 44780 Bochum, Germany
[2] Institute Laue-Langevin, BP 156, 38042 Grenoble, France
[3] Joint Institute for Nuclear research, 141980, Dubna, Russia
[4] Petersburg Nuclear Physics Institute, 188300, Gatchina, Russia

E-mail: Kirill.Zhernenkov@rub.de

Abstract. We report on the first application of polarized neutron reflectometry (PNR) to probe the re-magnetization kinetics in thin films subjected to periodic magnetic fields. Specular PNR and off- specular scattering were recorded from a 100 nm thick MBE grown Fe film under the application of AC magnetic fields with amplitudes up to 60 Oe and frequencies in the range of 0.3 - 1.8 MHz. It was found that up to 0.4 MHz a complete magnetization reversal of the Fe film takes place during each AC field cycle. Higher frequencies cause only a partial re-magnetization and above 0.8 MHz the AC field no longer alters the sample magnetization. High frequencies suppress magnetization fluctuations on a micrometer scale which are otherwise seen via off-specular scattering.

1. Introduction

Domain wall (DW) propagation in magnetic films and nanowires is presently a highly active field due to potential applications in magnetic storage media, such as the concept of a "Racetrack Memory" [1]. Three regimes can usually be distinguished. For low fields up to a critical value, the DW velocity increases linearly with the applied field. In this region the DW motion is viscous and the type of DW does not change during propagation. However, beyond a critical magnetic field, which is known as the Walker break down field [2], the DW speed rapidly decreases because of a new damping mechanisms and the transformation into different types of domain walls. DW motion in films and nanowires has been studied by various methods in recent years, such as by magnetoresistance [3], magnetic force microscopy [4], spin scanning electron microscopy (S-SEM) [5], and magneto-optics [6, 7]. Here we present a new method to probe the dynamics of magnetic domains in an AC magnetic field covering the range of magnetization reversal times down to fraction of a microsecond using polarized neutron reflectometry (AC-PNR). Most of the methods mentioned above work in direct space and time, providing images of propagating domain walls in magnetic nanowires. AC-PNR is a reciprocal space method, which has the advantage that it can measure the time evolution of magnetization depth profiles averaged over the sample surface.

2. Experimental procedures

The experimental geometry realized with the polarized neutron reflectometer ADAM at the Institut Laue Langevin[8](ILL, France) is sketched in Fig. 1. An AC magnetic field $H_{AC}(t) =$

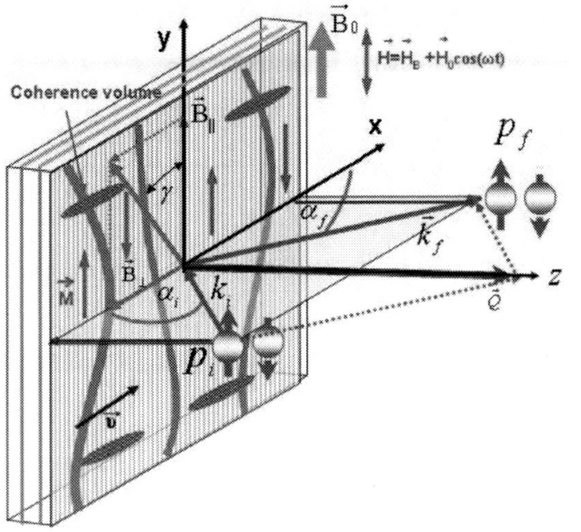

Figure 1. Scheme of the experiment. AC and bias DC field, as well as domain magnetization (red arrows), are collinear with polarization direction (blue arrows through "balls") which is normal to the scattering plane. The latter contains incoming, $\overrightarrow{\mathbf{k}}_i$, outgoing, $\overrightarrow{\mathbf{k}}_f$, wave vectors and the wave vector transfer $\overrightarrow{\mathbf{Q}}$. Red ellipses denote cross sections of the coherence ellipsoids with the sample surface. Specular reflection and off specular scattering intensity is a result of incoherent averaging over a number of those ellipsoids covering the whole surface (see for details Ref.[9]).

$H_0 \cos(2\pi f t)$ with frequencies $f = 0.3 - 1.8$ MHz was applied along one of the easy axis (Y-axis) within the film plane and perpendicular to the specular wave vector transfer $\overrightarrow{\mathbf{Q}}$. AC field amplitudes H_0 up to 60 Oe were applied to surely overcome the coercive field $H_c = 21$ Oe. In order to fix the initial saturated state of the sample, a bias field of $H_b = 25$ Oe was applied in addition to the AC field.

Two non-spin-flip (NSF) $R^{\pm\pm}$, and two spin-flip (SF), $R^{\pm\mp}$, reflectivities were recorded for $f = 0.3, 0.6$ and 1.5 MHz. A representative example for a set of data taken at f=0.3 MHz together with results of a model fit is depicted in Fig. 2. Data on specular reflection and off-specular diffuse scattering were collected at $f = 0$, 0.3, 0.45, 0.6, 0.8, 1.0 and 1.8 MHz with a position sensitive detector(PSD) for a polarized incident neutron beam but without polarization analysis of the scattered beam. Intensity maps arranged into R^+ and R^- panels are shown in Fig. 4(a,b), where the specular reflection ridge along the diagonal and diffuse scattering spots beside the ridge can be recognized. Recording AC-PNR data requires a special set-up, including a RF generator, amplifier, wide band transformer, and a resonance circuit allowing to vary amplitude and frequency of the periodic magnetic field over a wide range. The details of the set-up will be published elsewhere [10].

3. Experimental results

According to Fig. 2, in an AC field of 0.3 MHz the NSF reflectivities R^{++} and R^{--} are split above the angle of the total reflection. Hence, the sample magnetization averaged over the AC field period $T = 0.3$ microseconds is not zero. At the same time, each of NSF reflectivities reveal critical edges of total reflection for both spin components and hence the sample is not saturated. Weak SF reflectivity in Fig. 2 indicates small deviations of the magnetization vector from the applied field direction. Splitting between R^+ and R^- NSF reflectivities is also seen in Fig. 4. The R^+, R^--splitting significantly increases with increasing frequency and reaches a maximum value above $f = 0.8$ MHz. Simultaneously, the off-specular scattering becomes strongly asymmetric and then vanishes at 1.8 MHz. No splitting in the NSF reflectivities (see Fig. 3) and no asymmetry in the off-specular scattering is present if bias field is $H_b = 0$.

4. Discussion and conclusion

The experimental observations may quantitatively be described in a simplified scenario of re-magnetization driven by 180° DW propagation: DWs nucleated at the sample edges and/or

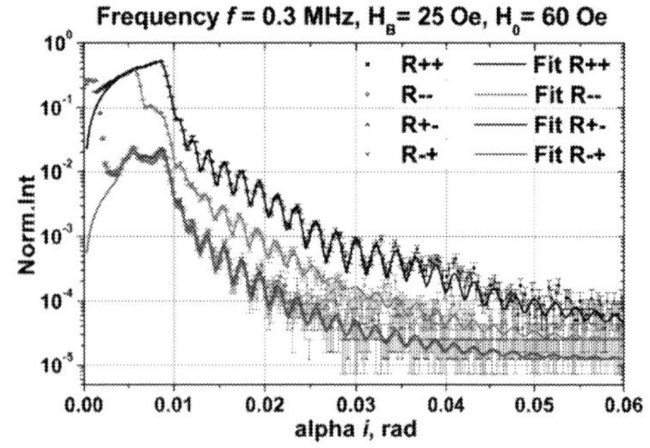

Figure 2. SF and NSF specular reflectivities measured (points) at f=0.3 MHz and H_b = 25 Oe fitted (lines) to the model of a partially magnetized state.

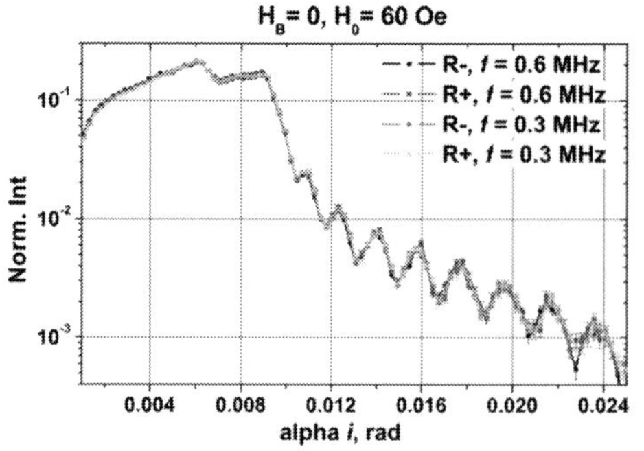

Figure 3. Specular reflectivities $R^+ = R^-$ at f=0.3 and 0.6 MHz, $H_0 = 60$ Oe and $H_b = 0$: The sample is in the totally demagnetized state.

in the bulk propagate along the X-axis against each other, erasing unfavorable domains. The DW's may start to move at the time moment $t = t_0$ (see, Fig. 5) at which $H(t_0) \approx -H_c$. If a field variation is so slow that $H(t) \leq -H_c$ during DWs propagation till they merge somewhere in the sample and annihilate, then reversal takes place. After reversal, the sample may stay in a negative saturation till $t = t_2$ at which $H(t_2) \approx +H_c$. Then the process is reversed: DWs are again nucleated and propagate back completing the re-magnetization cycle. At $H_b = 0$ re-magnetization in both directions should proceed over the same time span, while for $H_b \neq 0$ the sample may, on average, be preferentially magnetized in one direction. The fraction of the period spent in one of the saturated states can readily be deduced via fitting the data to the incoherent sum of reflection coefficients for two magnetized states weighted with the corresponding fraction of the period. The latter is plotted in Fig. 6 against the frequency at two values of the AC field amplitudes $H_0 = 60$ Oe and $H_0 = 55$ Oe. From Fig. 6 it follows that for $f < 0.4$ MHz the fraction of the period well coincides with that expected from the ratio between H_b and H_0. However, at elevated frequencies this fraction decreases and finally vanishes at $f \sim 1.5$ MHz. At high frequencies DWs do not succeed to pass all the way till annihilation before the moment of time $t = t_1$ at which the increasing external field reaches the value $H(t_1) \approx -H_c$. Then the reversal process is terminated until the moment of time $t = t_2$. After that the DWs propagate

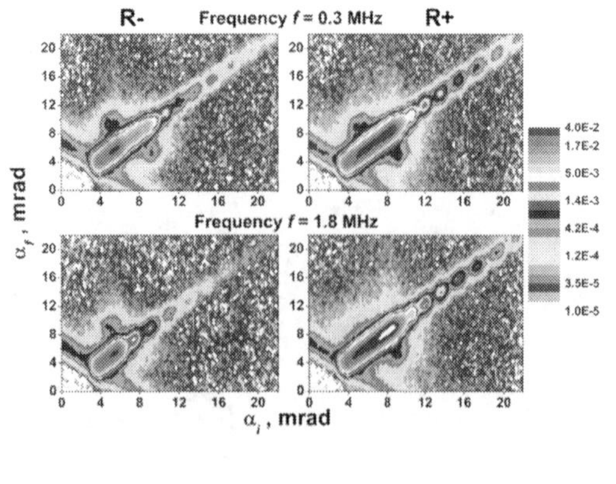

Figure 4. Scattering maps R^- (left) and R^+ (right)recorded for 0.3 MHz (top panel) and 1.8 MHz (bottom panel). Here the AC field amplitude is $H_0 = 60$ Oe, the bias field is $H_b = 25$ Oe. Specular reflection is represented by ridges running along the lines with equal angles $\alpha_i = \alpha_f$ of incidence and scattering. The off-specular scattering seen bulging out near specular ridges is strongly asymmetric at high frequency.

back restoring initial saturation state where the sample spends a certain amount of time before the next cycle starts. The lowest frequency at which the magnetization reversal is no longer complete allows to estimate the time for DW nucleation. For the present sample this is on the order of 0.5 microseconds. So far we can only determine the re-magnetization rate of the sample, but not the DW velocity. This requires to know the distance that the DWs travel in one cycle. Our results strongly indicate that the distance is bigger than the neutron coherence length which is estimated to be of about 100 micrometers. This gives a lower limit for the mean DW velocity of 30 m/sec. More accurate determination of the DW velocities will, however, be possible in forthcoming experiments with time resolved PNR analysis synchronized with the AC field. Such experiments should also shed more light on the nature of the off-specular scattering seen in Fig. 4, which may result from magnetic inhomogeneities (DW nucleation centers, residual domains, spin waves, etc.). At present we can only infer from our simulations of the off-specular maps that the residual magnetic domains are about 1500 nm in size and that the magnetization direction is randomly tilted by a few degrees against the applied field. Those magnetic inhomogeneities are sill detectable above the saturation field, but vanish at high frequencies, or in static fields $H \geq 120$ Oe.

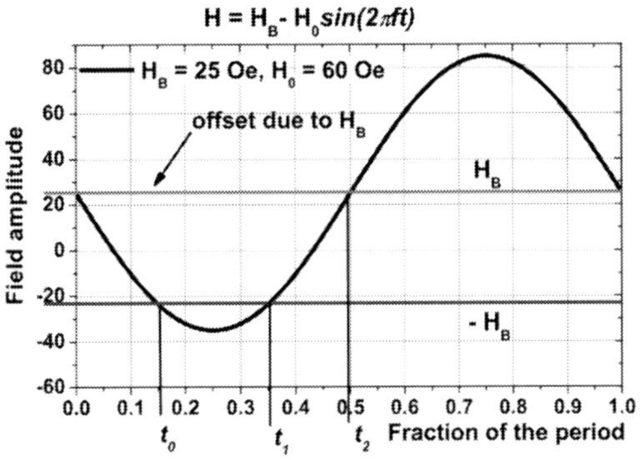

Figure 5. Variation of the AC field amplitude during one period.

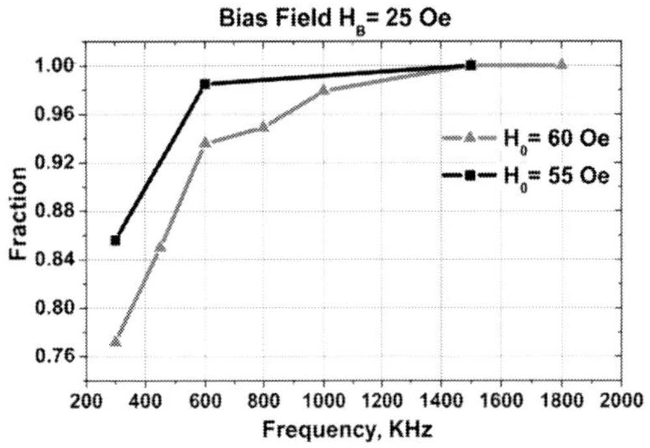

Figure 6. Dependence of the fraction of the initial state magnetization on the frequency for field amplitudes $H_0 = 55$ and $H_0 = 60$ Oe.

Acknowledgments
This work was funded by BMBF Verbundforschung Grant No. 05KN7PC1.

References
[1] Parkin S S P, Hayashi M and Thomaset L 2008 *Science* **320** 190
[2] Schryer N L and Walker L R 1974 *J. Appl. Phys.* **45** 5406
[3] Ono T, Miyajima H, Shigeto K, Mibu K, Hosoito N and Shinjo T 1999 *Science* **284** 468
[4] Yamaguchi A, Ono T, Nasu S, Miyake K, Mibu K and Shinjo T 2004 *Phys. Rev. Lett.* **92** 077205
[5] Kläui M et al 2005 *Phys. Rev. Lett.* **95** 026601
[6] Yamanouchi M, Chiba D, Matsukura F, Dietl T and Ohno H 2006 *Phys. Rev. Lett.* **96** 096601
[7] Nistor C, Beach G S D and Erskine J L 2006 *Rev Sci. Inst.* **77** 103901
[8] Wolff M, Zhernenkov K and Zabel H 2007 *Thin Solid Films* **515** 5712
[9] Zabel H, Theis-Bröhl K and Toperverg B P 2007 *Handbook of Magnetism and Advanced Magnetic Materials* vol.12 ed H Kronmüller and S Parkin (New–York: Wiley) p 1237
[10] Klimko S, Zhernenkov K, Toperverg B and Zabel H to be published

Polarized Neutrons and Synchrotron X-rays for Magnetism Conference 2009 IOP Publishing
Journal of Physics: Conference Series **211** (2010) 012017 doi:10.1088/1742-6596/211/1/012017

Supported and embedded Fe nanoparticles: influence of the environment on shape and interface contributions to the magnetic anisotropy

A Kleibert[1,2], F Bulut[3,4], W Rosellen[3], K H Meiwes-Broer[1], J Bansmann[4] and M Getzlaff[3]

[1] Institute of Physics, University of Rostock, Universitätsplatz 3, D-18051 Rostock, Germany
[2] Swiss Light Source, Paul Scherrer Institute, CH-5232 Villigen, Switzerland
[3] Institute of Applied Physics, University of Düsseldorf, Universitätsstr. 1, D-40225 Düsseldorf, Germany
[4] Institute of Surface Chemistry and Catalysis, Ulm University, Albert-Einstein-Allee 47, D-89069 Ulm, Germany

E-mail: getzlaff@uni-duesseldorf.de

Abstract. A continuously working arc cluster ion source was used to prepare mass-filtered Fe nanoparticles with a mean size of 6 to 10 nm. Their structure was determined by HRTEM. The nanoparticles were deposited into an Al matrix and additionally on a W(110) substrate for comparison. Within the matrix they maintain their spherical shape and do not show any magnetic anisotropic effect. For uncapped Fe nanoparticles on a tungsten surface, a flattening is observed by STM and the nanoparticles exhibit a distinct magnetic anisotropy with the easy magnetization axis being in the surface plane. Correlating both investigations we conclude that this observation is due to shape and interface-induced contributions to the magnetic anisotropy.

1. Introduction

Ferromagnetic nanoparticles with a tunable size of up to 20 nm and a narrow size distribution have attracted considerable interest for future applications ranging from medicine and chemistry to ultrahigh density magnetic storage devices [1]. Many different routes, especially chemistry-based methods suited for large scale applications, have been proposed and partially successfully tested for the production of nanoparticles with desired physical properties [2, 3]. The shape and the crystalline structure of such particles (e.g., [4]) are often influenced by kinetic barriers during the growth process and thus these systems might deviate from thermal equilibrium behavior [5]. As a consequence this class of materials opens the possibility to realize systems with tunable interesting magnetic, structural, and electronic properties. From a technological point of view it is essential to keep in mind that each realistic application requires supported and often embedded nanoparticles in order to guarantee functionality over a sufficiently long time period. However, many studies showed that the particle properties may significantly change when being in contact with a support or a matrix, respectively, particularly when magnetic properties are considered [6]. In order to distinguish pure geometrical (e.g. symmetry) and interface-related effects from additional preparation-induced phenomena it is necessary to investigate the same particles but in different environments. Experimentally, this can be achieved when studying nanoparticles preformed in the gas phase and subsequently deposited onto different substrates or by codeposition into respective matrices [3, 7–10].

In the present contribution we focus on preformed, mass-filtered Fe nanoparticles being deposited under soft-landing conditions onto (i) a W(110) substrate and (ii) by codeposition into an Al matrix. Combining high resolution transmission electron microscopy (HRTEM), scanning tunneling microscopy (STM), and X-ray magnetic circular dichroism (XMCD) enables us to study the correlation of structure, shape and the magnetic anisotropy [11] in both environments.

2. Experimental details

The Fe particles have been prepared by means of an arc cluster ion source (ACIS) [12]. This source provides mass-filtered pure and alloyed *3d* transition metal nanoparticles with a kinetic energy below 0.1 eV per atom ensuring fragmentation-free deposition conditions [13]. The particle diameters are tuned by applying a dc voltage U_{quad} to an electrostatic quadrupole filter. Using this experimental setup, particles in a size range between 4 and 20 nm are available [14]. XMCD and STM investigations have been carried out *in situ* under UHV conditions (base pressure $< 1 \times 10^{-9}$ mbar). The W(110) crystal was prepared by cycles of heating in oxygen atmosphere and flashing as described in the literature [15, 16]. The surface cleanliness and quality were checked by means of low energy electron diffraction prior to the cluster deposition. In order to avoid agglomeration as well as magnetic dipolar interactions particle ensembles with low density ($<$ 200 nanoparticles per μm^2) were prepared in all experiments described here. For the matrix sample first a thick layer of Al was evaporated from a crucible onto a native Si(111) wafer. Subsequently, the Fe particle deposition was started simultaneously to the Al evaporation. Finally, a thick capping layer of Al was prepared to ensure that the particles are fully covered by Al. The XMCD experiments were carried out using circularly polarized synchrotron radiation provided by the PM3 beamline at BESSY (Berlin). Details on the XMCD effect and its applications can be found in the literature [17–19]. Here, magnetization curves were obtained recording the sample drain current (total electron yield detection) with a fixed polarization and the light at an angle of incidence of 30° relative to the surface when applying an external magnetic field in the surface plane and 60° when applying the magnetic field along the perpendicular direction. The photon energy was set to the L_3 edge of Fe. A magnetic field $\vec{B}$ up to 30 mT was applied at variable angle with respect to the sample surface normal allowing to probe anisotropic magnetic properties. For the STM experiment a commercial MicroSTM (Omicron) was used. The probe tips are prepared by chemical etching of a tungsten wire. All measurements were carried out at room temperature in the constant current mode.

3. Results and discussion

3.1. Structure and shape of the Fe particles

Fe particles prepared by the ACIS cluster source have already been characterized in some of our previous studies [14, 20]. Similarly to other experiments a bulk-like bcc lattice in the size range from 6 to about 10 nm was found by means of *ex situ* TEM [21, 22]. The respective particle shape in thermal equilibrium is given by six (001) and twelve (110) surface facets forming a truncated dodecahedron as expected from the Wulff theorem [20, 22, 23]. Two high resolution TEM images are here given as examples in Figs. 1 (a) and (b). The particle in Fig. 1 (a) was deposited onto a carbon-coated TEM grid and subsequently exposed to ambient air for the transport to the microscope. A fourier transform of the image (shown in the inset) reveals a lattice parameter of 0.290 ± 0.005 nm being practically identical with the bulk value of 0.287 nm. After correction for the presence of the native oxide layer we can deduce an initial diameter of the pure metal particle of about 9 nm. We may note that the particle shape in Fig. 1 (a) is largely determined by the oxide shell consisting of irregularly grown oxide grains [24] and thus does not reflect the shape of the pure Fe particle prior to deposition. Fig. 1 (b) shows a particle of similar size but

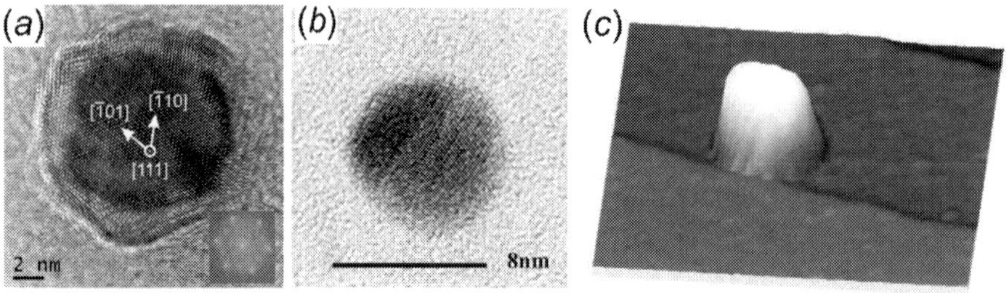

Figure 1. (a) HRTEM image of an Fe nanoparticle deposited on a carbon-coated TEM grid. An oxide shell was formed when exposing the particle to ambient air. A fourier transform of the image (shown in the inset) indicates a bcc lattice with bulk-like lattice parameter of 0.290 ± 0.005 nm in the metallic core. (b) A pure Fe nanoparticle of similar size but embedded in an aluminum matrix. (c) STM image of an individual cluster near a step edge on W(110). The height of the particle amounts to about 7 nm. The image size is 73 nm $\times$ 67 nm and the tunneling parameters are $I = 0.5$ nA, $U = +1.0$ V.

deposited into an Al matrix. As discussed in Ref. [20] the embedded particle is free of oxide and its apparent morphology can be attributed to the above described equilibrium shape.

The actual particle morphology in a deposition experiment may significantly depend on the chosen substrate and on the landing conditions. The equilibrium shape of a supported nanoparticle is in general modified by the interface energy [25]. The high surface energy of W(110) (4 $\mathrm{Jm^{-2}}$) relative to that of Fe (2.4 $\mathrm{Jm^{-2}}$) results in a distinct wetting behavior in respective heteroepitaxy experiments [26, 27]. Accordingly, one may expect that under equilibrium conditions Fe particles are smeared out on W(110) resulting in flat islands of monatomic height as discussed in [28]. However, kinetic barriers can hinder the preformed particles to adopt equilibrium conditions upon deposition. In molecular dynamics simulations it was found that for sufficiently small kinetic and adsorption energy released upon contact to the substrate the initial particle shape can be conserved [13]. Using *in situ* STM we find that under the present conditions the Fe nanoparticles clearly retain a three-dimensional shape upon deposition on W(110) as shown in Fig. 1 (c). Moreover, the particles are randomly distributed on the surface and no signature of diffusion to step edges or surface defects is found (not shown here). Since the lateral dimensions in STM images are affected by convolution effects when tip radius and particles are of comparable size [29, 30] one cannot directly infer the particle morphology from images as given in Fig. 1 (c). However, the particle height can unambiguously be measured [being 7 nm in Fig. 1 (c)] and can be related to the particle diameter determined by means of TEM.

A respective comparison of STM heights and TEM diameters is shown in the upper panel of Fig. 2 for different particle sizes being determined by the setting of U_{quad} in the mass-filtering unit of the ACIS. In each investigation more than 250 particles were measured concerning the diameter and the height, respectively, resulting in an error of about ± 0.5 nm. Obviously, the particles appear partially flattened on the W(110) surface. Thereby, the decrease in the height of the size-selected clusters on W(110) is about 1 nm when compared to the TEM images. A similar flattening has also been found for FeCo alloy clusters on W(110). As discussed in Ref. [31] we ascribe this to thermally activated surface diffusion initiated immediately upon the

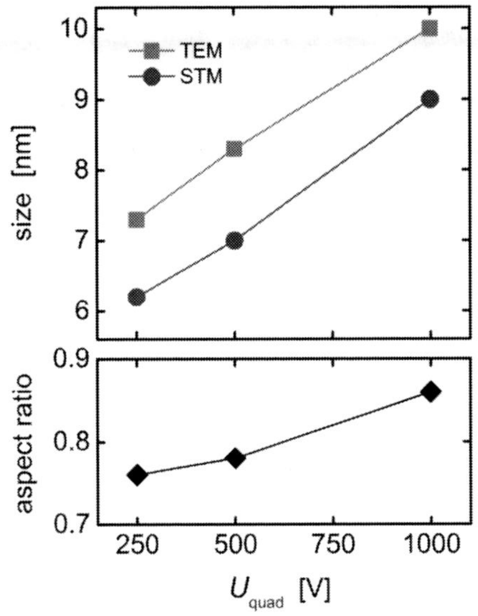

Figure 2. Upper panel: Comparison between the lateral size of size-selected clusters (TEM: red squares) and the height of clusters on W(110) (STM: blue circles) for different cluster sizes. The error amounts to ±0.5 nm. They were produced with variation of the voltages at the electrostatic quadrupole U_{quad}. Lower panel: Corresponding size-dependent aspect ratio. The error amounts to ±0.2.

impact on the substrate, i.e., before the released energy is dissipated. After thermalization no further changes are observed at room temperature and on the time scale of our experiments (a few hours). This also nicely demonstrates how structures far from equilibrium conditions can be achieved and stabilized by deposition of preformed nanoparticles. A detailed discussion concerning the equilibrium shape of crystalline nanoparticles deposited onto a lattice mismatched planar substrate can be found in [32]. Assuming a constant volume before and after deposition and a shape of a spherical cap of the particles in contact with the substrate we can calculate the size-dependent aspect ratio as shown in the lower panel of Fig. 2. The deviation from a spherical shape (aspect ratio = 1) clearly decreases with the particle size. This can also be understood, when considering thermally activated diffusion upon impact. Thereby the length of the diffusion paths being required for an effective material transport increases with the particle size, while the diffusion constant stays constant.

3.2. Magnetic behavior

In situ recorded magnetization curves of Fe nanoparticles on W(110) are significantly affected by the observed shape changes. As shown in Fig. 3 particles deposited with a diameter of about 10 nm are easily magnetized in the surface plane (solid circles), but significantly higher magnetic fields are required for out-of-plane magnetization (open circles). Our data suggest an uniaxial magnetic anisotropy with a hard axis perpendicular to the sample surface. However, the magnetic moments of the particles are not ferromagnetically blocked but thermally fluctuating in the sample plane. The respective magnetization curves can be described by the statistical physics partition function applied to a macrospin as discussed, e.g., in Refs. [33, 34]. The magnetic anisotropy energy of such a sample is then extracted by fitting the partition function to the datasets. The black lines in Fig. 3 are obtained by a simultaneous fit to both, the in-plane and out-of-plane data and reveal an anisotropy energy of $K = (5.7 \pm 1.7)$ μeV/atom. This value can be compared to calculated shape and interface anisotropy contributions. Based on the observed aspect ratio of about 0.9, respective calculations yield $K_{\text{shape}} = 1.3$ μeV/atom and $K_{\text{interface}} = 6.4$ μeV/atom, cf. Ref. [31], being close to the experimental value. However, the magnitude of the interface term may depend on the orientation of the particles on the W(110) surface. In particular, $K_{\text{interface}}$ might be different for particles resting with their (001) or their (110) facets on the substrate, respectively. As discussed above tip convolution effects prevent us to obtain the orientation of the deposited particles from the STM data. Thus, additional efforts are required to unambiguously determine the origin of the observed anisotropy [35]. Finally, we may note that the fits shown in Fig. 3 are obtained by numerical integration of the partition function. Recently, an analytical expression was derived, which applies when the magnetic field is parallel to the anisotropy axis [36]. This formula might be alternatively used to determine the anisotropy energy from respective data. A further possibility is given by analyzing the initial susceptibility of the magnetization curves with expressions given in Ref. [33].

Magnetization curves were also measured for Fe nanoparticles (with a diameter of 8 nm) embedded in the Al matrix as shown in Fig. 3. The full diamonds (open diamonds) correspond

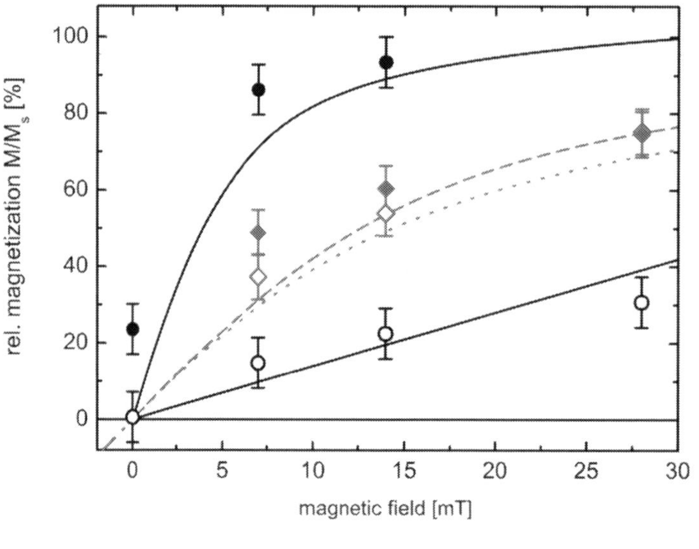

Figure 3. Relative magnetization of 10 nm Fe clusters on W(110) with the magnetic field applied in the surface plane (solid circles) and out-of-plane (open circles) as measured at room temperature. The solid lines are fits to the data as discussed in Ref. [31]. Respective data for 8 nm Fe nanoparticles embedded in an Al matrix with the magnetic field applied in the surface plane (solid diamonds) and out-of-plane (open diamonds). The dashed line depicts a corresponding Langevin curve and the dotted line shows the effect on the magnetization curve when the particles have a relatively weak uniaxial anisotropy with randomly oriented easy axes.

to the magnetization measured with the magnetic field applied parallel (perpendicular) to the sample surface. For this situation there is no evidence for an anisotropic behavior. We may note that the cubic magneto-crystalline anisotropy contribution of bcc Fe amounts to $K_{\text{Fe}} = 3.3$ μeV/atom [37] and the onset of superparamagnetism for particles in the present size range is therefore expected to be far below room temperature (< 25 K [38]). Although the temperature in the present experiments is more than 10 times larger than the blocking temperature, the magnetization curves may still deviate from a respective Langevin curve [36].

To evaluate the effect of the anisotropy we have calculated a Langevin curve given by $M/M_s = coth(NmB/k_BT) - k_BT/NmB$, with m being the magnetic moment per atom, N the number of atoms per particle, k_B the Boltzmann constant, T the temperature, and B the magnitude of the magnetic field. Using $m = 2.2\ \mu_B$ for Fe, $N = 28,000$ corresponding to a particle size of 8 nm, and $T = 290K$ we obtain the dashed curve in Fig. 3 showing a good agreement with the experimental data. For comparison we have computed the magnetization curve of particles with the same size, magnetic moments and at the same temperature, but additionally including the effect of a *uniaxial* anisotropy with randomly distributed easy axes in a respective particle ensemble. Assuming K_{Fe} the statistical partition function yields the dotted curve in Fig. 3 which shows noticeable deviations from the Langevin curve. However, we may note that the magneto-crystalline anisotropy of iron is of *cubic* nature with fourfold symmetry. This may lead to a more Langevin-like behavior of the present particles as suggested by the data in Fig. 3.

Independently of the above discussion, the impact of the environment on the magnetic properties of Fe nanoparticles becomes directly obvious from Fig. 3. The deposition of Fe particles onto a flat W(110) substrate leads to a partial flattening compared to the equilibrium shape. First, this reduces the symmetry of the particle shape. In addition to this the contact interface is different from the remaining particle surface. Both effects lead to an anisotropic magnetic behavior as discussed in detail in Ref. [31]. In contrast, embedding similar particles in an Al matrix retains the high symmetry of the preformed particles as shown in Fig. 1 (b). In fact, the lower surface energy of Al (about 1.1 Jm^{-2} [27]) relative to that of Fe leads to a dewetting behavior thus supporting a spherical particle geometry. As a consequence no shape anisotropy is developed and interface contributions cancel out resulting in an isotropic superparamagnetic behavior.

4. Conclusion

A continuously working arc cluster ion source was used to prepare mass-filtered Fe nanoparticles with sizes ranging from 6 to 10 nm. Their structure was determined by HRTEM. The nanoparticles were deposited on a W(110) substrate and embedded into an Al matrix. In the matrix they maintain their spherical shape, after deposition on a plane substrate without capping a flattening is found by STM, respectively. The magnetic properties were investigated by XMCD. Whereas the deposited nanoparticles exhibit a distinct magnetic anisotropy with the hard magnetization axis being perpendicular to the surface plane the embedded nanoparticles do not show any anisotropic effect. Correlating these investigations we conclude that this observation is due to distinct differences of the respective magnetic shape and interface anisotropy contributions in both systems.

Acknowledgments

We thank J. Carrey (Thomson-CSF, Orsay, France) for TEM investigations, the staff of BESSY for technical assistance, and the DFG for financial support within the priority call 1153 (clusters in contact with surfaces) via DFG BA 1612/3-3, DFG KL 2188/1-3, and DFG GE 1026/4-3.

References

[1] Huber D L 2005 *Small* **1** 482
[2] Sun S H, Murray C B, Weller D, Folks L and Moser A 2000 *Science* **287** 1989
[3] Palmer R E, Pratontep S and Boyen H G 2003 *Nat. Mater.* **2** 443
[4] Antoniak C, Trunova A, Spasova M, Farle M, Wende H, Wilhelm F and Rogalev A 2008 *Phys. Rev. B* **78** 041406
[5] Puntes V F, Krishnan K M and Alivisatos A P 2001 *Science* **291** 2115
[6] Bansmann J *et al.* 2005 *Surf. Sci. Rep.* **56** 189
[7] Bromann K, Felix C, Brune H, Harbich W, Monot R, Buttet J and Kern K 1996 *Science* **274** 956
[8] Getzlaff M, Kleibert A, Methling R, Bansmann J and Meiwes-Broer K H 2004 *Surf. Sci.* **566** 332

[9] Getzlaff M, Bansmann J, Bulut F, Gebhardt R K, Kleibert A and Meiwes-Broer K H 2006 *Appl. Phys.* A **82** 95
[10] Bansmann J, Getzlaff M, Kleibert A, Bulut F, Gebhardt R K and Meiwes-Broer K H 2006 *Appl. Phys.* A **82** 73
[11] Getzlaff M 2008 *Fundamentals of Magnetism* (Berlin Heidelberg: Springer)
[12] Methling R P, Send V, Klinkenberg E D, Diederich T, Tiggesbäumker J, Holzhüter G, Bansmann J and Meiwes-Broer K H 2001 *Eur. Phys. J.* D **16** 173
[13] Haberland H, Insepov Z and Moseler M 2009 *Phys. Rev.* B **79** 125423
[14] Kleibert A, Passig J, Meiwes-Broer K H, Getzlaff M and Bansmann J 2007 *J. Appl. Phys.* **101** 114318
[15] Musket R G, McLean W, Colmenares C A, Makowiecki D M and Siekhaus W J 1982 *Appl. Surf. Sci.* **10** 143
[16] Bode M, Pascal R and Wiesendanger R 1995 *Surf. Sci.* **344** 185
[17] Stöhr J 1999 *J. Magn. Magn. Mat.* **200** 470
[18] Stöhr J and Siegmann H C 2006 *Magnetism* (Berlin Heidelberg: Springer)
[19] Kortright J B, Awschalon D D, Stöhr J, Bader S D, Idzerda Y U, Parkin S S P, Schuller I K and Siegmann H C 1999 *J. Magn. Magn. Mat.* **207** 7
[20] Kleibert A, Meiwes-Broer K H and Bansmann J 2009 *Phys. Rev.* B **79** 125423
[21] Jamet M, Wernsdorfer W, Thirion C, Dupuis V, Mélinon P, Pérez A and Mailly D 2004 *Phys. Rev. B* **69** 024401
[22] Vystavel T, Palasantzas G, Koch S A and De Hosson J T M 2003 *Appl. Phys. Lett.* **82** 197
[23] Wulff G 1901 *Z. Kristallogr.* **34** 449
[24] Fauth K, Goering E, Schütz G and Kuhn L T 2004 *Journal of Applied Physics* **96** 399
[25] Henry C R 2005 *Prog. Surf. Sci.* **80** 92
[26] Gradmann U, Liu G, Elmers H J and Przybylski M 1990 *Hyperfine Interactions* **57** 1845
[27] Vitos L, Ruban A V, Skriver H L and Kollár J 1998 *Surf. Sci.* **411** 186
[28] Reuter D, Gerth G and Kirschner J 1997 *J. Appl. Phys.* **82** 5374
[29] Hoevel H and Barke I 2006 *Prog. Surf. Sci.* **81** 53
[30] Sell K, Kleibert A, von Oeynhausen V and Meiwes-Broer K H 2007 *Eur. Phys. J. D* **45** 433
[31] Kleibert A, Bulut F, Gebhardt R K, Rosellen W, Sudfeld D, Passig J, Bansmann J, Meiwes-Broer K H and Getzlaff M 2008 *J. Phys.: Condens. Matter* **20** 445005
[32] Müller P and Kern R 2000 *Surf. Sci.* **457** 229
[33] Chantrell R, Ayoub N and Popplewell J 1985 *J. Magn. Magn. Mat.* **53** 199
[34] Gambardella P *et al.* 2003 *Science* **300** 1130
[35] Kleibert A, Voitkans A and Meiwes-Broer K H 2009 *submitted*
[36] Fruchart O, Jubert P O, Meyer C, Klaua M, Barthel J and Kirschner J 2002 *J. Magn. Magn. Mat.* **239** 224
[37] Klein H P and Kneller E 1966 *Phys. Rev.* **144** 372–374
[38] Aharoni A 2000 *Introduction to the Theory of Ferromagnetism* (Oxford: Oxford University Press)

Journal of Physics: Conference Series **211** (2010) 012018 doi:10.1088/1742-6596/211/1/012018

Hard x-ray resonant scattering study of $Ni_{81}Fe_{19}(111)/CoO(111)$ exchange biased bilayer

R Feyerherm, E Dudzik, S Valencia and F Radu
Helmholtz-Zentrum Berlin für Materialien und Energie, BESSY, 12489 Berlin, Germany
E-mail: feyerherm@helmholtz-berlin.de

Abstract. We have employed hard x-ray resonant magnetic scattering to study the bulk antiferromagnetic domain state of the archetypal epitaxial $Ni_{81}Fe_{19}(111)/CoO(111)$ exchange biased bilayer ($Ni_{81}Fe_{19}$ = Permalloy = Py). Tuning the x-ray energy to the Co K-edge, we were able to detect the $(\frac{1}{2}\frac{1}{2}\frac{1}{2})$ antiferromagnetic magnetic Bragg reflection from a 2000 Å thick CoO film covered by a 120 Å Py layer and capped by 50 Å gold. The temperature dependence of the $(\frac{1}{2}\frac{1}{2}\frac{1}{2})$ Bragg intensity and the rocking-width agrees well with previous neutron diffraction data. The average lateral size of antiferromagnetic domains is about 40 nm close to $T_N = 291$ K.

1. Introduction

The exchange bias effect is of great importance for basic research and numerous device applications. It manifests itself in a shift of the hysteresis loop in the negative or the positive direction with respect to the applied field. Its origin is related to the magnetic coupling across the common interface shared by a ferromagnetic (FM) and an antiferromagnetic (AF) layer. The particular domain state of the AF layer has a significant effect on the magnitude of the exchange bias, which usually is much smaller that predicted by idealized models. Recently, the origin of the reduced exchange bias in epitaxial $Ni_{81}Fe_{19}(111)/CoO(111)$ bilayer has been investigated by combined soft and hard x-ray resonant magnetic scattering (XRMS) and polarised neutron diffraction [1]. Here, we provide a detailed description of the XRMS results, including new data.

2. Experimental

A 2000 Å thick CoO film was grown on an $Al_2O_3(11\bar{2}0)$ substrate, covered by a 120 Å Py layer and capped by 50 Å gold at the Helmholtz-Zentrum Berlin (HZB) by dc-magnetron sputtering as reported in [1]. The excellent smoothness of the Py/CoO interface has been proven by x-ray diffraction, showing the presence of several Laue oscillations around the Py(111) Bragg reflection. Polarized neutron reflectivity data result in a roughness parameter of ≤ 7Å for the interfaces. The AF CoO layer was studied using XRMS close to the Co K-edge (7.71 keV). The measurements were performed in zero magnetic field at the beamline MAGS, operated by the HZB at the synchrotron source BESSY [2]. The sample was mounted on a closed-cycle cryostat equipped with Be domes. Cooling the sample from 300 K to 10 K in a 800 Oe magnetic field applied parallel to the sample surface, i.e., saturating the ferromagnetic Py layer, showed no significant effect on the widths or intensities of magnetic superlattice reflections compared to zero-field cooling data. In order to separate magnetic from structural contributions, linear polarisation analysis of the scattered beam was carried out using the (006) reflection of a highly

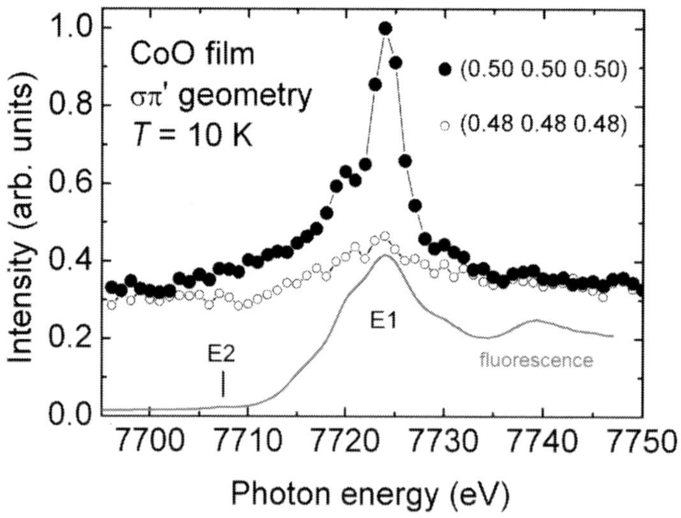

Figure 1. X-ray energy dependence of the $(\frac{1}{2}\frac{1}{2}\frac{1}{2})$ Bragg reflection intensity through the Co K-edge measured for the 2000 Å thick CoO film with linear polarization analysis ($\sigma\pi'$ channel). For comparison, the energy dependence at an off-peak reciprocal space position has also been measured, reflecting the background from fluorescence and ($\pi\pi'$ cross-talk of) diffuse scattering. In resonance, the absolute peak intensity of the $(\frac{1}{2}\frac{1}{2}\frac{1}{2})$ Bragg reflection was about 40 cts./sec above background at full ring current. The fluorescence signal is also plotted which gives a measure of the energy dependent absorption cross section.

oriented pyrolytic graphite (HOPG) crystal, which gives an analyser Bragg angle of 91.6°, close to the ideal value of 90°. The degree of linear polarization of the primary beam was about 90%.

3. Results and Discussion

The energy dependence of the intensity at the CoO $(\frac{1}{2}\frac{1}{2}\frac{1}{2})$ position (Figure 1) shows a significant resonance enhancement at the Co K edge, indicating that the corresponding Bragg reflection is of magnetic origin. The observed resonance is dominated by the dipolar transition $1s \rightarrow 4p$ (E1) with a maximum intensity at 7724 eV. The quadrupolar transition $1s \rightarrow 3d$ (E2) at 7707 eV causes a weak anomaly in the fluorescence spectrum but is invisible in the Bragg intensity data. This constrasts the behavior observed for a bulk sample, where the E2 transition is prominent [3].

This discrepancy is easily understood taking into account the limited sample thickness and the variation of the x-ray penetration depth μ across the Co K edge. In CoO, for energies closely below the edge we get approximately $\mu_\downarrow \approx 29$ μm, while $\mu_\uparrow \approx 4$ μm above. The $(\frac{1}{2}\frac{1}{2}\frac{1}{2})$ reflection appears at a Bragg angle $\theta = 9.4°$ and the effective sample thickness τ is $\tau = 2000$ Å$/\sin\theta = 1.2$ μm, which in any case is smaller than the penetration depth. Therefore, the scattered intensity is limited by the sample thickness and scales with the ratio τ/μ. With respect to the scattering from a bulk sample this leads to a modification of the energy dependence of the scattered intensity. For the following experiments the x-ray energy was set to 7724 eV.

Longitudinal (θ–2θ) scans of the (111) and $(\frac{1}{2}\frac{1}{2}\frac{1}{2})$ reflections exhibit significant line broadening related to the formation of small structural grains and AF domains [1]. The size L of these domains can be estimated using $L = 2\pi/\text{FWHM}$, where FWHM is the full width at half

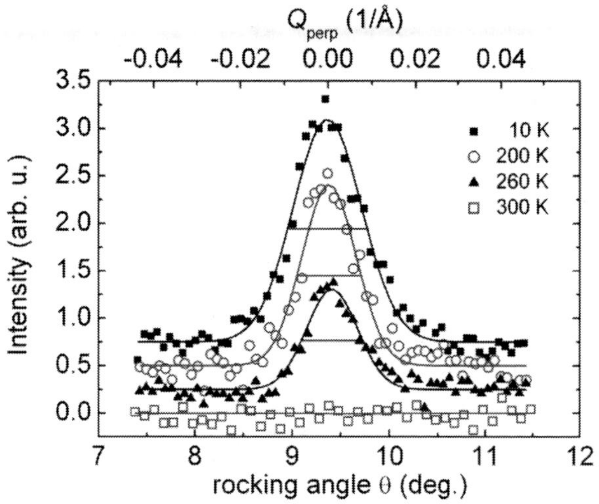

Figure 2. Transverse (θ rocking) scans of the $(\frac{1}{2}\frac{1}{2}\frac{1}{2})$ Bragg reflection for the 2000 Å CoO film. A diffuse background was determined from data taken at 300 K and subtracted. The transverse instrumental resolution was 0.02° and thus negligible. For clarity, the data are offset along the vertical axis.

maximum in reciprocal space. From the FWHM values of 0.011 $Å^{-1}$ and 0.013 $Å^{-1}$, respectively, we infer an average grain size of 57 nm and an average AF domain size of 48 nm for the out-of-plane direction [4]. This suggests the formation of AF domains within structural grains.

In order to get information about the in-plane AF domain size, the AF $(\frac{1}{2}\frac{1}{2}\frac{1}{2})$ Bragg reflection was measured by transverse (θ rocking) scans for various temperatures (Figure 2). The temperature dependence of the $(\frac{1}{2}\frac{1}{2}\frac{1}{2})$ Bragg intensity (Figure 3) reflects the square of the magnetic order parameter. The overall T-dependence and the observed transition temperature $T_N = 291$ K are in excellent agreement with previous neutron scattering data.

The temperature dependence of the FWHM of the $(\frac{1}{2}\frac{1}{2}\frac{1}{2})$ reflection is shown in Figure 4. Again, the XRMS data agree well with the results from neutron diffraction experiments. The quantitative analysis yields a 40 nm average lateral AF domain size close to $T_N = 291$ K. Interestingly, the line width is temperature dependent, suggesting a significant decrease of the AF correlation length on cooling. In the previous report we argued that the averaged sizes of AF domains and domain walls vary as function of temperature. This can be understood as an effect of the interplay between AF anisotropy and stiffness strength [1]. Quantitatively, however, an accurate estimate of the correlation length from the rocking width may be hampered by the structural changes of the CoO lattice below T_N. To this end we have measured rocking widths also of the (111) reflection (not shown). The preliminary data analysis shows that the temperature dependence of the (111) rocking width exhibits a similar temperature dependence as that of the $(\frac{1}{2}\frac{1}{2}\frac{1}{2})$ reflection, suggesting that the known monoclinic distortion in the magnetically ordered phase of CoO [5] may contribute to the observed temperature dependences by mosaic broadening. We have also measured the azimuth dependence of the $(\frac{1}{2}\frac{1}{2}\frac{1}{2})$ line width (not shown) and found no significant variation of the FWHM value as function of the azimuth angle, suggesting isotropic in-plane average domain sizes.

In conclusion, we have determined the temperature dependence of the intensity and the rocking-width of the $(\frac{1}{2}\frac{1}{2}\frac{1}{2})$ Bragg reflection from a 2000 Å CoO film on sapphire by XRMS. The results correspond well with previous neutron diffraction data.

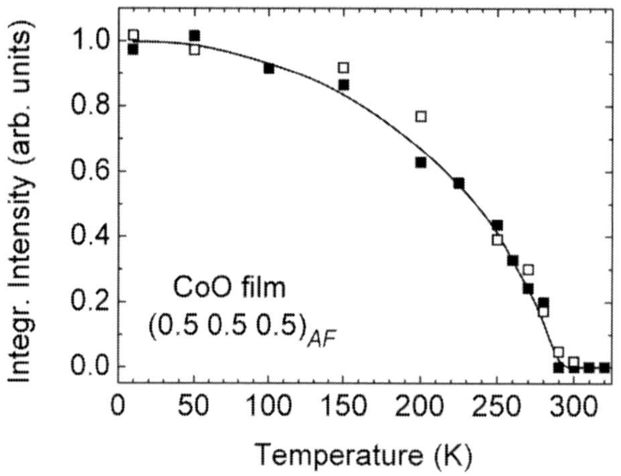

Figure 3. Temperature dependence of the $(\frac{1}{2}\frac{1}{2}\frac{1}{2})$ integrated Bragg intensity for the 2000 Å CoO film measured by XRMS (solid symbols). Corresponding neutron diffraction data (open symbols), taken from [1], are shown for comparison. The line is a guide to the eye.

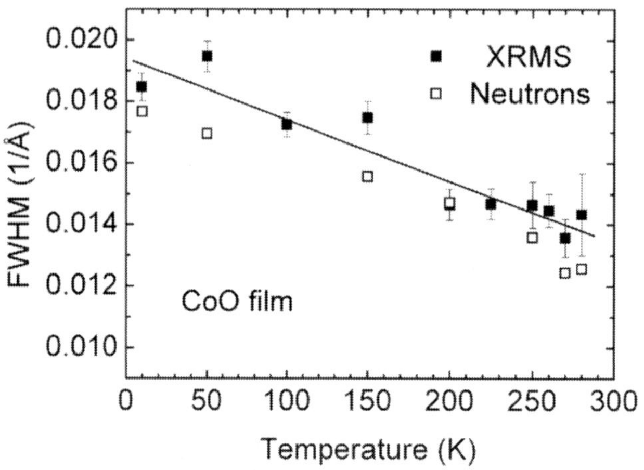

Figure 4. Temperature dependence of the FWHM line width of the $(\frac{1}{2}\frac{1}{2}\frac{1}{2})$ Bragg reflection from the XRMS data (closed symbols). The corresponding neutron diffraction data (open squares), taken from [1], are shown for comparison. The line is a guide to the eye.

References

[1] Radu F *et al* 2009 *Phys. Rev.* B **79** 184425
[2] Dudzik E, Feyerherm R, Signorato R and Zilkens C 2006 *J. Synchrotron Rad.* **13** 421
[3] Neubeck W, Vettier C, Lee K B and de Bergevin F 1999 *Phys. Rev.* B **60** R9912
[4] With the experimental resolution $\Delta Q_{\|} = 1.2 \times 10^{-3}$ Å^{-1} any instrumental line broadening can be neglected
[5] Jauch W, Reehuis M, Bleif H J and Kubanek F 2001 *Phys. Rev.* B **64** 052102

Interlayer exchange coupling in Er|Tb superlattices mediated by short range incommensurate Er order

E Pfuhl[1], J Voigt[2], S Mattauch[2], D Korolkov[2] and Th Brückel[1,2]

[1] Institut für Festkörperforschung, IFF, Forschungszentrum Jülich, D-52425 Jülich,
[2] Jülich Center for Neutron Science, IFF, Forschungszentrum Jülich, Outstation at FRM II, Lichtenbergstrasse 1, D-85747 Garching

E-mail: e.pfuhl@fz-juelich.de

Abstract. We study the magnetic correlations in Er|Tb superlattices by means of off-specular scattering of polarized neutrons. We show here the co-existence of inhomogeneous magnetic states: i) ferromagnetic order of moments within the Tb layers below 230 K (FM), correlation length of about 10 bilayer, ii) an incommensurate modulated magnetic order, restricted to single Er layers and iii) antiferromagnetic coupling of ferromagnetic layers below 70K (AFC). Polarised off-specular neutron scattering under grazing incidence reveals that i) magnetic fluctuations appear when the sample is cooled below 70 K, ii) these fluctuations lead to AFC, when the sample is cooled to 10 K, which iii) persists, when the sample is subsequently heated up to 45 K, while the order is not present during the cooling cycle. Also the short range incommensurate order changes accordingly, implying that the magnetic order in the Er layers mediates the interlayer coupling between ferromagnetic Tb layers.

1. Introduction

The discovery of magnetic interlayer exchange coupling initiated intense research activities concerning rare-earth superlattices (overview in [1,2]). While earlier studies focused on systems featuring non-magnetic spacers (e.g. consisting Y layers [3,4]) more recent work was devoted to systems containing two magnetic layers [5-11]. The latter cases reveal a variety of novel magnetic phases due to the competing anisotropies, proximity effects and epitaxial strain. In Er|Tb superlattices the competing anisotropies prevent the formation of long range incommensurate structures as found in the bulk materials, except for the thinnest Tb layer thickness ([Er_{21}|Tb_5], indices denoting the number of atomic layers) [7]. The presence of epitaxial strain stabilizes the ferromagnetic order in the Tb layers for Tb layer thicknesses larger than 10 atomic layers below T_C = 230 K. Wide angle neutron diffraction and x-ray resonance exchange scattering revealed an antiferromagnetic coupling (AFC) between adjacent layers in contrast to the macroscopic magnetization that has been observed by SQUID measurements [10]. Furthermore short range incommensurate magnetic correlations appear below 80 K, which seem to be restricted to single Er layers.

Here we present off-specular scattering of polarized neutrons to resolve the nature of the AFC in an [Er_{21}|Tb_{10}] superlattice containing 61 bilayers and to probe the range of the incommensurate correlations perpendicular and parallel to the surface plane. The sample was grown with the hexagonal [0001] direction parallel to the surface normal, which is also parallel to the propagation vector of the incommensurate bulk magnetic structures. The high epitaxial

Journal of Physics: Conference Series **211** (2010) 012019 doi:10.1088/1742-6596/211/1/012019

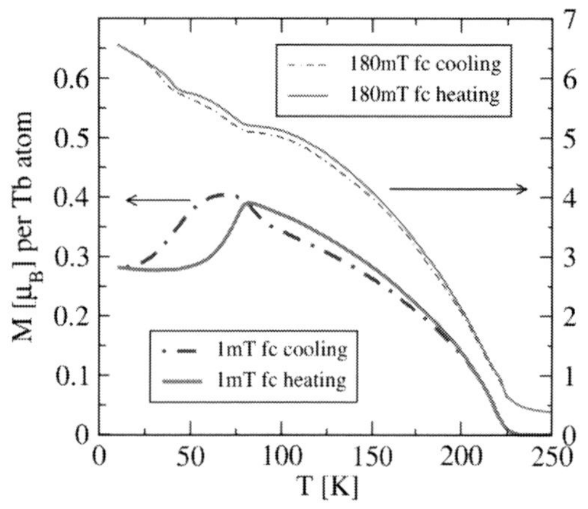

Figure 1. (Color online) Temperature dependence of the in-plane magnetization of an $Er_{21}|Tb_{10}$ superlattice in two different magnetic fields. Blue curve measured during cooling, red curve measured during heating. The diamagnetic contribution from sample container and substrate was corrected for. The applied magnetic fields are representative for the field conditions during the neutron scattering experiment.

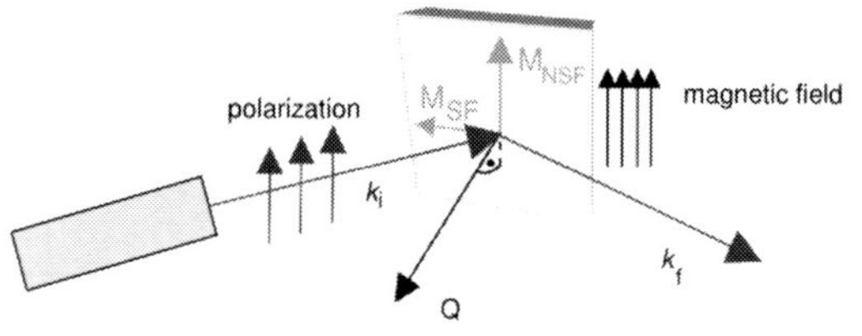

Figure 2. (Color online) Schematic experimental setup for the neutron scattering. Magnetic field and neutron polarization aligned parallel to the basal plane. Magnetization parallel to this contributes to NSF channel and perpendicular to SF channel.

quality was confirmed by low energy electron diffraction and x-ray diffraction. Details of the preparation and characterization can be found elsewhere [7].

2. Macroscopic magnetization

For the macroscopic characterization we used a SQUID magnetometer (Quantum Design MPMS) in a temperature range from 10 K to 250 K. For these measurements a small piece was cut from the sample later used for neutron scattering. Subsequent x-ray diffraction assured that the sample quality was not deteriorated. The piece was aligned with the surface normal perpendicular to the applied magnetic field. The temperature dependence of the in-plane magnetization at two different applied fields is shown in Fig. 1. The curves have been measured by cooling the sample in the respective magnetic field and subsequent heating to avoid a reduction of the measured magnetization by the formation of randomly oriented domains, when cooled in zero field. We tentatively normalize the magnetization to the number of Tb atoms within the sample, since the basal plane ferromagnetism is an original phase of bulk Tb. Hence we may compare the magnetization to the saturation value of 9.72 μ_B for metallic Tb. The number of Tb atoms was determined from the area of the substrate, the number of Tb layers as known from the preparation, and the unit cell area in the epitactically strained sample. As the latter can not be quantified exactly we estimate the error on the Tb atom number to 4%.

Below $T_C \approx 230$ K the magnetization steeply increases indicating a ferromagnetic order of the Tb moments throughout all the applied field. Below 90 K the qualitative behavior is different for the 180 mT data and the 1 mT data. Upon cooling we observe a kink, that has been attributed to an additional contribution from the Er moments to the magnetization [10]. Below the maximum at 60 K the magnetization decreases significantly when the sample is cooled in an applied field $\mu_0 H = 1$ mT. For this magnetic field, we observe a clear thermal hysteresis upon heating, as the

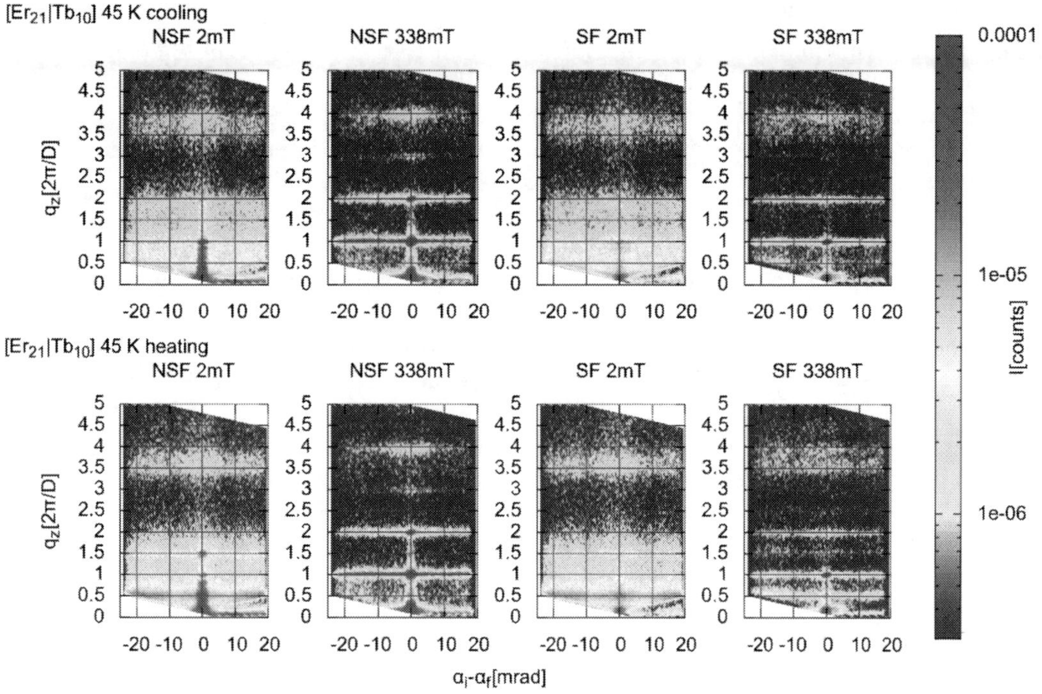

Figure 3. (Color online) NSF and SF maps of scattering intensity as a function of q_z in units of the superlattice periodicity $2\pi/D$ and $\alpha_i - \alpha_f$, yielding the information about correlations parallel to the growth direction vertically and the lateral correlations horizontally. Color coded intensities are normalized to monitor counts. Both rows were measured at 45 K. The upper row shows intensity maps measured during the cooling cycle, the lower row was recorded upon heating after the sample had been cooled to 10 K in the respective applied magnetic field.

magnetization starts to increase only above 50 K. The heating curve crosses the cooling curve at ≈ 80 K and remains slightly above the cooling curve up to T_C. For larger applied fields, the thermal loop is also not completely closed. We observe an additional kink at a slightly different temperature for the cooling and the heating cycle, respectively. For comparison the macroscopic magnetization was measured at the same values of applied field and temperature as the neutron measurement. For this being only a few points this data is not displayed in Fig. 1. Nevertheless there is still a small thermal hysteresis to be observed under these conditions and the magnetization curve at 180 mT is representative.

3. Neutron Measurements

To get an insight into the magnetization of a single layer and the range of the magnetic coupling we measured the polarized off-specular neutron scattering along the specular rod at the TREFF instrument ($\lambda = 4.73$ Å) at FRM II (formerly HADAS at FRJ-2 [13]). For this we again applied the magnetic field along the in-plane direction and the scattering vector Q was parallel to the out-of-plane direction. As the scattering vector has a large component parallel to the surface normal and only a small component within the surface plane, we probe mainly the magnetization within the surface plane, which coincides with the ab plane of the hexagonal lattice. The external magnetic field and the neutron polarization are aligned perpendicular to the scattering plane, i.e. within the superlattice surface. Magnetic moments parallel to the guide field cause non-spin-flip scattering (NSF), while magnetic moments perpendicular to the field flip the spin (spin-flip, SF). We have measured all 4 channels of incoming and outgoing polarization. Since the differences between the up-up and down-down states of the neutron polarization are negligible, we present only the sum of the respective intensities. The same applies for the up-down and down-up states. Due to the fact that the nuclear scattering lengths of Er and Tb are similar ($b_{Er} = 7.8$ fm and $b_{Tb} = 7.4$ fm) the nuclear contrast between the layers almost vanishes, leaving only a small nuclear contribution to the NSF. Fig. 3 shows the off-specularly scattered intensity maps as a function of $\alpha_i - \alpha_f \approx 2\frac{q_x}{q_y}$ and q_z in units of $\frac{2\pi}{D}$, with the bilayer thickness $D = 8.5$ nm. The former gives a measure of the lateral correlations on the length scale from several 1-100 μm. Parallel to q_z one can probe correlations on a length scale of hundrets of nm. We show representative scans at 2 mT and at 338 mT, which is the maximum field that can be applied.

Three types of features can be distinguished in Fig. 3: i) Reflections at integer values of q_z indicating FM order of Tb layers as anticipated from the SQUID measurements. They are represented by narrow Bragg sheets indicating long range vertical correlations spanning across some 10-15 bilayer. The sheets exist already at 150 K, the highest measured temperature. ii) Reflections at half integer values denoting a doubled magnetic unit cell and therefore indicating antiferromagnetic coupling of ferromagnetic layers. They appear only after the sample had been cooled to 10 K. Under small field conditions the NSF channel exhibits true specular reflectivity at these q_z values, while only the Bragg sheets are visible in the SF channel. Hence the AFC is long range only for the component parallel to the external field with a correlation length $> 1.5\mu$m. For the component perpendicular to the field we estimate the correlation length from the width of the Bragg sheet to be smaller than 500 nm. When the sample is cooled in a large field, the situation is different: The reflections at half integer values exist only in the SF channel, indicating that only the component perpendicular to the external field couples antiferromagnetically. Upon heating, these reflections still exist at $T = 45$ K. During the cooling cycle diffuse intensity is observed in the q_z range $0.3 < q_z(\frac{2\pi}{D}) < 1$ at this temperature. iii) Broad diffuse intensity at $3.5 < q_z(\frac{2\pi}{D}) < 4$. For this feature the thermal history has two effects. Comparing the data measured at low field for the cooling and heating cycle, the center of the intensity is shifted towards smaller q_z in the heating run. But the center and the magnitude of the intensity in the SF and NSF channel remain similar. In combination with earlier magnetic resonant x-ray scattering results [7], this intensity could be attributed to short range helix-like magnetic correlation reminiscent of the magnetic structure found in bulk Er. It has to be noted that the intensity at $q_z = 3.5\frac{2\pi}{D}$ is enhanced in the heating run in the NSF and the SF channel. The situation is different, when the sample is cooled in a strong field: upon cooling, the intensity is centered around $q_z = 4\frac{2\pi}{D}$ and it is broader in the SF than in the NSF channel. Upon heating the NSF intensity remains centered around $q_z = 4\frac{2\pi}{D}$. In the SF channel the diffuse intensity is found for $3.5 < q_z(\frac{2\pi}{D}) < 4$ with a maximum around $q_z = 3.5$. So far it is not clear, how the intensity at strong magnetic fields has to be interpreted. The experimental result indicates different Fourier components for the parallel and perpendicular components of the magnetic moments. It seems plausible that the coupling of ferromagnetic layers is connected with the short range correlations, particularly the antiferromagnetic coupling of the perpendicular component.

4. Discussion and Outlook

The macroscopic data and the neutron reflectivity show that the $[Er_{21}|Tb_{10}]$ superlattice develops ferromagnetic order below $T_C = 230$ K. The intriguing features of the superlattice are revealed by the off-specular scattering. The surprising suppression of the macroscopic magnetization at low temperature can be linked to the appearance of magnetic fluctuations causing diffuse scattering. The temperatures at which we observe anomalies in the magnetization coincide roughly with the phase transition temperatures in bulk Er. The magnetization starts decreasing at temperatures, at which the basal plane order changes in bulk Er [12]. The off-specular scattering reveals the fluctuations associated with the suppression of the magnetization. While upon cooling the diffuse scattering at low q_z is rather unstructured, clear AFC develops between neighboring layers once the sample is cooled to base temperature. Due to the limited neutron flux we were not able to determine exactly the temperature, where the AFC becomes long range. It is intriguing to realize that the turn angle accumulated across 21 atomic layers of Er approximates $\Phi = 5\pi$ for the ground state wave vector in bulk Er being $5/21\frac{\pi}{c}$ [15], which favours the AFC between neighbouring Tb layers. The application of a magnetic field leads to a ferromagnetic alignement of the magnetization component parallel to the field, while the AFC survives for the component perpendicurlar. So far we cannot explain the origin of the thermal hysteresis, but we speculate that magneto-elastic interactions are involved, as they drive the lock-in of the propagation vector in bulk Er [14]. To conclude we have shown, that ferromagnetic order appears below a transition temperature $T_C = 230$ K. The ferromagnetic alignement is disturbed at low temperatures, leading first to magnetic fluctuations and finally to AFC. This might be attributed to the propagation vector in the Er layers becoming commensurate with the superlattice periodicity. Further refinement of the presented data is ongoing and will give quantitative information about the layer magnetization in Er and Tb layers, respectively. To fully understand the complex magnetic structure of the $[Er_{21}|Tb_{10}]$ superlattice it is necessary to develop a model that takes the different energy contributions and disorder at the interfaces into account. A systematic study of the composition dependence is ongoing and will reveal new hints to solve this problem.

References

[1] Majkrzak C F *et al* 1991 *Adv. Phys.* **40** 99
[2] Bruno P 1995 *Phys. Rev.* B **52** 411
[3] Jehan D A *et al* 1993 *Phys. Rev.* B **48** 5594
[4] Yafet Y 1987 *J. Appl. Phys.* **61** 4058
[5] Simpson J A *et al* 1996 *Z. Phys.* B **101** 35
[6] Bryn-Jacobson C *et al* 1997 *Phys. Rev.* B **55** 14360
[7] Voigt J *et al* 2006 *Eur. Phys. J.* B **49** 441
[8] Voigt J *et al* 2004 *Europhys. Lett.* **65** 560
[9] Cowley R A 1998 *JMMM* **177-181** 1156
[10] Yamazaki H *et al* 1998 *JMMM* **177-181** 1207
[11] Yamazaki H *et al* 2001 *Phys. Stat. Sol.* (b) **228** 741
[12] Koehler W C 1972 *Magnetic Properties of Rare Earth Metals* ed Elliot R J (New York: Plenum Press)
[13] Rücker U *et al* 2000 *Physica* B **276-278** 95
[14] Gibbs D *et al* 1986 *Phys. Rev.* B **34** 8182
[15] Jensen J and Mackintosh A R 1991 *Rare Earth Magnetism* (New York: Oxford University Press)

Effect of Ag as a surfactant on the thermal stability in Cu/Co multilayers

M Gupta,[1] A Gupta,[1] S M Amir,[1] J Stahn[2] and J Zegenhagen[3]
[1] UGC-DAE Consortium for Scientific Research, University Campus, Khandwa Road, Indore 452 001, India
[2] Laboratory for Neutron Scattering, ETH Zürich and Paul Scherrer Institut, 5232 Villigen PSI, Switzerland
[3] ESRF, B. P. 220, F-38043 Grenoble Cedex, France

E-mail: mgupta@csr.ernet.in

Abstract. In the present work we studied the effect of Ag as a surfactant in Cu/Co multilayers prepared by ion beam sputtering. Two identical samples of Cu/Co multilayers with 0.2 nm Ag on a Cu buffer layer and without this Ag layer were deposited on Si substrates. It is known that Cu has a lower free energy than Co, and therefore, the Cu/Co interfaces are not symmetric. Addition of Ag alters the kinetics and thermodynamics of the growth and triggers the layer by layer growth as Ag floats on the growing front balancing the surface free energy of Cu and Co. It was found that with addition of Ag surfactant the thermal stability of Cu/Co multilayer improves.

1. Introduction

Ferromagnetic layers separated by non-ferromagnetic layers in a multilayer arrangement are known to exhibit the giant magneto resistance (GMR) [1] through spin-dependent scattering taking place at the interfaces. Therefore the microstructure of the layers plays an important role in determining the GMR in such multilayers. The intermixing between layers due to chemical diffusion, the roughness or the layer curvature, crystallinity of the individual layers strongly influence the electrical and magnetic properties [2, 3]. The Cu/Co multilayers (with layer thickness few nm) are known for high GMR due to the immiscibility of Cu and Co. Though the volume diffusion between Cu and Co is negligible, the grain boundary diffusion of Cu and Co atoms may play a significant role in determining the GMR properties. Further, Cu has a lower surface free energy of 2300 mJ/m^2 than Co (2700 mJ/m^2). Therefore wetting of Cu layer on Co takes place but agglomeration of Co layer on Cu occurs [5]. This means that Cu is energetically preferred on the growth surface during deposition of Co on Cu. This leads to asymmetric interfaces in Cu/Co multilayers. Recently, it was suggested that by adding a 'surfactant' during the thin film growth, sharp interfaces can be obtained in a multilayer [4, 5, 6, 7, 8]. A surfactant is chosen in such a way that its free energy allows a balance between the components of multilayers whose surface free energy differs significantly. Generally elements with large atomic volume are chosen as surfactant as it favors their floating out to the surface during growth of overlayer, leaving behind smooth interfaces that are less prone to intermixing. Metals such as Pb, In and Ag have been used as a surfactant to achieve sharp interfaces. However very limited work has been done to quantify the effect of surfactant in terms of inter-diffusion and its ability to

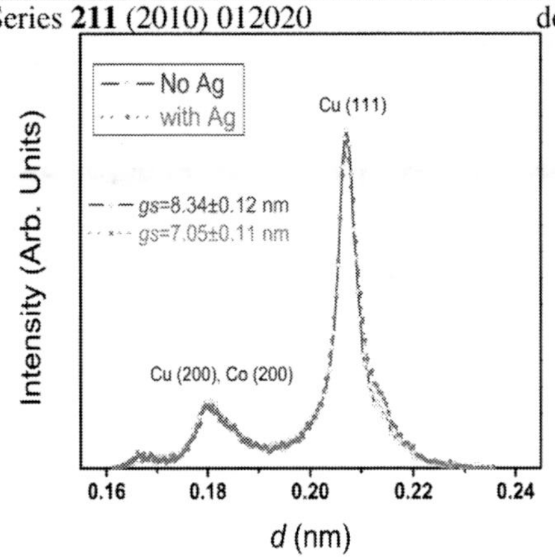

Figure 1. X-ray diffraction pattern of the the Si/[Cu 10nm/Ag 0.2 nm[Cu 3 nm/ Co 2nm]$_{10}$ and Si/[Cu 10 nm/Ag 0.2 nm[Cu 3 nm/ Co 2 nm]$_{10}$ samples. The grain size (gs) is slightly smaller for the sample prepared using Ag.

increase the thermal stability. Our aim is to quantify the inter-diffusion in nanometer range thin films using a precise technique like neutron reflectivity. It may be noted that reflectometry or grazing incidence reflection offer a depth resolution down to sub nm level [9]. Therefore precise measurements of inter-diffusion can be done using neutron reflectivity. Here in this work we have used Ag as a surfactant during the deposition of Cu/Co multilayer. The preliminary results obtained in Cu/Co multilayers by using a 0.2 nm Ag layer on a buffer layer of 10 nm Cu layer are reported in this work.

2. Experimental

The samples were prepared using ion beam sputtering technique. The ion beam source used to produce Ar^+ ions was a Kaufman type hot cathode ion gun installed in a vacuum chamber at an angle of 45°. The Ar^+ ion were allowed to fall alternatively on Cu and Co targets to deposit a multilayer structure [Cu 3 nm/Co 2 nm]$_{10}$ on a Si(100) substrate. The Ag target was also mounted on a specially designed target holder which can accommodate four different targets. Prior to the deposition of multilayer, a Cu buffer layer of thickness 10 nm was deposited on the Si substrate. In case of sample prepared with Ag surfactant a 0.2 nm thin layer of Ag was deposited on Cu buffer layer. The pressure during the deposition of the films was of the order of 10^{-3} mbar while the pressure obtained before inserting Ar gas in the chamber was about 1×10^{-6} mbar. More details about sample preparation are given in ref. [10].

The samples were measured using laboratory and synchrotron x-rays. The synchrotron measurements were carried out at ID32 beamline of ESRF, France. The x-ray diffraction (XRD) and x-ray reflectivity (XRR) measurements were performed at different energies. The neutron reflectivity (NR) measurements were performed at SINQ/PSI, Switzerland using neutrons of wavelength 0.5 nm. The NR measurements were performed at room temperature in the as-deposited and after annealing the samples in a vacuum furnace at different temperatures for 1 h. In addition to NR measurements, the polarized neutron reflectivity (PNR) measurements were also performed. The PNR measurements were performed with an applied magnetic field of 800 G at the samples. The direction of magnetic field was kept parallel to the surface of the samples which is transverse to the direction of the neutron beam.

3. Results and discussions

Fig. 1 shows the grazing incidence x-ray diffraction pattern of the samples prepared with and without Ag surfactant. It was found that the XRD pattern of the sample prepared with Ag

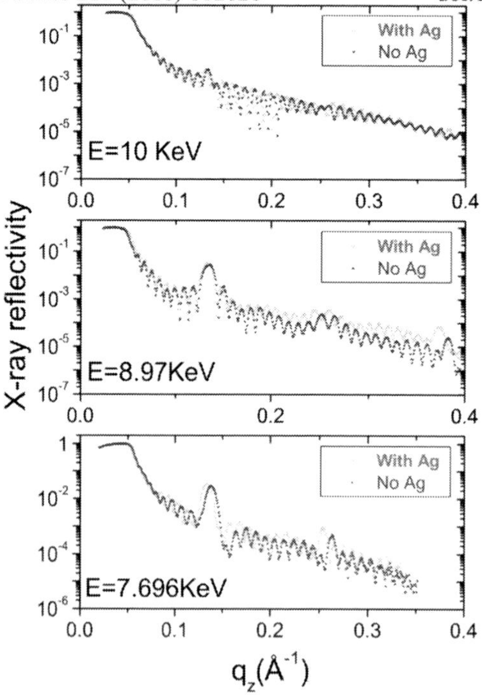

Figure 2. X-ray reflectivity pattern of the Si/[Cu 10nm/Ag 0.2 nm[Cu 3 nm/ Co 2nm]$_{10}$ and Si/[Cu 10 nm/Ag 0.2 nm[Cu 3 nm/ Co 2 nm]$_{10}$ samples. The measurements were performed in the as-deposited sample at different energies at ESRF. The energy of the x-rays were tuned near Cu and Co absorption edges and at 10 KeV.

surfactant was slightly broader compared to the sample prepared without Ag. The grain size (gs) for the sample prepared with Ag was about 7 nm while for the sample prepared without Ag it was 8.3 nm. This indicates that in presence of Ag, the grain growth of Cu is some what restricted.

In order to observe the effect of Ag layer on the interfaces of the multilayer, the x-ray reflectivity measurements were performed near the absorption edges of Cu (8.97 KeV) and Co (7.696 KeV) and at a higher energy of 10 KeV. Fig. 2 show the measured reflectivity for both samples at different energies. The choice of these energies was made to increase the scattering contrast between Cu and Co. In both cases near the absorption edges of Cu and Co, the XRR pattern of sample without Ag showed a higher roughness compared to the sample prepared with Ag. However at the higher energy of 10 KeV, the reflectivity of both samples appears together. The positions of the Bragg peaks in case of samples prepared with and without Ag surfactant are slightly shifted relative to each other. This may happen due to error in thickness caused during the deposition process. The fitting of the measured data taken at the absorption edges of Cu and Co was done using Parratts formalism [11]. From the fitting of the data it was observed that the roughness of the Cu buffer layer is the same in both cases however the interface roughness of Cu-on-Co is 0.5 nm while that of Co-on-Cu is 0.3 nm for the sample prepared without surfactant. For the sample prepared using Ag surfactant, the interface roughness of both interfaces is almost equal to 0.4 nm. This indicates the surfactant yields interfaces with similar roughness amplitude. The surfactant effect allows a more homogeneous distribution of the interfacial defects.

The neutron reflectivity pattern of the samples prepared with and without Ag as a surfactant are shown in Fig 3. The measurements were performed in the as-deposited samples and after annealing the samples at 373 K, 473 K, 573 K and at 673 K for 1 hour at each temperature. The annealing were performed in a vacuum furnace with a base pressure of the order of 1×10^{-6} mbar.

Table 1. The inter-diffusion length (L_d) and magnetic moment per Co atom obtained from the neutron data.

Sample	Co magnetic Moment with Ag (μ_B)	Co magnetic Moment without Ag (μ_B)	L_d with Ag (nm)	L_d without Ag (nm)
As-deposited	1.7	1.5	-	-
373 K	1.7	1.6	0.8	1.05
473 K	1.7	1.6	1.05	2.05
573 K	0.6	0.25	2.5	-
673 K	0	0	-	-

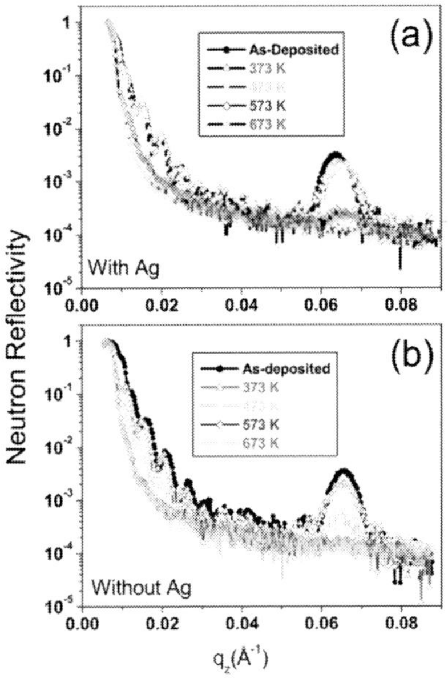

Figure 3. Neutron reflectivity pattern of the Si/[Cu 10 nm/Ag 0.2 nm[Cu 3 nm/ Co 2 nm]$_{10}$ samples (a) and Si/[Cu 10nm/[Cu 3nm/ Co 2nm]$_{10}$ samples (b). The measurements were performed in the as-deposited sample and after annealing the samples in the temperature range of 373 K to 673 K with a step of 100 K. The annealing of the samples was performed in a vacuum furnace.

The Bragg peak corresponding to the multilayer period appears at nearly the same angles in both samples. The intensity of the Bragg peak is expected to decay due to the inter-diffusion as the annealing temperature is raised. This decay is seen to be faster in the samples when no surfactant was used.

In case of the sample prepared using Ag surfactant the Bragg peak intensity decreases marginally up to an annealing temperature of 473 K. However in case of sample where no Ag surfactant was used even at 473 K there is significant decrease of Bragg peak intensity. At 573 K the Bragg peak completely disappears for the sample where no surfactant was used while for the sample with 0.2 nm Ag the Bragg peak could still be seen however its intensity is significantly reduced. At 673 K the Bragg peak completely disappears in both cases. This result

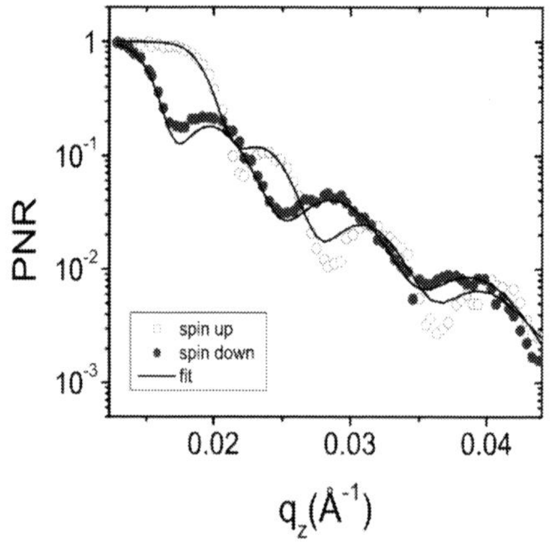

Figure 4. Polarized neutron reflectivity pattern of the as-deposited Si/[Cu 10 nm/Ag 0.2 nm[Cu 3 nm/ Co 2 nm]$_{10}$ sample.

clearly demonstrates that the thermal stability of the Cu/Co multilayer is improved with Ag surfactant. From the measured neutron reflectivity data the inter-diffusion in both samples can be quantified and the decay of the Bragg peak intensity can be used to calculate the diffusion coefficient using the expression [9, 12, 13]

$$I(t) = I(0)\exp\left(-\frac{8\pi^2 n^2 D}{\ell^2}t\right), \tag{1}$$

where $I(0)$ is the intensity before annealing and $I(t)$ is the intensity after annealing time t at temperature T, ℓ is the bilayer periodicity n is the order of reflection which is equal to unity for 1st order Bragg peak. With known diffusion coefficient (D) calculated using eq 1 the inter diffusion length (L_d) can be calculated with the expression $L_d^2 = 6Dt$ in the direction normal to the multilayer [14]. The inter-diffusion lengths obtained in this way are given in table 1. As can be seen from the table the L_d is significatively smaller for the sample prepared with Ag surfactant as compared to the sample prepared without any surfactant. This result clearly shows that by using Ag surfactant the inter-diffusion in Cu/Co multilayers can be reduced.

In addition, the PNR measurements yield the magnetic moment of Co in these samples. PNR is a technique which is able to yield the absolute value of a magnetic moment per atom in a magnetic thin film with high accuracy [15]. In contrast to the bulk magnetization magnetometer techniques, e.g. dc extraction, VSM, or SQUID, no correction due to the magnetic signal from the substrate has to be applied in PNR. Further, the sample dimensions and mass do not play a crucial role in the determination of the magnetic moment. During the experiment, polarized neutrons with spin parallel or anti-parallel (also termed as spin + / up, or spin − / down) to the direction of magnetization on the sample are reflected off the surface of a sample at grazing incidence. The measurements were performed with an applied field of 800 G, which is sufficient to reach the saturation magnetization in all the layered samples.

Fig. 4 show the PNR pattern of the sample prepared using Ag surfactant. Since the PNR pattern of a sample prepared with or without surfactant is similar in shape, therefore pattern corresponding to one sample is shown. The measured PNR pattern were fitted using SimulReflec software [16]. Table 1 show the obtained values of the magnetic moment per Co atom in units μ_B obtained after fitting the PNR data in both cases. In the fitting routine the magnetic profile of Co is assumed to be constant throughout the layer. It was found that in the Co magnetic moment was slightly smaller in case of sample prepared without Ag surfactant. Also the decay of Co magnetic moment was faster as the annealing temperature was increased in case of sample prepared without Ag. This could be understood from the results obtained with NR data and it was found that inter-diffusion is faster as the annealing temperature is increased in case of sample prepared without Ag. A faster inter-diffusion indicates an intermixed layer of Cu-Co resulting in a decreased value of Co magnetic moment. The obtained result demonstrates that the thermal stability of Cu/Co multilayers can be improved using Ag as a surfactant.

Acknowledgments

Authors would like to acknowledge the help received from Mr. S. Potdar in sample preparation. This work is based on experiments performed at the Swiss spallation neutron source SINQ, Paul Scherrer Institut, Villigen, Switzerland. Financial support received from Department of Science and Technology, Government of India under its scheme 'Utilization of International Synchrotron Radiation and Neutron Scattering facilities' for carrying out experiments is gratefully acknowledged. A part of this work was performed under the Indo Swiss Joint Research Programme with grant no. INT/SWISS/JUAF(9)/2009.

References

[1] Baibich M N, Broto J M, Fert A, Van Dau F Nguyen, Petroff F, Etienne P, Creuzet G, Friederich A and Chazelas J 1998 *Phys. Rev. Lett.* **61** 2472
[2] Nagamine L C C M, Biondo A, Pereira L G, Mello A, Schmidt J E, Chimendes T W, Cunha J B M and Saitovitch E B 2003 *J. Appl. Phys.* **94** 5881
[3] An Y, Zhang H, Dai B, Mai Z, Cai J and Wu Z 2004 *J. Appl. Phys.* **100** 023516
[4] Steigerwald D A and Egelhoff W F Jr 1989 *J. Vac. Sci. Technol.* A **7** 2167
[5] Chopra H D, Yang D X, Chen P J and Egelhoff W F 2002 *Phys. Rev.* B **65** 094433
[6] Yang D X, Shashishekar B, Chopra H D, Chen P J and Egelhoff W F 2001 *J. Appl. Phys.* **89** 7121
[7] Camarero J, Ferron J, Cros V, Gomez L, De Parga A L V, Gallego J M, Prieto J E, De Miguel J J and Miranda R 1998 *Phys. Rev. Lett.* **81** 850
[8] Larson D J, Petford-Long A K, Cerezo A, Bozeman S P, Morrone A, Ma Y Q, Georgalakis A and Clifton P H 2003 *Phys. Rev.* B **67** 144420
[9] Gupta M, Gupta A, Stahn J, Horisberger M, Gutberlet T, and Allenspach P 2005 *Phys. Rev.* B **70** 184206
[10] Gupta M, Gupta A, Phase D M, Chaudhari S M, and Dasannacharya B A 2003 *Appl. Sur. Sci.* **205** 309
[11] Parratt L G 1954 *Phys. Rev.* **95** 359
[12] Rosenblum MP, Spaepen F, Turnbull D. 1980 *Appl. Phys. Lett.* **37** 184
[13] Speakman J, Rose P, Hunt JA, Cowlam N, Somekh RE, Greer A L 1996 *J. Magn. Magn. Mat.* **156** 411
[14] Schmidt H, Gupta M, Gutberlet T, Stahn J and Bruns M 2008 *Acta Materialia* **56** 464
[15] Blundell S J and Bland J A C 1992 *Phys. Rev.* B **46** 3391
[16] SimulReflec: Reflectivity curve simulations and fitting, online at http://www-llb.cea.fr/prism/programs/simulreflec/simulreflec.html

Preparation and analysis of epitaxial Fe monolayers buried in Pd

D Schumacher[1], S Mattauch[2], U Rücker[1] and Th Brückel[1,2]

[1] Institut für Festkörperforschung, Forschungszentrum Jülich GmbH, 52425 Jülich, Germany
[2] Jülich Centre for Neutron Science, Forschungszentrum Jülich GmbH, 85747 Garching, Germany

E-mail: d.schumacher@fz-juelich.de

Abstract. Pd almost fulfils the Stoner criterion to be ferromagnetic. In order to understand effects of ferromagnetic inclusions in Pd it is of great interest to study in detail the influence of an ultrathin Fe layer in Pd. Here we report on the preparation and characterization of epitaxial Pd containing Fe layers of a few monolayer thickness. The growth procedure on Ag buffered GaAs(001) substrates was optimized by in-situ Low Energy Electron Diffraction (LEED) and Auger Electron Spectroscopy (AES) to obtain chemical purity and monocrystallinity of the layers with smooth interfaces. By using a variable deposition temperature during the growth of the first Pd layer on the Ag buffer we were able to avoid interdiffusion between Ag and Pd. X-ray- and polarized neutron reflectometry revealed a very well defined interface between the Pd and Fe on top, but an intermixing zone between Fe and Pd on top. With decreasing Fe layer thickness we have observed an increasing magnetization compared to bulk Fe moments.

1. Introduction

Proximity effects in magnetic multilayers are expected to be particularly pronounced for ultrathin ferromagnetic layers in direct contact to magnetically polarizable metals, namely Pd. The basic underlying principles are far from being understood, while these proximity effects might have important consequences for the design of certain spintronic devices. Therefore, we want to analyse both the influence of an ultrathin ferromagnetic layer on a polarizable environment and the transition of 3D to 2D behaviour of the ferromagnetic layer by reducing its thickness down to a few monolayers. The situation is different from ultrathin ferromagnetic layers deposited on isolating substrates, because here the Fe atoms are embedded in a metallic matrix, where the band structure extends over a 3-dimensional system, even if the chemical structure is 2-dimensional. Also the influence of the Pd environment back on the ferromagnetic Fe is subject of the investigation. The results of this experiment might ultimately be relevant for an improved design of spintronic devices, where stabilization of ferromagnetism in ultrathin ferromagnetic films is required.

It has been shown that concentrations of less than 2 % Fe in bulk Pd lead to an up to three times bigger macroscopic magnetic moment than one would expect resulting from the Fe alone [1]. In addition, the radius of influence of each single Fe atom in the Pd bulk can be as large as 50 Å. It was also investigated how the magnetization of a ferromagnetic substrate is changed, if Pd is deposited monolayer by monolayer [2][3][4]. These experiments revealed that Pd is polarized when deposited on Fe, Co or Ni. Even though these effects seem to be much smaller than reported on Fe diluted in bulk Pd, there are some hints that the effect depends a lot on the quality of the layer structure. The range of influence of the ferromagnetic substrate on Pd was biggest in the samples where layer by layer growth was observed. Beyond that, it has been shown that in better structured Pd/Fe multilayers up to 4 monolayers of Pd at the interface are magnetically influenced by Fe with a maximum moment of $0.4\,\mu_B$ per Pd atom in the monolayer directly in contact with the ferromagnet [5]. These experiments still do not reproduce effects of the order of the "Giant moments" in bulk Pd. It thus can be expected that even better structured multilayers lead to even bigger effects. In addition, by reducing the Fe layer thicknesses down to the monolayer regime one can investigate the coupling of the Fe to the Pd environment on both sides and - in case of roughness - 3-dimensionally, which should give a similar situation to the bulk Pd containing Fe impurities.

Thus we now want to explore the spatial and magnetic structure of an epitaxially grown well structured Pd/Fe multilayer system with best possible interfaces. In this paper we present an optimized growth procedure of epitaxial Pd containing Fe monolayers together with the structural analysis by LEED, AES and x-ray-reflectometry and the magnetic characterization by SQUID magnetometry and polarized neutron reflectometry.

2. Sample preparation and in-situ characterization

Samples were grown by MBE on GaAs(001) substrates covered by a 5 Å thick Fe seed layer (to determine the growth direction of Ag to be (001)) and a 1500 Å thick Ag buffer layer. In-situ characterization was done by LEED and AES. The substrate preparation procedure is well known and optimized [6][7]. LEED analysis confirmed epitaxial growth of Pd on Ag at all temperatures between room temperature and 200 °C. In addition, at higher temperatures sharper LEED spots indicate higher crystalline quality. In contrast, AES revealed interdiffusion of Ag into the deposited Pd layer at temperatures higher than 100 °C. To obtain both well structured and pure Pd layers, a variable deposition temperature was applied. It was shown that starting the deposition at room temperature, then after the first monolayers (about 10 Å) increasing the temperature to 120 °C for the rest of the layer leads to high quality monocrystalline and pure Pd thin films (see fig. 1). It was also shown that postannealing above 120 °C is not possible without interdiffusion of Ag and Pd. Therefore, all further layers must not be grown at a temperature above 120 °C, which nevertheless seems to be sufficient for growing monocrystalline multilayer stacks: Even very thin Fe layers (down to single monolayers) deposited at 120 °C on Pd showed monocrystallinity (LEED) and stability over long time scales (AES). We did not observe any hint for interdiffusion between Fe and Pd even after postannealing at 120 °C for 20 hours. We have grown $[Pd_{150\,Å}/Fe_{x\,Å}]_{10}/Pd_{150\,Å}$ multilayers with x=150,10,4,2,1 at a substrate temperature of 120 °C for all layers except the first one, where the variable temperature scheme described above was applied.

3. Ex-situ characterization

We have taken polarized neutron, x-ray-reflectometry and SQUID magnetometry data. The polarized neutron reflectometry (PNR) was performed at the TREFF reflectometer of the JCNS at the FRM II research reactor. The neutron wavelength was fixed at 4.73 Å. The efficiencies of the polarizer, analyzer and both spin flippers are above 98 %. Figure 2 shows as an example x-ray-reflectometry and PNR data for the multilayer and the simulations fitted to these data. Especially the analysis of the $[Pd_{150\,Å}/Fe_{150\,Å}]_{10}/Pd_{150\,Å}$ multilayer has shown that it is necessary to introduce a mixed Pd/Fe layer on top of every Fe layer to obtain a good agreement between measurement and simulation. On the other hand, between a Pd layer and a Fe layer on top a sharp interface is present as was also indicated by the AES in-situ characterization. The thickness of the mixed layer for the sample is about 10 Å. Therefore, for all the Fe layers of

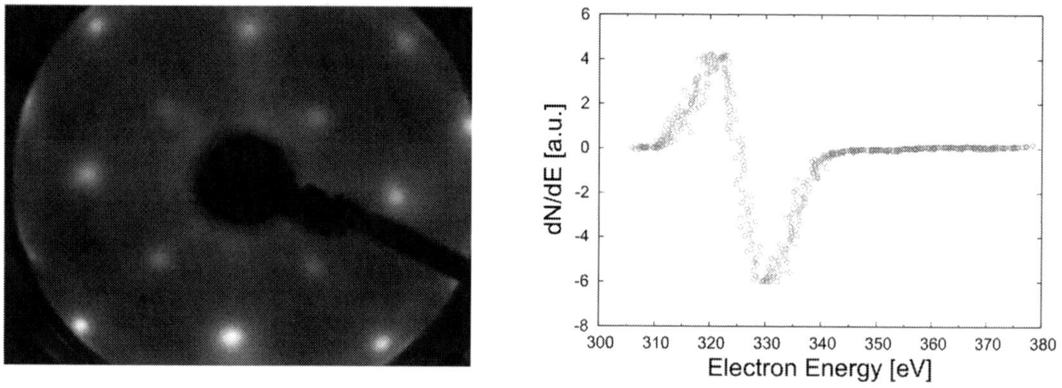

Figure 1. Left: LEED pattern of a 150 Å thick Pd layer deposited on a Ag buffered GaAs substrate with a variable deposition temperature; right: Auger electron spectrum of the same layer. The main peak of Pd at 331 eV is clearly visible, no peak is seen at 350 eV which is the energy of the main peak of Ag.

thickness smaller than 10 Å, we assume that the complete Fe layers are mixed with Pd with a gradually increase of the Pd concentration from the bottom to the top of the Fe layers.

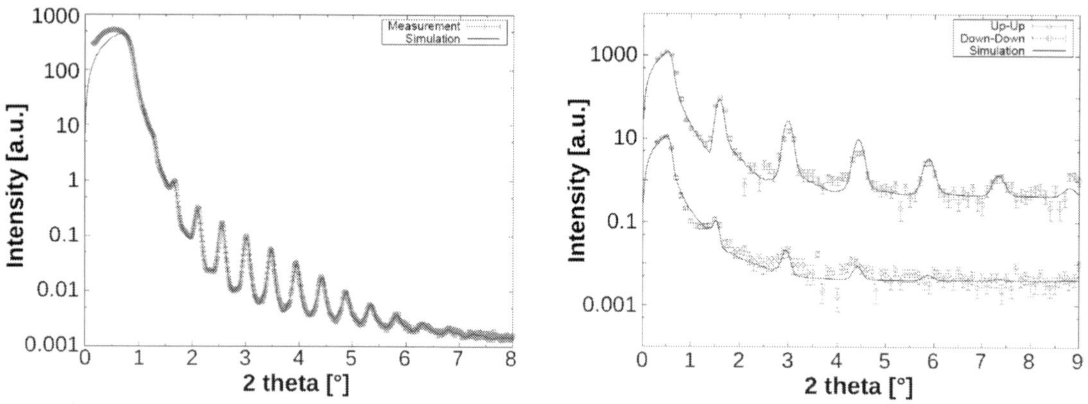

Figure 2. Left: X-ray reflectometry measurement and simulation of the $[Pd_{150\,Å}/Fe_{10\,Å}]_{10}/Pd_{150\,Å}$ multilayer; right: PNR measurement and simulation of the $[Pd_{150\,Å}/Fe_{10\,Å}]_{10}/Pd_{150\,Å}$ multilayer (Up/up channel is multiplied by 100).

By introducing a mixed layer very good agreement between experimental data and the simulated model can be obtained for all the samples. On the other hand this hampers the accurate determination of the amount of Fe in the samples. The superlattice period for all samples is higher than expected from the sample preparation which is just a constant calibration offset factor of the quartz crystal oscillator used in the MBE to control growth rate and layer thicknesses. As Pd and Fe are deposited from the same position in the MBE chamber this calibration factor is the same for both materials. Therefore, the assumption was made that the ratio of Pd to Fe amount in the samples are given by the nominal value, e.g. 15 to 1 for the $[Pd_{150\,Å}/Fe_{10\,Å}]_{10}/Pd_{150\,Å}$ multilayer. To estimate the Fe content in the multilayers the superlattice period that could be determined very accurately by reflectometry should also contain Fe and Pd of the same ratio. For the analysis of the SQUID data we used the resulting values for the Fe content to calculate a reference ferromagnetic moment (RFM) that these amounts of bulk Fe (magnetic moment per atom: $2.2\,\mu_B = 2.0 \cdot 10^{-20}$ emu) would cause (see table 1).

The magnetic scattering length density $Nb_{m,Fe}$ for the Fe layers in the $[Pd_{150\,Å}/Fe_{10\,Å}]_{10}/Pd_{150\,Å}$ sample determined by PNR is very close to the bulk value of Fe $Nb_{m,Fe,bulk}$ ($Nb_{m,Fe}/Nb_{m,Fe,bulk} = 0.92 \pm 0.06$) and in very good agreement with the SQUID results. In addition, the mixed layers in this multilayer have scattering length densities below the Fe bulk value. However, this value does crucially depend on the choice of the position of the interface: Due to the gradual transition from Fe to Pd at the interface, the thickness of the mixed layer can be chosen in a range of several Å and similar good fits can be obtained by changing the other parameters accordingly. Therefore, it is very difficult to extract detailed quantitative information about the magnetic structure of the intermixing layer and compare them with the SQUID results. The same holds for the other samples with thinner Fe layers, where it must be assumed that the complete Fe layers are mixed with Pd. For these samples we also obtained magnetic scattering length densities below the Fe bulk values, but the big uncertanties caused by the intermixing hinder a sensible comparison with the SQUID data.

With the SQUID magnetometer we have measured hysteresis curves at different temperatures between 2 and 350 K. Saturated ferromagnetic moments and coercive fields are decreasing with increasing temperature as expected for all samples (see fig. 3). In addition, the RFM is in very good agreement with these data lying between the saturated moments measured at 2 K and 350 K (see table 1). The low temperature data taken at 2 K reveal an increasing ratio of measured moment to RFM from a value of approximately 1 to 2.6 with decreasing Fe thickness. In contrast, these data suggest that in the samples with relatively thick Fe layers the effect on Pd is much smaller compared to that reported for the bulk Pd with Fe impurities.

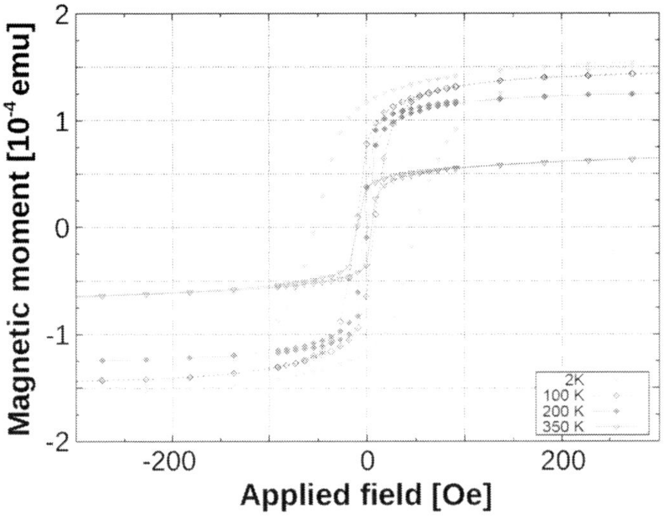

Figure 3. Temperature dependence of the hysteresis curves of the $[Pd_{150\,Å}/Fe_{2\,Å}]_{10}/Pd_{150\,Å}$ multilayer (substrat signal is substracted).

4. Conclusions and Outlook

We have developed a procedure to grow pure, monocrystalline Pd layers on Ag buffered GaAs substrates. A variable temperature during growth of the first Pd layer was applied to avoid interdiffusion with Ag. We were able to deposit very thin, monocrystalline and stable Fe films

Table 1. Results of the macroscopic SQUID measurements (FM = ferromagnetic). The reference ferromagnetic moments (RFM) were calculated by the atomic moment per Fe atom in bulk Fe multiplied by the number of Fe atoms.

	Pd_{150}/Fe_{10}	Pd_{150}/Fe_4	Pd_{150}/Fe_2	Pd_{150}/Fe_1
superlattice period [Å] (reflectometry)	183.4 (2)	179.8 (2)	164 (1)	155 (1)
expected Fe layer thickness [Å]	11.4 (3)	4.7 (3)	2 (1)	1 (1)
RFM μ_{Fe} [10^{-4} emu]	4.9 (1)	2.0 (1)	1.0 (4)	0.4 (4)
FM moment at 2 K μ_{2K} [10^{-4} emu]	5.4063 (4)	2.2407 (3)	1.6054 (4)	1.0458 (6)
μ_{2K}/μ_{Fe} [10^{-4} emu]	1.10	1.12	1.61	2.61
FM moment at 350 K μ_{350K} [10^{-4} emu]	4.5871 (2)	1.9624 (1)	0.7504 (1)	0
μ_{350K}/μ_{Fe} [10^{-4} emu]	0.94	0.98	0.75	0

on Pd. The preparation of Pd/Fe multilayers was also possible even though x-ray- and polarized neutron reflectometry revealed an intermixing at the interfaces of Fe and Pd on top. Surprisingly, in the samples with Fe layers thicker than 2 Å the effect on Pd seems to be very small even though due to the intermixing the situation should be more similar to the situation in dilute bulk Fe/Pd alloys, where the "Giant Moments" were found, than in a perfect multilayer system, where both materials are spatially separated from each other. Nevertheless, the SQUID data yield some indication for an increasing magnetization with decreasing Fe layer thickness compared to the bulk values of Fe. Whether this increase is due to an increased Fe magnetic moment per atom or to an induced magnetization in Pd, could not be clarified due to the intermixing. To determine the exact magnetic structure of the sample and the amount of polarized Pd, X-ray Magnetic Circular Dichroism (XMCD) measurements are planned. In addition, the growth procedure of the multilayers will be optimized. A variable deposition temperature for all Pd layers might also avoid intermixing of Pd and Fe. Furthermore, polarized neutron measurements will also be done at the new reflectometer MARIA at FRM II, which will yield much more detailed information about the multilayer systems due to the higher signal/noise ratio available.

References

[1] Verbeek B H, Nieuwenhuys G J and Mydosh J A 1980 *Phys. Rev.* B **22** 5426
[2] Gradmann U 1993 *Ultrathin transition metal films* (Amsterdam: Elsevier Science Publishers)
[3] Gradmann U, Dürkop T and Elmers H J 1997 *J. Magn. Mater.* **165** 56
[4] Kohlhepp J, Fritzsche H, Elmers H J and Gradmann U 1995 *J. Magn. Mater.* **148** 95
[5] Vogel J, Fontaine A, Cros V, Petroff F, Kappler J P, Krill G, Rogalev A, Goulon J 1997 *Phys. Rev.* B **55** 3663
[6] Bürgler D E, Schmidt C M, Wolf J A, Schaub T M and Güntherodt H J 1996 *Surf. Sc.* **366** 295
[7] Bürgler D E, Schmidt C M, Schaller D M, Meisinger F, Hofer R and Güntherodt H J 1997 *Phys Rev.* B **56** 4149

Journal of Physics: Conference Series **211** (2010) 012022 doi:10.1088/1742-6596/211/1/012022

Peculiarities of magnetic states in Ferromagnet/Superconductor heterostructures due to the proximity effects

V L Aksenov, Yu N Khaidukov and Yu V Nikitenko

Joint Institute for Nuclear Research, Dubna, Moscow Region, 141980, Russia

E-mail: aksenov@nf.jinr.ru

Abstract. The peculiarities of magnetic states in ferromagnet (F)/superconductor (S) heterostructures due to the proximity effects are investigated with the use of polarized neutron reflectometry. An anomalously strong reduction in the magnetization of the F layer due to superconductivity is observed in the S/F/S system. The dependence of the effects on an external magnetic field is studied for the first time. The enhanced neutron standing waves are used for registration of weak signals.

1. Introduction

The term "proximity effect" in heterostructures usually means the appearance of the properties of one phase in another, caused by the contact of these phases. In the physics of ferromagnet (F)/superconductor (S) proximity effects there are a number of intriguing predictions, which call for further clarification and experimental verification [1]. In particular, it was predicted that due to the proximity effects various scenarios of the influence of the superconducting phase transition on the magnetic properties of the layered F/S systems are possible: the formation of inhomogeneous ordered magnetic structure (cryptoferromagnetic), formation of the domain structure, magnetization "leakage" from F to S layer (inverse proximity effect). The magnetization of the system is reduced as a result. Experimentally the problem has been studied using the ferromagnetic resonance method [2, 3] and by the polarized neutron reflectometry (PNR) [4 – 7]. The effect of the reduction of the F layer magnetization and changes of magnetic profiles of the system below the superconducting transition were observed, but the understanding of physical mechanisms of these observations is not complete.

In this paper we present the results of investigations of F/S and S/F/S layered structures with the use of the polarized neutron reflectometry. We study the effects in dependence to an external magnetic field, which has not been done before. The due attention has been paid to the study of the quality of layers and interfaces, as it is known that considerable and developed interface roughness can disguise physical effects and change the interpretation. In the heterostructures with high structural quality the effects can be weak. We propose that in such a case the regime of enhanced neutron standing waves should be used in the polarized neutron reflectometry.

2. Field dependence of magnetization reduction

In our recent paper [7] we reported on an anomalously strong reduction of magnetization in the S/F/S structure Nb(50 nm)/^{57}Fe(3.9 nm)/[Si(3.4 nm)/Mo(3.4 nm)]$_{40}$/Si. Here the Nb layer and Si/Mo structure are two superconductors with transition temperatures $T_{C\text{-}Nb} = 9.2$ K and $T_{C\text{-}Si/Mo} = 4.5$ K, respectively. The F phase between them was presented by the iron layer with Curie temperature $T_K = 55$ K. Such a small temperature was explained by high intermixing with neighboring layers. The PNR measurements were performed with the time-of-flight polarized neutron reflectometer REMUR at the IBR-2 reactor (JINR, Dubna) at temperatures T = 15 K, 5 K and 2 K in the magnetic field H = 0.5 kOe. It was obtained that the magnetization decreases below $T_{C\text{-}Nb}$ and $T_{C\text{-}Si/Mo}$ from 1.9 kOe at 15K to 1.3 kOe (≈30%) at 5K and to 0.1 kOe (≈95%) at T = 2K. A row of Abrikosov vortices was observed in the center of the Nb layer at 2 K. A possible reason for the absence of vortices at 5 K can be the shunting of the magnetic flux by the F layer with non-zero magnetization, which leads to the reduction of intensity of the magnetic field in the Nb layer. The estimations give H ≈ 150 Oe, which is much less than $H_{C1} \approx 360$ Oe.

The experiments have been repeated recently on the reflectometer ADAM at ILL at H = 225, 750 and 970 Oe [8]. At H = 750 and 970 Oe we observed similar behavior as at H = 500 Oe. However, at H = 225 Oe no change of spin asymmetry has been observed. Here spin asymmetry $S \equiv [R^+ - R^-]/[R^+ + R^-]$ is the ratio of difference to the sum of two neutron reflectivities in different spin states. With the assumption that the magnetic part of the optical potential is much less than the nuclear one, it can be easily shown that the spin asymmetry is proportional to the magnetization profile of the structure. In Fig.1 the temperature evolution of the integrated absolute value of the spin asymmetry weighted on statistical error $f = \int[|S(Q)|/\delta S(Q)]dQ$ is shown. The normalization to the statistical error helps to avoid artificial effects in the case when the spin asymmetry is small. We use modulus of S(Q) because for this sample the spin asymmetry is an oscillating function and therefore simple integration would give a value close to zero. Thus, one can say that the parameter f is somehow proportional to the absolute value of magnetization of the whole sample.

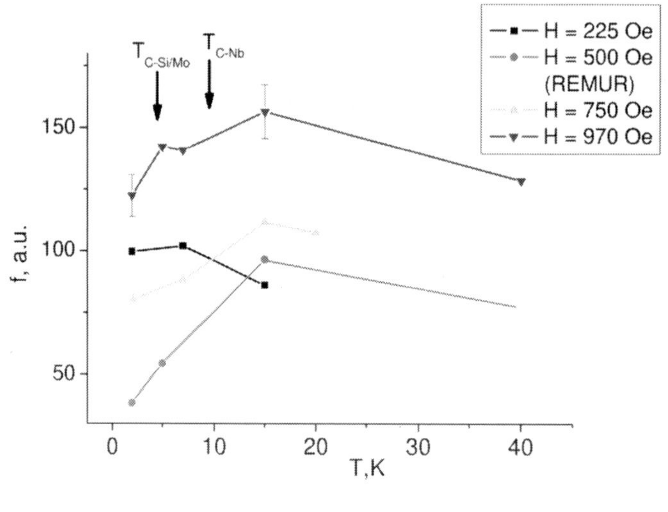

Figure 1. $f \equiv \int[|S(Q)|/\delta S(Q)]dQ$ at different magnetic fields. The details see in the text.

The performed measurements show that the reduction of the average magnetization of the F layer due to the influence of superconductivity depends on the external magnetic field. This process develops at the fields H ≥ 0.5 kOe in two stages: after the first superconducting transition in the Nb layer and then after the second transition in the [Si/Mo] layer. At the field H = 0.5 kOe we observe an anomalously strong reduction. The suppression of magnetism by superconductivity decreases with increasing H. The analysis of the diffuse scattering, field and temperature dependence of the magnetization leads us to the conclusion that there are two mechanisms of the magnetization reduction: the appearance of the domains oriented by the magnetic field and the suppression of the magnetization inside each domain by the superconducting exchange energy. At $T < T_C$ a part of electrons in the F layer due to the proximity effect forms Cooper pairs and therefore does not participate in the ferromagnetic alignment.

3. Field dependence of the magnetic structure at the S/F interface

In this section we consider the details of the magnetic structure at the S/F interface at different external magnetic fields. The peculiarities of the magnetic state can be clarified with the use of the regime of neutron standing waves in a reflectometry experiment. This regime allows us to scan the spatial vicinity of an interface to detect magnetic non-collinear or scattering regions such as domains, clusters, vortices, etc. [9]. The study of the structure Pd(1.5 nm)/V(39 nm)/Fe(3 nm)/ 20×[V(3 nm)/Fe(3 nm)]/MgO has been done on the reflectometer D-17 at ILL (Grenoble) at T = 3K and 7 K at magnetic fields H = 0.2, 0.7, 1.0, 1.5 and 4.5 kOe. The periodic structure 20×[V/Fe] (PS) plays a role of a standing wave generator. The superconducting critical temperature and magnetic field (applied parallel to the surface) in the S layer are $T_C(0) = 3.7$ K and $H^{\|}_{C2}(0) \approx 22$ kOe respectively. In Fig. 2 the spin asymmetries at magnetic fields H = 0.7 and 1.5 kOe are shown. The peak at $Q = Q_B = 1\ nm^{-1}$ corresponds to the 1st order Bragg reflection from the PS. The oscillations at $Q = [0.1 \div 1]\ nm^{-1}$ are mainly defined by the magnetic state at the S/F interface. At $H \leq 1$ kOe no significant changes of the spin asymmetry at the superconducting transition have been observed. From the spin asymmetries at T = 7 K it follows that the growth of peak at $Q = Q_B$ with increasing magnetic field takes place. This growth points to the process of magnetization of the PS. Starting from H = 1.5 kOe the magnetization of the S/F interface also takes place. When the V(39nm) layer goes into the superconducting state (see curves at T = 3 K) spin asymmetry S(Q) at 1.5 kOe looks the same as spin asymmetries below 1 kOe. The same effect has been observed at higher field H = 4.5 kOe. Thus, one can draw a conclusion that the magnetization of the S/F interface is suppressed during the transition into the superconducting state of V(39 nm). The magnetic state of a periodic structure is not changed meanwhile.

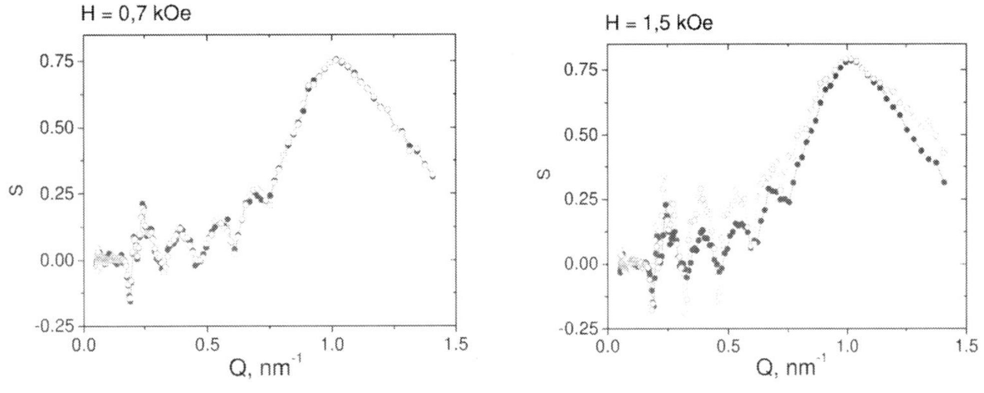

Figure 2. Spin asymmetries at T = 7 K (open circles) and T = 3 K (solid circles) at different magnetic fields.

The intensity of diffuse scattering from magnetic inhomogeneities in the regime of standing waves (i.e. at $Q = Q_B$) can be written as: $J_{diff}(Q) \sim \int n(z, Q)\, \sigma_{diff}(z)\, dz$ [9]. Here $\sigma_{diff}(z)$ is the cross-section of diffuse scattering at a depth z, n(z, Q) is the neutron density at a given depth and Q. For the latter we can write an approximate form: $n(z, Q) = (1 + R^{1/2})^2 - 4\sin^2[(Q\Delta z + \alpha(Q - Q_B))/2]$. Here R is the reflectivity at Q_B, the factor α depends on the number of repetitions N and thickness of the period D of the PS as $\alpha = ND/2$. The expression for n(z, Q) shows that at a given Q the neutron density has its maximum at a certain distance $\Delta z = \alpha(Q_B/Q - 1)$ from the S/F interface. Thus, the diffuse scattering intensity provides information on the distribution of inhomogeneities at a depth Δz. Its maximum at certain Q corresponding to the maximum of the diffuse scattering cross-section is situated at Δz. The expression for the diffuse scattering can be simplified for the case when the function $\sigma_{diff}(z)$ is locally limited around the S/F interface at a distance much smaller than a period of the standing wave. For this case we can omit integration. In Fig.3 the normalized z-distribution of inhomogeneities $\kappa_D = \sigma_{diff}(\Delta z)/\sigma_{diff}(0)$ at different magnetic fields is presented. From Fig. 3 we can conclude that at H = 1.5 kOe the distribution maximum is closer to the F layer while at H = 4.5 kOe it is closer to the S layer. This observation can be interpreted in the following way. At H = 1.5 kOe there is an inhomogeneous magnetic state of the F layer, which becomes homogeneous at higher fields. At H = 4.5 kOe ($\geq H_{C1}$) we have a homogeneous magnetic state in the F layer and a developed vortex state in the S layer.

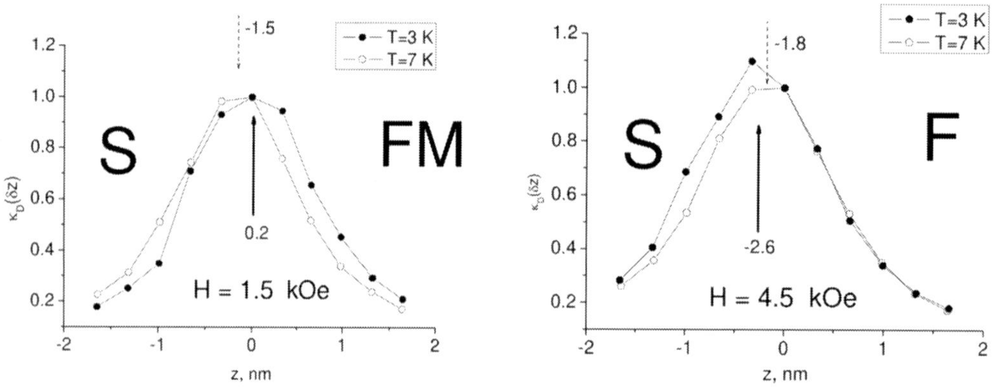

Figure 3. Depth distribution of the inhomogeneities in the vicinity of the S/F interface at T = 7 K (open circles) and T = 3 K (solid circles). Solid and dashed lines show the position of the mass center above and below T_C.

4. New approach in reflectometry of proximity effects

The influence of superconductivity on ferromagnetism in S/F heterostructures is usually pronounced for weak ferromagnets. This weakness can be achieved when a strong ferromagnet in the F layer is diluted or the thickness of the F layer is small (in the order of 1 nm). In both cases we have a weak magnetic signal in a reflectometry experiment to detect it by usual method. There is a possibility to enhance the signal in this case by using enhanced neutron standing waves [9].

For the experimental study we used a structure Cu(33 nm)/V(40 nm)/Fe(1 nm)/MgO, prepared by molecular beam epitaxy in RMKI KFKI (Budapest, Hungary) [10] on the basis of theoretical calculations [11]. Here S/F bilayer V(40 nm)/Fe(1 nm) is placed between Cu and MgO with high neutron scattering length densities to create regime of resonance enhanced neutron standing waves. The evaporation of iron directly onto the substrate allowed us to obtain structurally homogeneous F layer with extremely low roughness ($\sigma \sim$ 1-2 monolayers). Superconducting transition temperature of the vanadium layer and Curie temperature of the iron layer are $T_C = 3.5$ K and $T_K \geq 300$ K, respectively.

The preliminary investigation of the sample was performed at the reflectometer N-REX+ (FRM2, Münich) at room temperature [10]. The neutron wave function in the structure was calculated with the use of the neutron scattering length density profiles obtained in this experiment. It was obtained that the neutron density around the S/F interface was $n_{S/F} = 50$ in comparison with the density $n_0 = 1$ in the incident beam.

Figure 4 presents the results of the reflectivity measurements in spin-flip and non-spin-flip channels at T = 10 K and H = 20 Oe performed at the reflectometer ADAM (ILL). The magnetic field was in the sample plane and perpendicular to the magnetization of the F layer as shown in the inset to Fig. 4. The value of the resonance-enhanced spin-flip signal over the initial beam was 10%. This made it possible to measure the temperature dependence of the resonance peak with an accuracy of around 1% (Fig. 5a) and the diffuse scattering during 8-12 hours.

The decrease in the spin-flip intensity at $T < T_C$ (Fig. 5a) can be connected with the reduction of magnetization in the F layer. The increase of the diffuse scattering around T = 5 K just before T_C (Fig. 5b) may be due to the appearance of short-range order clusters, which destroy the long-range magnetic order. The subsequent increase in the spin-flip intensity at T < 2.5 K ($\approx$0.6 T_C) can be related to the appearance of induced magnetization in the S layer, which is antiparallel to the magnetization of the F layer and therefore noncollinear to the external magnetic field. The increase in the diffuse scattering around T = 2 K can be connected with the appearance of magnetic clusters in the S layer. Another possible mechanism of the presented behavior at T = 2 K consists in increasing of the reverse magnetization due to the increase in the exchange interaction below 5 K.

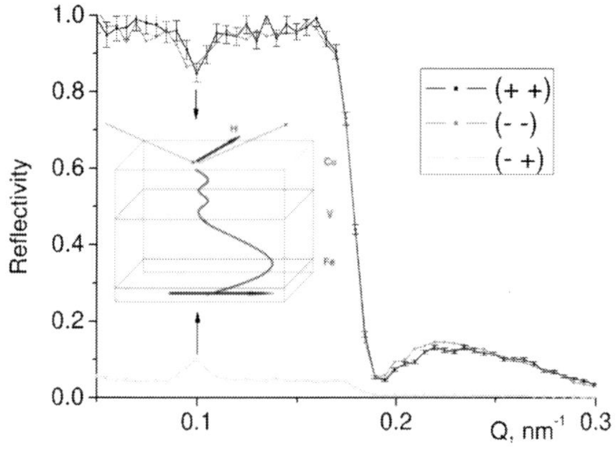

Figure 4. Spin-flip and non-spin-flip reflectivity at T = 10 K. In the inset – scheme of the experiment: **H** and **M** – vectors of external field and magnetization of the F layer, solid line inside the structure shows neutron density inside the structure at resonance.

The presented results show the efficiency of the use of the regime of enhanced neutron standing waves for investigations of weak changes of magnetic states at S/F interfaces: in the given example we were able to detect a 4% decrease in the spin-flip intensity, which corresponds to a 2% decrease in the magnetization of the F layer.

5. Conclusions

The peculiarities of magnetic states in the F/S heterostructures due to the superconducting phase transition in the S layer depending on the external magnetic field have been investigated for the first time. In the S/F/S structure an anomalously strong reduction in the magnetization of the F layer up to 95% from M_{sat} has been revealed. Both in the S/F and S/F/S structures a change in the nonhomogeneous domain state is observed below the superconducting temperature T_C. The reduction in the mean magnetization of a domain due to the superconducting exchange energy is suggested. Our results do not support the model of cryptoferromagnetic state.

It has been demonstrated that the details of the magnetic structure at a given spatial position, for example at the S/F interface, can be clarified with the help of the regime of neutron standing waves in the polarized neutron reflectometry. The regime of enhanced neutron standing waves is proposed for investigations of weak effects.

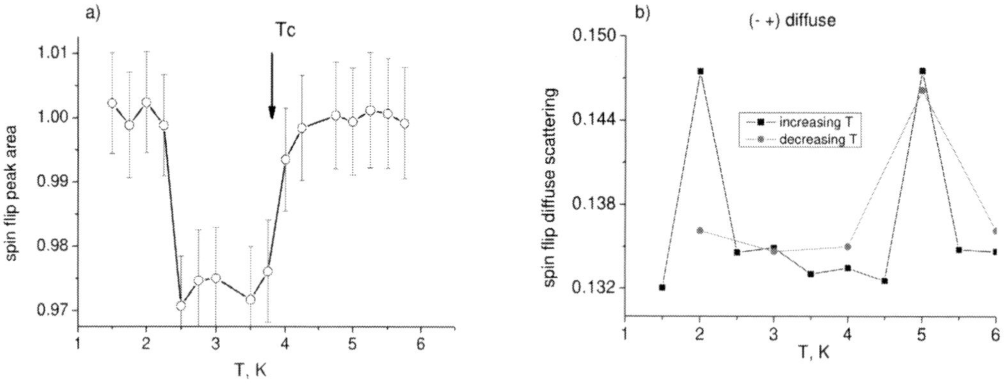

Figure 5. Temperature dependence of specular spin-flip (a) and diffuse (b) scattering around the resonance.

Acknowledgements
The authors wish to thank L.Bottyan (RMKI KFKI, Hungary), B.Nagy (RMKI KFKI, Hungary), K.Zhernenkov (University of Bochum, Germany), A.Rühm (MPI, Germany) for their assistance in the measurements. The work has been done with the financial support of the Russian Foundation for Basic Research (grant № 09-02-00566). One of the authors (Yu.Khaidukov) would like to acknowledge the support of the Foundation for Assistance to Small Innovative Enterprises of the Russian Federation (grant № 8455 UMNIK-08-3).

References

[1] Bergeret F S, Volkov A F and Efetov K B 2005 *Rev. Mod. Phys.* **77** 1321
[2] Mühge Th, Garif'yanov N N, Goryunov Yu V, Theis-Bröhl K, Westerholt K, Garifullin I A and Zabel H 1998 *Physica* C **296** 325
[3] Garifullin I A, Tikhonov D A,Garif'yanov N N,Fattakhov M Z, Theis-Bröhl K, Westerholt K, and Zabel H 2002 *Appl. Magn. Reson.* **22** 439
[4] Stahn J, Chakhalian J, Niedermayer Ch, Hoppler J, Gutberlet T, Voigt J, Treubel F, Habermeier H-U, Cristiani G, Keimer B and Bernhard C 2005 *Phys. Rev.* B **71** 140509 (R)
[5] Aksenov V L, Jernenkov K N, Khaidukov Yu N, Nikitenko Yu V, Petrenko A V, Proglyado V V, Andersson G and Wäppling R 2005 *Physica* B **356** 9
[6] Aksenov V L, Nikitenko Yu V, Petrenko A V, Uzdin V M, Khaidukov Yu N and Zabel H 2007 *Cryst. Rep.* **52** 381
[7] Aksenov V L, Nikitenko Yu V, Khaidukov Yu N, Vdovichev S N, Borisov M M, Morkovin A N and Mukhamedzhanov E Kh 2009 *J. Surf. Invest.* **3** 495
[8] Khaidukov Yu 2008 *report on experiment # 5-54-38 at ILL*
[9] Aksenov V L, Ignatovich V K and Nikitenko Yu V 2006 *Cryst. Rep.* **51** 734
[10] Khaidukov Yu N, Zhernenkov K N, Nikitenko Yu V, Nagy B, Bottyan L, Rühm A, Teichert A, Steitz R and Aksenov V L submitted to *Cryst. Rep., e-preprint: arXiv:1001.2895v1*
[11] Khaidukov Yu N, Nikitenko Yu V and Aksenov V L arXiv:0801.3967v1
[12] Löfwander T, Champel T, Durst J and Eschrig M 2005 *Phys. Rev. Lett.* **95** 187003

Journal of Physics: Conference Series **211** (2010) 012023 doi:10.1088/1742-6596/211/1/012023

Magnetization Flop in Fe/Cr GMR Multilayers

E Josten[1], U Rücker[1], S Mattauch[2], D Korolkov[2], A Glavic[1] and Th Brückel[1,2]

[1] Institut für Festkörperforschung - Forschungszentrum Jülich GmbH, 52425 Jülich, Germany
[2] Jülich Centre for Neutron Science - Forschungszentrum Jülich GmbH, 85747 Garching, Germany

E-mail: e.josten@fz-juelich.de

Abstract.
We study the magnetization of antiferromagnetically (AF) coupled Fe/Cr multilayers, with 7 and 19 periods respectively, under the influence of an external magnetic field. Due to the odd number of Fe layers the uncompensated magnetization of the multilayer at remanence is equal to the magnetization of a single Fe layer. Measurements have been performed using the magnetooptical Kerr effect (MOKE) for an integral overview and polarized neutron scattering under grazing incidence for a depth resolved information about the magnetization orientation of the individual Fe layers.

We find a dependence of the orientation of the remanent magnetization on the number of AF coupled Fe double layers. For a small number of periods (here: 7). The uncompensated magnetic moment is sufficient to align the magnetization of all layers parallel or antiparallel to the applied magnetic field. On the other hand, in the multilayer with 19 AF-coupled Fe layers all Fe layers are magnetized perpendicular to the field.

1. Introduction

Layered magnetic structures are main elements of today's spintronic information technology [1]. A prominent example is the applications of the giant magnetoresistance effect (GMR) [2] in hard disk drives or magnetic field sensors. The complexity of physical properties of magnetic multilayers (ML), due to the competition between interlayer (IL) exchange coupling, the different anisotropies and external influences is not yet fully understood.

In this paper we report on the remagnetization behaviour of AF-coupled multilayer structures as a function of the number of multilayer periods. To address this issue a model system is needed, which shows reproducible concurrent influences. This is given in epitaxial Fe/Cr-MLs. The coupling phenomena in Fe/Cr layered structures have been discovered in the 1980's as the first IL exchange coupled system of transition metals [3, 4, 5]. The AF coupled Fe/Cr-MLs are grown with molecular beam epitaxy (MBE), macroscopically characterized by MOKE and measured with neutron scattering with full polarization analysis under grazing incidence. This method allows one to investigate the magnetic behaviour layer by layer as a function of the applied magnetic field and provides additional lateral information on roughnesses and magnetic domains.

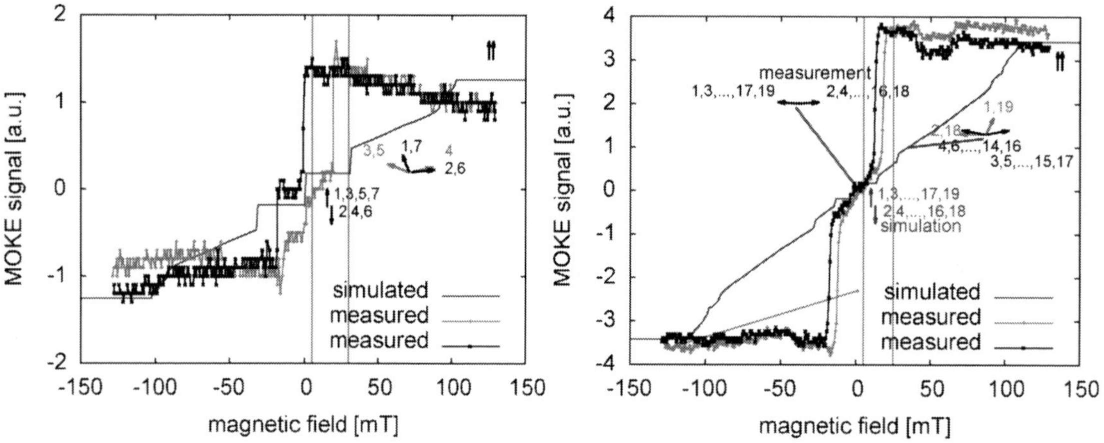

(a) $[Fe/Cr]_7$ (b) $[Fe/Cr]_{19}$

Figure 1. MOKE measurements (red: increasing, black: decreasing field) and simulations (blue) of AF coupled [Fe/Cr] MLs. The vertical pale lines indicate fields selected for neutron measurements. The numbered arrows show the in-plane orientation of the magnetization. The main difference between the two measurements is the behaviour at remanence. The $[Fe/Cr]_7$ system shows a jump at $H = 0$ while the $[Fe/Cr]_{19}$ system smoothly crosses the zero. In the intermediate field range, there appears to be a strong difference between measurement and simulation, because of the transverse Kerr effect that distorts the curve where the magnetization is oriented perpendicular to the field, and because of the limited penetration depth, which emphasizes the top layer, while the simulation uses equal weights for all layers. (color online)

2. Sample preparation and macroscopic characterization

Two samples with the layer compositions $[Fe_{10\,nm}/Cr_{1\,nm}]_{x6}/Fe_{10\,nm}$ and $[Fe_{10\,nm}/Cr_{1\,nm}]_{x18}/Fe_{10\,nm}$ on a GaAs(001) substrate with a $Ag_{150\,nm}$ buffer layer have been grown epitaxially by MBE using the standard procedure described in reference [6, 7]. In the rest of the text, we use the abbreviations $[Fe/Cr]_7$ and $[Fe/Cr]_{19}$ for the two samples, respectively. Because of the single crystalline growth the magnetic easy axes due to crystalline and shape anisotropy are known to be parallel to the four in-plane [100] directions of the substrate.

The magnetization as a function of the applied magnetic field is precharacterized by MOKE. This method enables a qualitative measurement of the magnetization as a function of the magnetic field. The magnetization signal is distorted due to the limited penetration depth of the light, due to contributions of the transversal Kerr effect during the rotational transition of the magnetization and due to thermal drift and noise. Nevertheless, it allows the determination of the type and strength of the IL coupling by fitting the transition fields to a model based on the total energy.

Figure 1 shows the MOKE measurements of both samples with the magnetic field aligned parallel to one of the easy axes. One can easily identify the AF character of the interlayer coupling from the reduced magnetization at small fields. To be able to understand the MOKE measurement of the complex multilayer, we have performed simulations based on a simple model of minimizing the total magnetic energy with respect to the angle between the magnetic moments of the individual layers and the direction of the applied field. The AF coupling, Zeeman energy and the cubic crystalline anisotropy have been taken into account. The interplay of these three contributions determine the alignment of the magnetic moments in the layers and thus the magnetization component parallel to the field. The model is based on the assumption that there is a homogeneous magnetization in every layer (single domain state) and neglects potential

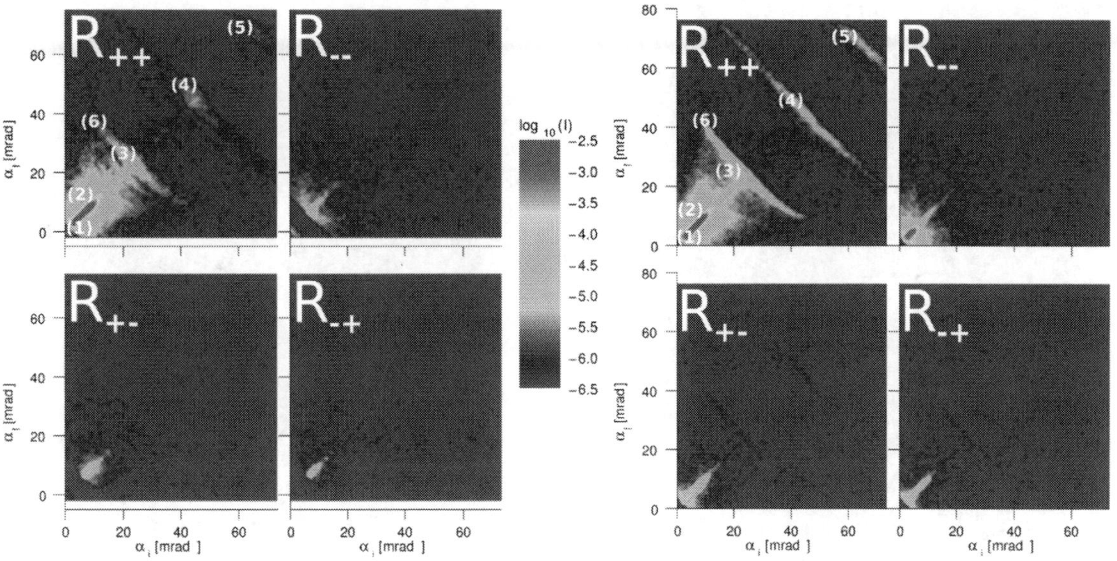

(a) [Fe/Cr]$_7$ (b) [Fe/Cr]$_{19}$

Figure 2. Neutron measurement in saturation field of 300 mT. Plotted is the logarithm of the scattering intensity vs. incident and outgoing angles α_i, α_f. The four polarization channels are the NSF channels (R_{++} and R_{--}) and the SF channels (R_{+-} und R_{-+}). Indicated are the primary beam blocked by the beamstop (1), the plateau of total reflection (2), the first (3), second (4) and third order (5) Bragg-peak (giving information about the layer structure) and the Bragg-sheets (information about correlated roughness)(6). (color online)

walls, so that the hysteresis effects cannot be reproduced. The values of the constants of bilinear coupling J_1 and biquadratic coupling J_2 have been adjusted by a reference measurement of an [Fe/Cr/Fe] sample, where extinction effects can be neglected.

For the [Fe/Cr]$_7$ sample we find a reasonable agreement of measurement and simulation. The net magnetization of the ML at remanence is the magnetization of a single layer. This is found to be sufficient for the alignment of the magnetization of all layers alternatingly parallel and antiparallel to the field. When crossing zero field, the magnetization of all layers is reversed, resulting in a jump in the MOKE curve. With increasing field the influence of the AF coupling energy is reduced compared to the Zeeman energy which tends to magnetize all layers along the field. The compromise is then a state, where all magnetizations are somewhat aligned perpendicular to the field, alternatingly left and right, with a canting towards the field direction. The canting induces a magnetization component in every layer parallel to the field in favour of the Zeeman energy, while the alternation satisfies the AF coupling. The magnetic moments cant more and more until they finally reach saturation, when they are all aligned parallel to the field.

The MOKE measurement of the [Fe/Cr]$_{19}$ sample shows a smooth transition at zero field. This is contradictory to the previously described measurement and contradictory to the simulation. The transition towards the state, where the net magnetization is aligned with the field, is not observed experimentally. Together with the neutron data, it became clear that the coercivity effects, which are not included in the model calculation, inhibit this transition, as the energy gain is not sufficient to cross the anisotropy barrier.

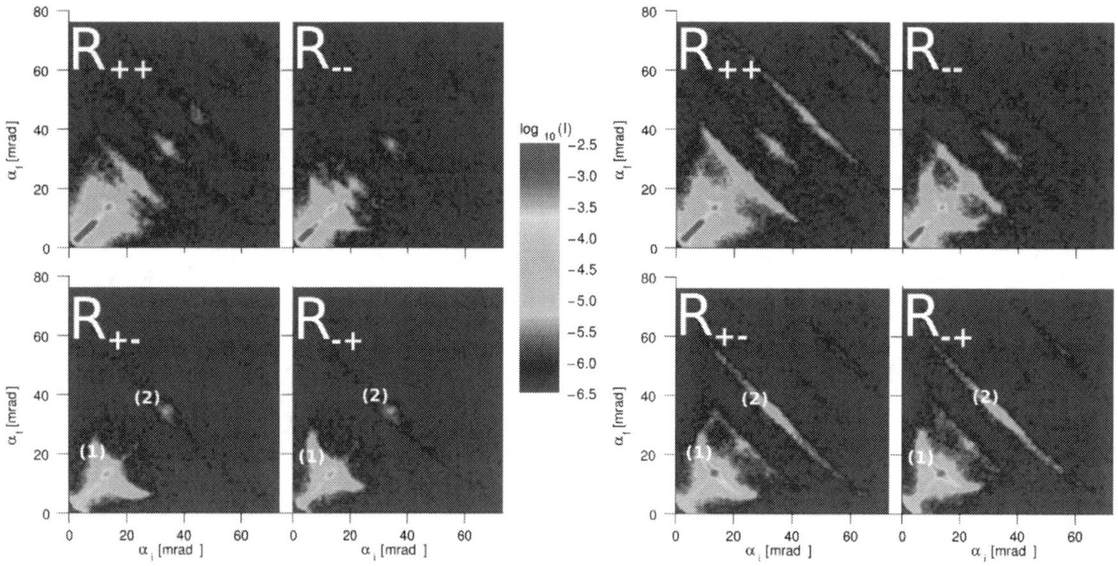

(a) $[Fe/Cr]_7$ (30 mT)

(b) $[Fe/Cr]_{19}$ (25 mT)

Figure 3. Neutron measurement in intermediate field. Indicated are the AF superstructure Bragg peaks of the order $\frac{1}{2}$ (1) and $1\frac{1}{2}$ (2). (color online)

3. Polarized neutron reflectometry and off-specular scattering

Polarized neutron reflectometry (PNR) and off-specular scattering reveal information about the in-plane magnetization vector of individual layers and their domain structure. The experiments have been performed at the reflectometer TREFF of the JCNS at the FRM II research reactor in Garching. The setup allows to simultaneously measure specular and off-specular scattering with full polarization analysis. The neutron wavelength was fixed at 4.73 Å. The efficiencies of the polarizer, analyzer and both spin flippers are above 98%.

An external magnetic field has been applied in-plane parallel to one of the easy axes. Based on the MOKE measurements, we have defined the magnetic fields suitable for the neutron measurements. We have decided to perform three measurements on one branch of the hysteresis curve, from saturation at 300 mT down to remanence.

The figures 2 - 4 show the scattered intensity in a logarithmic colour plot as a function of the angle of incidence α_i and the outgoing angle α_f. The four graphs R_{++}, R_{--},R_{+-}, and R_{-+} show the different combinations of incoming and outgoing polarization of the neutron beam. The incident and outgoing angles α_i and α_f are defined as the angles between the incoming and outgoing wave vectors $\vec{k_i}$ and $\vec{k_f}$ and the sample surface, respectively. If $\alpha_i = \alpha_f$, we observe specular reflectivity. This part contains only information on the out-of-plane structure because the scattering vector $\vec{Q} = \vec{k_f} - \vec{k_i}$ is perpendicular to the surface. In off-specular geometry ($\alpha_i \neq \alpha_f$) the scattering vector has a component in plane which contains information on lateral correlations e.g. from roughness and magnetic domains with a size smaller than the in-plane coherence length ($\approx$ 20 μm)[8, 9]. Scattering in the non-spin-flip (NSF) channels (R_{++} and R_{--}) are due to the correlations of the structure and the magnetization components parallel to the external field. The information in the spin-flip (SF) channels (R_{+-} and R_{-+}) has its origin only in the magnetization component perpendicular to the external field.

In saturation (fig. 2) all magnetic moments are aligned with the field, so no SF scattering

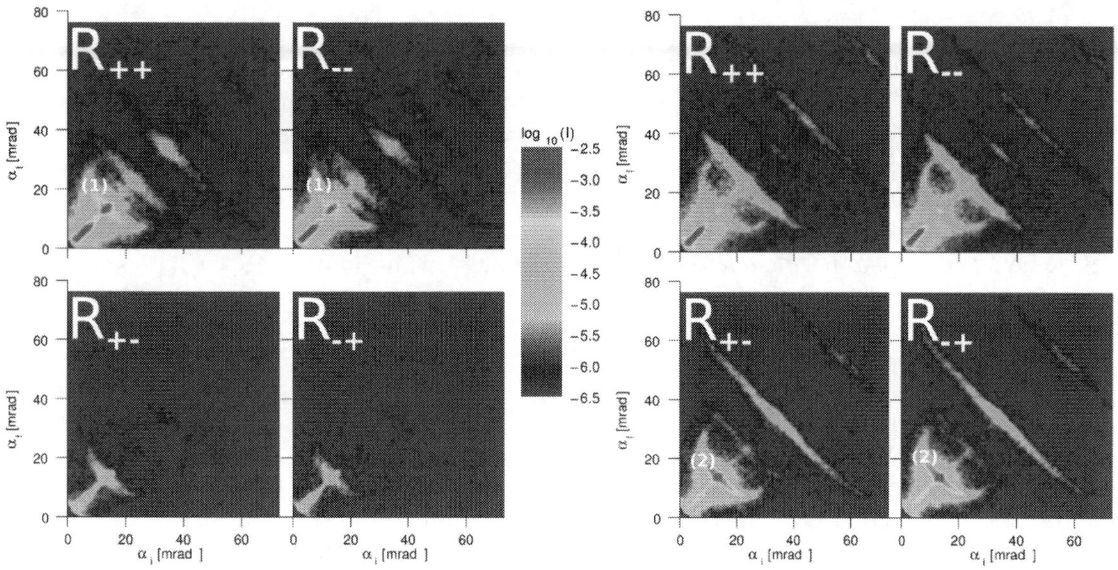

(a) $[Fe/Cr]_7$ (b) $[Fe/Cr]_{19}$

Figure 4. Neutron measurement in remanence field of 5 mT. Indicated are the AF superstructure Bragg peaks of order $\frac{1}{2}$ in the NSF channel of the $[Fe/Cr]_7$ system (1) and in the SF channel of the $[Fe/Cr]_{19}$ system (2). (color online)

is observed. The intensity in the SF channels is purely parasitic due to the limited flipping ratio of the instrument. As the nuclear scattering length density of Cr is almost equal to the difference between nuclear and the magnetic scattering length density of Fe, there is no contrast for spin-down neutrons. Therefore, in R_{--} only the total reflection is observed.

At the intermediate field of 30 mT or 25 mT, resp., (fig. 3) we observe intensity in the SF-channels (Bragg peaks of half orders). This is induced by an additional magnetization component perpendicular to the external field, because of the stronger influence of the AF coupling compared to the Zeeman energy at reduced magnetic field. A magnetic contrast between adjacent Fe layers has evolved, so the magnetic unit cell is twice as large as the structural unit cell. The behavior of both systems in this field range is the same.

At remanence (fig. 4) there is obviously a big difference between both samples. In the $[Fe/Cr]_7$ system there is magnetic information (doubled unit cell because of AF alignment) only in the NSF-channel, because all moments are (anti-)parallel to the quantization axis. The remaining intensity difference between R_{++} and R_{--} is a signature of the remaining uncompensated magnetic moment in field direction. This is in agreement with the MOKE data and its simulation. The $[Fe/Cr]_{19}$ system shows the half order AF Bragg peaks only in the SF channels while only the structural integer order Bragg peaks remain in the NSF channels (the faint intensity in the other channels is a parasitic effect due to the limited flipping ratio). This shows that the moments are all aligned perpendicular to the small external field. The intense wide Bragg sheets in the SF channels indicate a magnetic domain pattern with domain sizes in the μm regime.

For a quantitative analysis we have modeled the specular part of the intensity measured (fig. 5). With a fit of the model parameters to the data, we were able to retrieve the directions of the magnetic moments with respect to the external field (tab. 1). In saturation field all moments align parallel to the external field, so that we could use it for fixing all structural parameters.

In an intermediate field of 30 mT ($[Fe/Cr]_7$) or 25 mT ($[Fe/Cr]_{19}$) we find a more complex magnetic structure. The competing influences of Zeeman-, anisotropy- and AF coupling energy lead to differently aligned outer and inner layers because the inner layers feel the coupling from both sides. The results from the PNR fit are in good agreement with the simulation results describing the MOKE measurements as can be seen in table 1.

In remanence the statements of the qualitative analysis hold. In the $[Fe/Cr]_7$ system the moments align parallel and antiparallel to the field while in the $[Fe/Cr]_{19}$ system the magnetization is aligned perpendicular to the field with a slight canting in field direction. This is consistent with the MOKE measurement, but only in the case of $[Fe/Cr]_7$ it is also consistent with the model calculation.

The difference to the result of the simulation in the case of $[Fe/Cr]_{19}$ can be explained as follows: The model only minimizes the total energy of the system without taking into account potential walls for domain wall movement or activiation energies to cross a hard anisotropy axis. Effects of coercivity, which make a system remain in a state it had previously, can not be described. The difference between the systems can be attributed to the increase of energy which is necessary to flip the magnetization of the whole ML across a hard anisotropy axis. All layers have to be flipped at once, so the activation energy is proportional to the number of Fe layers. In contrast, the magnetic field only works on the magnetic moment of the single layer, whose magnetization is not compensated by the AF coupling. In the $[Fe/Cr]_7$ system it is energetically favorable to flip all 7 Fe layers to gain the Zeeman energy of the uncompensated layer. For 19 Fe layers the Zeeman energy of the uncompensated layer is the same while flipping the whole system needs much more energy. Thus for the $[Fe/Cr]_{19}$ system the gain in Zeeman energy is not sufficient to overcome the anisotropy barrier, so that the system remains with the magnetic moments aligned perpendicular, slightly canted to the external field direction.

4. Conclusion

We carried out MOKE measurements, specular and off-specular polarized neutron scattering of AF coupled Fe/Cr MLs. The magnetic behavior of the systems shows a dependence on the number of Fe layers. In the intermediate and high field range, the systems show identical behaviour as it is governed by the competition between AF coupling and Zeeman energy. At

(a) $[Fe/Cr]_7$ system

Fe-layer	**Angle to external field**			
1 and 7	0	8,1	21	0
2 and 6	180	−104,3	−91	0
3 and 5	0	16,5	73	0
4	180	−76,4	−79	0
	PNR fit 5 mT	**PNR fit 30 mT**	**MOKE-Sim. 32 mT**	**PNR fit 300 mT**

(b) $[Fe/Cr]_{19}$ system

Fe-layer	**Angle to external field**			
1 and 19	−83,86	−54,5	−72	0
2 and 18	86,55	87,9	82	0
3,5,7,...,15,17	−83,86	−76,3	−80	0
4,6,8,...,14,16	86,55	81,5	80	0
	PNR fit 5 mT	**PNR fit 25 mT**	**MOKE-Sim. 25 mT**	**PNR fit 300 mT**

Table 1. Angles of the magnetization vectors of the individual Fe layers as retrieved from the fit of the specular neutron reflectivity or the MOKE simulation, resp.

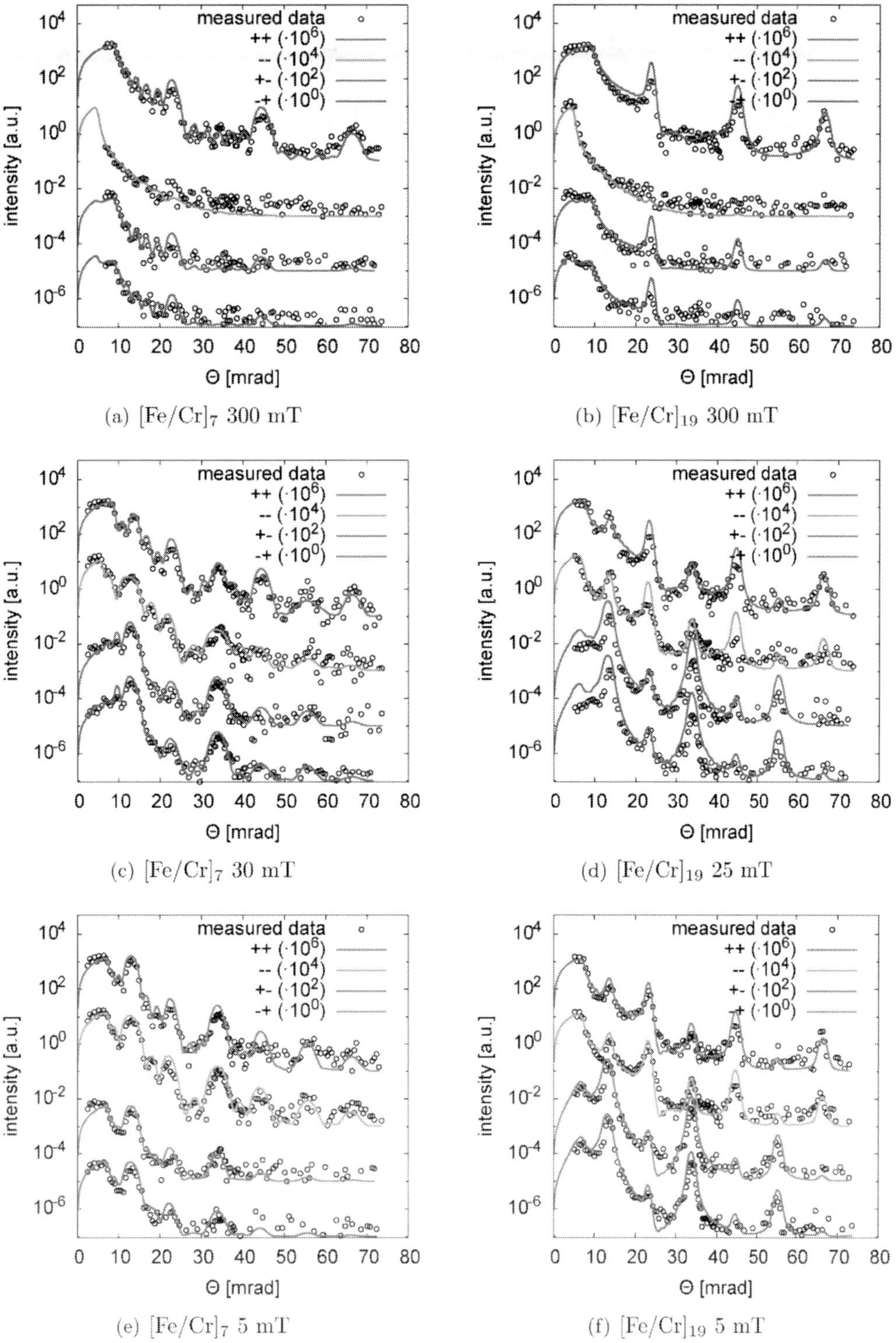

(a) $[Fe/Cr]_7$ 300 mT

(b) $[Fe/Cr]_{19}$ 300 mT

(c) $[Fe/Cr]_7$ 30 mT

(d) $[Fe/Cr]_{19}$ 25 mT

(e) $[Fe/Cr]_7$ 5 mT

(f) $[Fe/Cr]_{19}$ 5 mT

Figure 5. Fit of the specular reflectivity of both samples at all magnetic fields. The different polarization channels are scaled for better perceptibility, the symbols indicate the experimental points, the lines show the fits. (color online) remanence, the alignment is driven by the Zeeman energy of the uncompensated magnetic moment working against the coercivity. In the $[Fe/Cr]_7$ system the Zeeman term is sufficient to align the magnetization of all layers along or opposite to the applied field. With increasing number, the situation changes. In a ML with 19 AF-coupled Fe layers, we have found the entire magnetic signal in the spin-flip channels, which means that all Fe layers remain magnetized perpendicular to the field.

We have been able to model the magnetic behavior of AF-coupled systems with a simple approach. The simulations of the magnetic behavior of these MLs are consistent with our MOKE and neutron measurements within the limitations of the model.

References

[1] Grünberg P 2008 *Rev. Mod. Phys.* **80** 1531
[2] Fert A, Grünberg P, Barthélémy A, Petroff F and Zinn W 1995 *J. Magn. Magn. Mater.* **140** 1
[3] Grünberg P, Schreiber R, Pang Y, Brodsky M B and Sowers H 1986 *Phys. Rev. Lett.* **57** 2442
[4] Binasch G, Grünberg P, Saurenbach F and Zinn W 1989 *Phys. Rev.* B **39** 4828
[5] Baibich M, Broto J, Fert A, Nguyen Van Dau F, Petroff F, Etienne P, Reuzet G, Friedrich A and Chazelas J 1988 *Phys. Rev. Lett.* **61** 2472
[6] Bürgler D E, Schmidt C M, Schaller D M, Meisinger F, Hofer R and Güntherodt H J 1997 *Phys. Rev.* B **56** 4149
[7] Bürgler D E, Schmidt C M, Wolf J A, Schaub T M and Güntherodt H J 1996 *Surface Science* **366** 295
[8] Toperverg B P 2002 *Matter and Materials* **12** 247
[9] Anker J F and Felcher G P 1999 *J. Magn. Magn. Mater.* **200** 698

Polarized Neutrons and Synchrotron X-rays for Magnetism Conference 2009 IOP Publishing
Journal of Physics: Conference Series **211** (2010) 012024 doi:10.1088/1742-6596/211/1/012024

Overcoming the field-of-view restrictions in soft x-ray holographic imaging

C Tieg[1], R Frömter[2], D Stickler[2], H Stillrich[2], C Menk[2], S Streit-Nierobisch[3], L-M Stadler[3], C Gutt[3], O Leupold[3], M Sprung[3], G Grübel[3] and H P Oepen[2]

[1]European Synchrotron Radiation Facility (ESRF), B.P. 220, F-38043 Grenoble Cedex, France
[2]Institut für Angewandte Physik, Universität Hamburg, Jungiusstrasse 11, D-20355 Hamburg, Germany
[3]Deutsches Elektronen-Synchrotron (DESY), Notkestrasse 85, D-22607 Hamburg, Germany

E-mail: carsten.tieg@esrf.fr

Abstract. We present a new concept for imaging by soft x-ray holography. Microscopy-like imaging capabilities were achieved by the separation of mask and sample. The use of two independent silicon nitride membranes, one for the field-of-view-defining mask and the reference beam, and the other for the sample, allows to image different areas on the sample. The movement of the field-of-view across the sample is realized by a piezomotor-driven sample stage that permits relative and stable positioning with nm-precision. We demonstrate the capabilities of the x-ray holographic microscopy (XHM) technique by showing images with 60 nm spatial resolution of an artificially structured 100 nm thick gold film with a lateral size of $19 \times 4\,\mu m^2$.

1. Introduction

Very few imaging techniques feature at the same time element selectivity, nm resolution, magnetization sensitivity, flexible sample environment, and the capability to control external parameters like temperature and magnetic field. A recent photon-based imaging technique that combines these characteristics is soft x-ray holography. Since its first demonstration in 2004 [1], soft x-ray holography has proven to be a powerful tool for the study of magnetic systems, though the first images of non-magnetic specimen were already taken in the early 90s [2]. The method's inherent magnetization sensitivity, which is based on the x-ray magnetic circular dichroism effect at the $L_{2,3}$ absorption edges, was readily exploited for the study of layered systems with perpendicular magnetic anisotropies, like [Co/Pt] [1, 3, 4, 5] and [Co/Pd] [3, 4, 6, 7] multilayer systems. In addition, the potential for biological imaging was recently demonstrated by a study of diatoms and fibroblast cells [8]. In the hard x-ray regime it was shown that holographic imaging can be combined with phase retrieval methods in order to achieve diffraction-limited spatial resolution [9]. There are some advantages of soft x-ray holography in comparison to transmission soft x-ray microscopy, which offers similar key features and a spatial resolution of 12 nm [10]. First, with soft x-ray holography one can achieve nm spatial resolution with μm stability of the setup when using a combined mask-sample structure. And second, there is more room available in the vicinity of the sample for its manipulation.

Synchrotrons are the common x-ray sources for this imaging technique. However, it was recently demonstrated that tabletop soft x-ray sources based on pulsed lasers can be used as well [11]. Soft x-ray holography is a promising tool for the study of ultrafast magnetization processes using the temporal beam structure of the upcoming free-electron laser sources [12]. However, up to now a widespread use of this new technique has been hindered by certain limitations imposed on the field-of-view (FOV) and the sample.

In soft x-ray holography, samples are commonly deposited on a transparent silicon nitride membrane. On the side facing the beam the membrane accommodates an opaque metal film with apertures for the FOV-defining object hole and one or more holes to define the reference waves [13]. The back-side of the same membrane is used as a substrate for the sample. The object beam, which traverses the sample, interferes with the unperturbed reference beams and produces a holographic image on the sensor of a CCD camera. The intensity of the hologram carries the phase relation between object and reference beam. The real-space image of the non-periodic sample structure is reconstructed by a fast Fourier transform (FFT) of the hologram. This can be done in real-time during the experiment. Magnetic contrast is achieved by taking the difference between holograms recorded with opposite photon helicity at the L_3 absorption edge. Tuning the photon energy to the low-energy side of the resonance allows utilizing the phase contrast. This can be advantageous for radiation sensitive samples [6].

Though the ultimate spatial resolution of the reconstructed image is in principle limited by the wavelength, it is in practice the size of the reference hole that matters [1]. The best resolution reached up to now is $30-50$ nm for a wavelength of 1.59 nm [1]. Besides the technical difficulties to mill narrow reference holes with high aspect ratios into the opaque film, it is not advantageous in terms of the signal-to-noise ratio to aim for reference holes with sizes as small as 10 nm and below. To achieve a reasonable contrast in the reconstructed image, the intensity of the reference beam should be sufficiently high in comparison to the object beam. For a given resolution, the upper limit for the size of the FOV-defining object hole is therefore determined by the size of the reference hole. For a typical mask with an object and reference hole diameter of $2\,\mu$m and $0.1\,\mu$m, respectively, the intensity ratio of object to reference beam before traversing the sample is 400:1. This ratio is slightly lowered by the attenuation of the object beam intensity as a result of the absorption by the membrane and a thin metallic sample. For example, the x-ray transmission at the energy of the Co L_3 edge of a 200 nm thick membrane and a 10 nm thick Co sample is 0.75 and 0.6 [14], respectively. This gives an absorption-corrected intensity ratio of 180:1. A second reason for the size limitation of the FOV is the degree of coherence of the beam. To obtain an interference pattern, the transverse coherence length has to be larger than the distance between object and reference hole. Otherwise the contrast decreases in the reconstructed images. This effect has been recently demonstrated by comparing images of the same sample recorded with different transverse coherence lengths which were obtained by varying the wavelength [8]. A coherent beam is commonly prepared by the use of a pinhole. Placing a pinhole with a diameter of $d = 50\,\mu$m for soft x-rays with a wavelength of $\lambda = 1.5$ nm at a distance of $z = 1$ m in front the sample yields a transverse coherence length of $\xi_t = \lambda z/2\pi d = 5\,\mu$m only. Hence, images with a large FOV of about $50\,\mu$m and a spatial resolution better than 100 nm are therefore not feasible with this conventional mask-sample design. Finally, we also want to mention the limitation imposed by the CCD detector properties on spatial resolution and the maximal FOV that can be imaged. While the former requires a large chip size, the latter needs a small pixel size, which is technologically challenging to achieve with a single device.

In this work we show that the limitations on the FOV can be circumvented by separating the mask from the sample. This is realized by using two membranes. The separated mask allows to move the FOV together with the reference holes across the sample and thus to image different areas on the sample. Like in microscopy, a series of overlapping real-space images can be matched together to give a large-scale image of the sample. We demonstrate the potential of this new concept, for which we use the term x-ray holographic microscopy (XHM), by presenting images of an artificially structured Au film.

Journal of Physics: Conference Series **211** (2010) 012024 doi:10.1088/1742-6596/211/1/012024

2. Experimental Setup

The key to overcome the restriction of a position-fixed and size-limited FOV in soft x-ray holography is to separate the imaging elements from the actual sample by using two silicon nitride membranes. The principal setup is illustrated in Fig. 1. We have used commercial $0.5 \times 0.5\,\mathrm{mm}^2$ Si_3N_4 membranes with a thickness of 200 nm supported by a $5 \times 5\,\mathrm{mm}^2$ silicon frame. The upstream membrane acts as support for the opaque 800 nm thick Au film. We used focussed ion beam milling (FIB) to mill a $2\,\mu$m aperture for the object beam and three holes with diameters of 100, 320 and 350 nm for the reference beams into the Au film. The milling process of these high-aspect-ratio holes was performed from the metal side. This gives rise to a conical shape with the smaller hole diameter on the membrane side, as indicated in Fig. 1. The downstream membrane, which is mounted face-to-face to the mask-carrying membrane, supports the sample. It is fixed to a piezomotor-driven in-vacuum x/y–stage. The stage allows with nm precision and stability that both membranes can glide smoothly with respect to each other. In order to guide the reference beams through the sample membrane without being absorbed, we have milled a trench into it. Width and length of the trench were adjusted to accommodate the required lateral travel range of the sample.

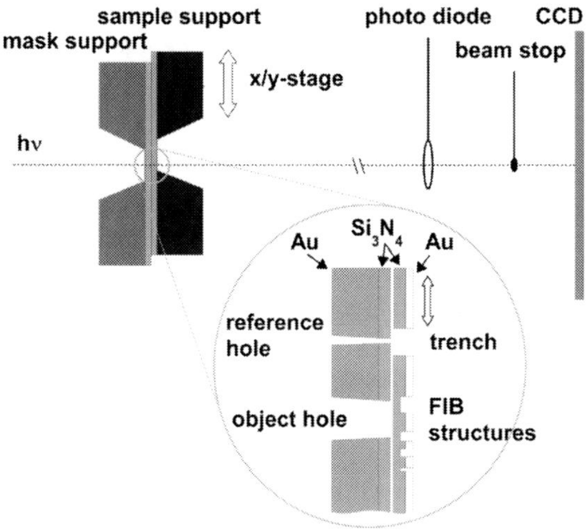

Figure 1. Schematic side view of the imaging setup used for x-ray holographic microscopy (XHM). Mask and sample are supported by two separate Si_3N_4 membranes. The mask support is fixed while the sample support can be moved in the plane perpendicular to the beam. By the relative movement of both membranes, the position of the field-of-view, which is defined by the object hole in the mask, can be changed. A trench in the sample membrane allows within a certain travel range an unperturbed passage of the reference beam through the sample.

For aligning mask and sample in the beam we have used a retractable photo diode. The measurements were performed at the soft x-ray beamline ID08 of the ESRF. A description of the setup can be found elsewhere [5]. The energy was tuned to 778 eV, corresponding to the Co L_3 absorption edge. The intense direct beam is blocked by a 0.7 mm diameter beam stop situated in front of a 16 bit CCD camera with 1300×1340 pixels and a pixel size of $20 \times 20\,\mu\mathrm{m}^2$. The camera was placed 36 cm behind the sample, allowing to record a maximum in-plane momentum transfer from the center ($q = 0$) to the edge of the chip of $q_{\max} = \frac{4\pi}{\lambda} \sin\theta_{\max} = 0.14\,\mathrm{nm}^{-1}$. Here, $2\theta_{\max}$ denotes the maximal scattering angle. This gives a spatial resolution $\Delta x = 2\pi/2q_{\max}$ imposed by the measurement geometry in the reconstructed images of about 20 nm. This value is below the size of the reference holes that defines the actual resolution limit [1].

3. Results and discussion

For this proof-of-principle experiment we have used a sample based on a FIB-structured Au film. Mask and sample are shown in Fig. 2. These scanning electron microscope (SEM) images were taken after the holography measurements. The mask was imaged from the membrane side while the artificially structured sample is seen from the metal film side (cp. Fig. 1). The mask hosts the FOV-defining object hole with a diameter of $2\,\mu$m, and three reference holes, drilled at a distance of 7.5 μm from the object hole center. They are arranged in a three-fold symmetry. The reference holes have a conical shape with the smaller diameter on the membrane side as a result of the milling process from the Au side. The upper and bottom left reference hole have a diameter of 320 nm and 350 nm, respectively. They allow high-contrast imaging while the bottom right one with a diameter of only 100 nm is optimized for high-resolution imaging.

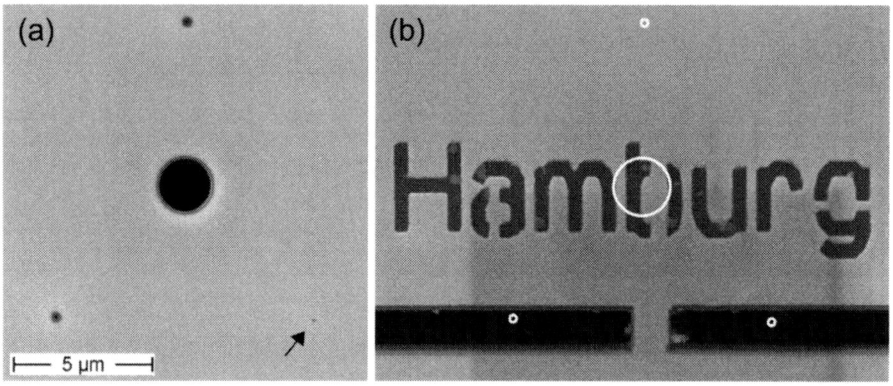

Figure 2. SEM image of mask (a) and sample (b) (same length scale). The 800 nm thick Au mask accommodates a 2 μm object hole, and a reference hole with 350 nm diameter at the bottom left, 320 nm at the top and 100 nm at the bottom right. Note that the membrane was completely removed in the object hole. The position of these four apertures on the FIB-structured sample is indicated in (b). A transparent trench (horizontal black bar at the bottom) assures for different imaging positions that at least one of the two lower reference beams can pass the 100 nm thick Au film of the sample without being obstructed. For XHM, a series of holograms at adjacent FOV positions is recorded, and the individually reconstructed images are merged together.

For the sample preparation, we deposited a continuous 100 nm Au film on the back-side of the downstream membrane and milled the word Hamburg into the film down to the Si_3N_4. The entire FIB structure spans an area of $17 \times 4\,\mu m^2$. In order to achieve a complete removal of the Au within the letters, the FIB milling process was extended until part of the membrane had been removed. This is indicated in the sample cross-section in Fig. 1. However, owing to an inhomogeneous milling yield, some Au remained on the membrane in the form of crystallites. These crystallites appear in Fig. 2 (b) in medium gray. In the bottom part of the sample we milled a trench through the Au film and the membrane. The complete removal of the membrane is obvious from the darker gray value as compared to the letters. The purpose of the trench's discontinuity below the letter b is to maintain mechanical stability of the sample membrane and the Au film. The vertical distance of the trench from the Hamburg-letters and its size were adjusted to the design of the mask in order to assure that at least one of the two bottom reference holes always lied above the trench while scanning the sample. The circles on the sample in Fig. 2 (b) replicate the mask. In principle, recording a hologram at two adjacent positions along the vertical direction is sufficient to image the whole vertical extent of a letter. Note that the upper reference hole is blocked. At a photon energy of 778 eV, the 100 nm Au film and the 200 nm sample membrane attenuate the intensity of this reference beam to about 15%.

Fig. 3 shows as an example the hologram and its FFT image of the sample region at the lower right part of the letter H. The SEM image in the inset of (a) gives the position of the object hole in front of the sample. The hologram was recorded with a total exposure time of 20 s. The beam stop blocks the direct beam in the center. Three point-symmetric pairs of this sample region appear in the FFT reconstruction. The sub-image α and its complex conjugate α^* have the

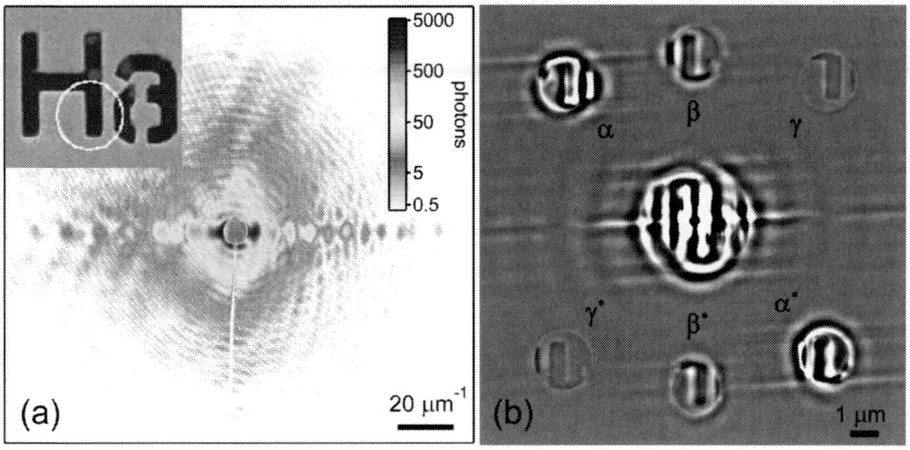

Figure 3. A hologram (a) and the real part of its Fourier transform (b). It was recorded while the bottom right part of the letter H was placed behind the object hole. The SEM image in (a) shows the position of the object hole on the sample. The FFT image contains the real-space image of the sample from all three reference holes, each producing a point-symmetric pair of the FOV. Sub-image γ and its complex conjugate γ^* correspond to the 100 nm reference hole.

lowest spatial resolution. They correspond to the 350 nm reference hole. Although blocked, the 320 nm reference hole gives rise to sub-image β, which has a weaker contrast as compared to α. We have used sub-image γ for reconstructing the entire sample. It corresponds to the 100 nm reference hole, and has a spatial resolution of 60 nm using the 20%–80% knife-edge definition.

As in optical microscopy, scanning the sample with respect to the optics allows to image samples that exceed the size of a single FOV. This capability is demonstrated in Fig. 4. The image, spanning the entire sample, is composed of 40 single FOVs. Contrast and spatial resolution are high enough to identify the Au crystallite on the membrane that result from an insufficient FIB milling process. For example, a Au crystallite is visible at the left and right end of the horizontal bar in the letter H. They have sizes of about 100 nm and 350 nm, respectively. It is important to stress again that it is not feasible to image such a large area with one exposure by conventional holography based on a position-fixed object hole of this size. In addition to the requirement of a much larger transverse coherence length, the relative intensity of the reference beam compared to the object beam would not be sufficient to obtain a reconstructable hologram if the size of the reference hole was not increased as well.

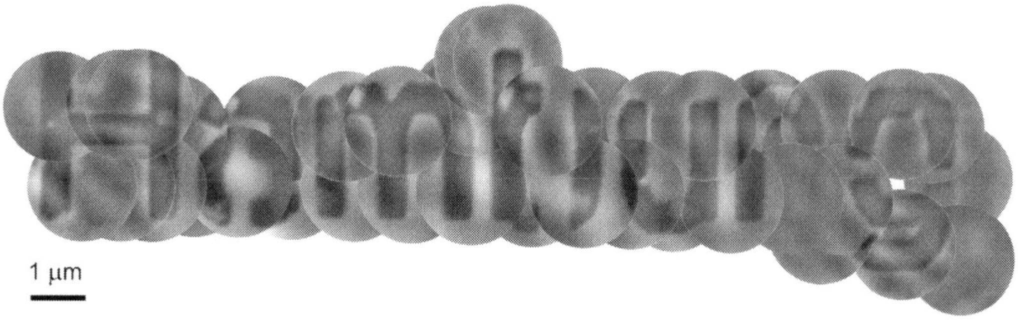

Figure 4. Image of a FIB-structured 100 nm thick Au film obtained from merging 40 single reconstructed holograms.

We have also used our x-ray holographic microscopy setup for magnetic domain imaging. We refer to our recent work where we demonstrate this capability by a study of the magnetic domain structure in a $[Co_{0.7\,nm}/Pt_{2.0\,nm}]_8$ multilayer with a laterally varying anisotropy as a result of the exchange coupling to an adjacent Fe wedge [15].

4. Summary

We have demonstrated a concept for x-ray holographic microscopy (XHM). It it based on the use of two separate silicon nitride membranes, one to accommodate the mask and the other to support the sample. The FOV-defining object hole, which is fixed in size and position on the mask membrane, can be moved across the sample with nm precision and stability by means of an x/y–stage. This design allows to record holograms at different sample positions. Hence, large sample areas can be imaged by combining individually reconstructed holograms. In comparison to conventional holography based on combined mask-sample structures [1], the preparation effort is much less. The technically demanding mask fabrication by FIB milling is now independent of the sample preparation. In addition, the mask can be used for several samples. Our XHM technique opens the door for the study of properties that vary along lateral length scales exceeding the size of a single FOV with a diameter of only $1 - 3\,\mu m$ [15]. We believe that this technique will lead to a much broader use of soft x-ray holography due to its flexibility and microscopy-like capabilities.

Acknowledgments

The authors thank R. Barrett for engineering support, G. Retout, R. Homs-Regojo, and T. Trenit for technical assistance. The group of Hamburg University acknowledges financial support by the German research foundation DFG within SFB 668, and the Free and Hanseatic City of Hamburg in the context of the "Landesexzellenzinitiative".

References

[1] Eisebitt S, Lüning J, Schlotter W F, Lörgen M, Hellwig O, Eberhardt W and Stöhr J 2004 *Nature* **432** 885
[2] McNulty I, Kirz J , Jacobsen C, Anderson E H, Howells M R and Kern D P 1992 *Science* **256** 1009
[3] Hellwig O, Eisebitt S, Eberhardt W, Schlotter W F, Lüning J and Stöhr J, 2006 *J. Appl. Phys.* **99** 08H307
[4] Hauet T, Günther C M, Pfau B, Schabes M E, Thiele J U, Rick R L, Fischer P, Eisebitt S and Hellwig O 2008 *Phys. Rev.* B **77** 184421
[5] Streit-Nierobisch S *et al.* 2009 *J. Appl. Phys.* **106** 083909
[6] Scherz A *et al.* 2007 *Phys. Rev.* B **76** 214410
[7] Günther C M, Radu F, Menzel A, Eisebitt S, Schlotter W F, Rick R, Lüning J and Hellwig O 2008 *Appl. Phys. Lett.* **93** 072505
[8] Guehrs E, Günther C M, Könnecke R, Pfau B and Eisebitt S 2009 *Opt. Express* **17** 6710
[9] Stadler L-M, Gutt C, Autenrieth T, Leupold O, Rehbein S, Chushkin Y and Grübel G 2008 *Phys. Rev. Lett.* **100** 245503
[10] Chao W, Kim J, Rekawa S, Fischer P and Anderson E H 2009 *Opt. Express* **17** 17669
[11] Sandberg R L, Raymondson D A, La-o-vorakiat C, Paul A, Raines K S, Miao J, Murnane M M, Kapteyn H C and Schlotter W F 2009 *Opt. Lett.* **34** 1618
[12] Dürr H A *et al.* 2009 *IEEE Transactions On Magnetics* **45** 15
[13] Schlotter W F *et al.* 2006 *Appl. Phys. Lett.* **89** 163112
[14] Nakajima R, Stöhr J and Idzerda Y U 1999 *Phys. Rev.* B **59** 6421
[15] Stickler D, Frömter R, Stillrich H, Menk C, Tieg C, Streit-Nierobisch S, Sprung M, Gutt C, Stadler L M, Leupold O, Grübel G and Oepen H P *accepted by Appl. Phys. Lett.*

Polarized Neutrons and Synchrotron X-rays for Magnetism Conference 2009 IOP Publishing
Journal of Physics: Conference Series **211** (2010) 012025 doi:10.1088/1742-6596/211/1/012025

Towards a Tomographic Reconstruction of Neutron Depolarization Data

Michael Schulz[1,2], Andreas Neubauer[1], Sergey Masalovich[2], Martin Mühlbauer[1,2], Elbio Calzada[2], Burkhard Schillinger[1,2], Christian Pfleiderer[1], Peter Böni[1]

[1]Technische Universität München, Physik Department, E21, James Franck Str, 85748 Garching, Germany

[2]Technische Universität München, Forschungsneutronenquelle Heinz Maier Leibnitz (FRM II), Lichtenbergstr. 1, 85748 Garching, Germany

E-mail: michael.schulz@frm2.tum.de

Abstract. In this paper we show a first approach to the three dimensional reconstruction of spatially resolved neutron depolarization data. We will show measurements with a position sensitive CCD detector on a longitudinal polarization analysis setup using ^{3}He polarizers and analyzers installed at the radiography beamline ANTARES at FRM II, Munich. A tomographic reconstruction of data acquired with an inhomogeneous $Pd_{1-x}Ni_x$ sample shows that this method is a powerful tool to identify regions of different magnetic properties inside the sample.

1. Introduction

The combination of neutron imaging with polarization analysis is a new and powerful method which has proven to be useful in the investigation of magnetic field distributions in complex solenoids [1]. It may additionally be used for visualisation of effects like flux-pinning in superconductors [1], variations of the magnetisation in thin magnetic films [2] or even visualisation of domain walls for samples with very large domains [3]. Another very promising field of application is the investigation of inhomogeneous ferromagnets [4]. In particular, the depolarization of the neutron beam may be used to map out variations of ferromagnetic properties as a function of external parameters such as stress, pressure, applied magnetic and electric field or temperature as well as internal characteristics such as chemical composition, defects and strain.

The possibility of reconstructing the orientation of magnetic domains from tomographic measurements has been extensively discussed (see [5] and references therein). However this is only possible for the subsequent transmission of very few domains. A further prerequisite for this technique is a spatial resolution, which must be better than the domain size. Since typical domain sizes are of the order of a few μm this method can currently not be used on standard neutron radiography setups.

In this paper we will present a 3D reconstruction of neutron depolarization data, which yields a volume model of the sample and its magnetic properties on a macroscopic scale. Depolarization tomography is based on the spatially resolved measurement of the polarisation of a neutron beam after transmission of a sample under different projection angles. The data are then reconstructed using a standard filtered backprojection algorithm.

2. Experiments

The setup used for the experiments described in this paper is shown in Fig. 1. It was installed at the radiography beam line ANTARES at FRM II and consists of a ^{3}He polarizer and analyzer, a precession coil spin flipper, a closed-cycle cryostat with a base temperature of 3.5 K, which holds the sample and a position sensitive CCD detector that records the image on a LiF/ZnS scintillator with a thickness of 200 μm. The sample was mounted in a container filled with He exchange gas reducing the temperature gradient to less than 0.1 K along the sample. A monochromatic neutron beam of 3.2 Å was obtained with a double crystal monochromator [6]. To minimize the effect of stray fields around the sample, a pair of Helmholtz coils was placed around the cryostat, which produced a field of 80 G parallel to the guide field. For the tomography, the cryostat was mounted on a rotary stage with its axis of rotation parallel to the magnetic field at the sample position.

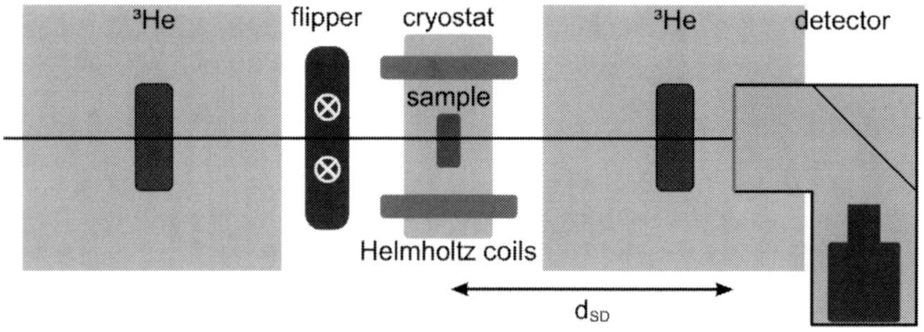

Figure 1. Schematic drawing of the setup used for the experiments described in this paper. The ^{3}He polarizer and analyzer are indicated with the magnetostatic cavities surrounding them. The sample was mounted in a closed-cycle cryostat around which a pair of Helmholtz coils was mounted. The detector is a position sensitive CCD with 2048 × 2048 pixels.

In order to increase the relaxation time of the polarised ^{3}He, it has to be kept in a very homogeneous magnetic field [7] with field gradients less than 10^{-4}. Thus, a bulky magnetostatic cavity containing the ^{3}He cell had to be placed between sample and detector and consequently limited the minimum sample to detector distance to $d_{SD} \approx 300$ mm, resulting in a spatial resolution of ≈ 0.75 mm. The initial polarisation of the ^{3}He gas was about 70% yielding a beam polarisation of $\approx 80\%$ with a relaxation time of the order of 150 h.

The sample we investigated was a $Pd_{1-x}Ni_x$ polycrystal, which was grown with the Czrochalski technique and has a nominal Ni concentration of $x = 2.67\%$. The sample has a cylindrical shape with a diameter of 11 mm and a length of 26 mm. $Pd_{1-x}Ni_x$ is a weak itinerant ferromagnet with a strong dependence of the Curie-temperature T_C on the Ni concentration x [8].

First, a radiographic temperature scan of the neutron depolarisation was made in a temperature range from 5 K to 40 K to find an appropriate temperature at which to record the tomography. Here, one has to avoid the condition of total depolarization of the beam after transmission of the sample, which is equivalent to non-penetration in absorption radiography leading to artifacts in the reconstruction. This can be done by setting a sufficiently high sample temperature, where the ordered moment M_s and consequently the depolarisation is lower. The Curie temperatures found for this sample range from below 5 K to 26 K. Based on the results of this measurement, a tomography of the sample was recorded at a temperature of 8 K. During the experiment, the sample was rotated over 180° with angular steps of 1°. At each angular position 5 spin-up and spin-down images were recorded with an exposure time of 60 s each and later summed up to obtain better neutron counting statistics. The total measurement time for the tomography was approx. 30 h.

3. Results and Discussion

With a spin flipper the beam polarisation after transmission of the sample can be determined as $P = 1/(PP_0)(I_\uparrow - I_\downarrow)/(I_\uparrow + I_\downarrow)$ and the absorption is calculated as $A = (I_\uparrow + I_\downarrow)/I_0$, where $I_\uparrow$ and $I_\downarrow$ are the spin-up and spin-down intensities, respectively and I_0 is the sum of these intensities measured without the sample in place. The term PP_0 is the product of the polarising and analysing efficiency, which can be measured outside of the sample region during the experiment and thus allows to correct for the relaxation of the ^{3}He polarisation with time. In the case of a random orientation of the magnetic domains, the beam polarisation after transmission of the sample follows an exponential decay [9] $P = P_0 e^{-\rho d}$ with the sample thickness d. Here $\rho = 1/3\gamma\delta B^2/v^2$ can be interpreted as a neutron depolarisation density, where γ is the neutron's gyromagnetic ratio, δ is the average domain size, B is the average magnetic field inside a domain and v is the neutron velocity. This law resembles the exponential attenuation law and the neutron depolarisation density can therefore be reconstructed with exactly the same filtered back projection algorithm [10]. Only the magnetic field and the domain size are variable in this expression and consequently ρ, which can also be interpreted as a spin-flip probability, contains information about the product of the domain size and the magnetisation of the domains. Further details on the reconstruction procedure will be discussed elsewhere [11].

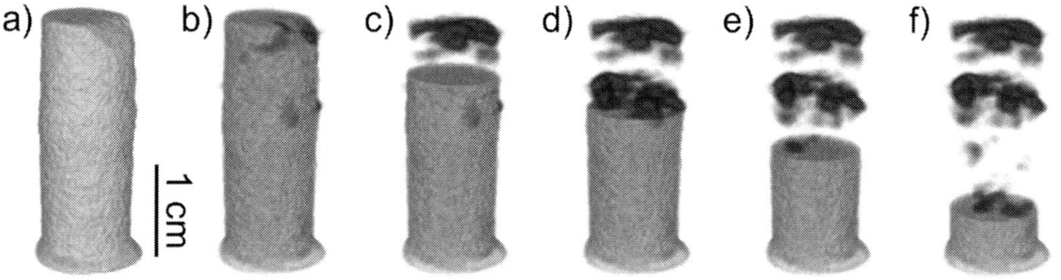

Figure 2. 3D Reconstruction of the ND data. a) View of the crystal; b) through f) Paramagnetic regions of the sample are shown in light grey. Ferromagnetic regions are displayed in blue. Several cuts through the sample are shown from c) to f).

Both the polarisation and the absorption data were reconstructed separately and then visualised with the volume rendering software VGStudio Max [12] as shown in Fig. 2. Here, the paramagnetic regions of the sample are displayed in grey, whereas the ferromagnetic regions are shown in blue. In order to reduce the noise in the reconstruction, the 3D data were median filtered. Fig. a) shows an outside view of the absorption data of the crystal as an overview. At the bottom of the sample the glue which was used to fix the sample on the sample holder is visible. Furthermore at the top of the sample one can clearly see an edge which was cut off the crystal for bulk measurements. For images b) through f) the paramagnetic regions of the sample were rendered more transparently to visualize the ferromagnetic parts inside the sample. Furthermore images c) through f) show horizontal cuts through the absorption data for better visibility of the depolarisation data. For all images the orientation of the 3D object was the same.

One can readily see that the sample is extremely inhomogeneous and has vast paramagnetic regions. Furthermore the ferromagnetic parts of the sample tend to arrange in horizontal layers, which are however, also not perfectly homogeneous. This might be due to the crystal growth process, which was perpendicular to these planes. A change in the growth parameters (i.e. the growth velocity) could be responsible for variations in the Ni concentration and consequently in the magnetic properties of the sample. The more detailed quantitative analysis of the distribution of the amplitude of the magnetisation in the sample will be discussed elsewhere.

4. Conclusion and Outlook

The experiment described in this paper shows that in addition to the results obtained in neutron depolarization radiography one can obtain 3D information on the distribution of magnetic properties of a sample by using a tomographic method. This could help to improve the understanding of substances showing a strong dependence of the Curie temperature on the composition or phase separation. Furthermore this method allows to locate regions of desired magnetic properties from a 3D model and later cut these out of the sample for further investigation with bulk or neutron scattering measurement methods.

Several other substances such as Ni_3Al and Heusler alloys have already been studied with this method and it turns out that many samples, which were thought to be of very high quality show drastic variations of their magnetic properties on a macroscopic scale. [11] This observation becomes especially important if neutron scattering studies are performed on such samples, since these use an large beam which performs an implicit average over the sample volume.

We are currently as well studying the influence of different crystal growth conditions on the resulting crystal quality and will try to identify the ideal parameters for high quality crystals using radiography and tomography with polarized neutrons.

Acknowledgments

We wish to thank A. Hilger and N. Kardjilov from CONRAD at HZB, Berlin for giving us the opportunity to do first tomography measurements at their beam line. Financial support through DFG Forschergruppe FOR 960 on Quantum Phase Transitions is gratefully acknowledged.

References

[1] Kardjilov N, Manke I, Strobl M, Hilger A, Treimer W, Meissner M, Krist T and Banhart J 2008 *Nature Physics* **4** 399
[2] Piegsa F, van den Brandt B, Hautle P, Kohlbrecher J and Konter J 2009 *Physical Review Letters* **102** 145501
[3] Grünzweig C, David C, Bunk O, Dierolf M, Frei G, Kühne G, Schäfer R, Pofahl S, Ronnow H and Pfeiffer F 2008 *Applied Physics Letters* **93** 112504
[4] Schulz M, Böni P, Franz C, Neubauer A, Calzada E, Mühlbauer M, Schillinger B, Pfleiderer C, Hilger A and Kardjilov N 2009 *Proceedings of the ICNS 2009, Journal of Physics - Conference Series* **accepted for publication**
[5] Jericha E, Szeywerth R, Leeb H and Badurek G 2007 *Physica B: Physics of Condensed Matter* **397** 159–161
[6] Schulz M, Boeni P, Calzada E, Muehlbauer M and Schillinger B 2009 *Nucl. Inst. and Meth. A. Spectrometers, Detectors and Associated Equipment* **605** 33 – 35
[7] Andersen K, Chung R, Guillard V, Humblot H, Jullien D, Lelièvre-Berna E, Petoukhov A and Tasset F 2005 *Physica B: Physics of Condensed Matter* **356** 103–108
[8] Nicklas M, Brando M, Knebel G, Mayr F, Trinkl W and Loidl A 1999 *Physical Review Letters* **82** 4268–4271
[9] Halpern O and Holstein T 1941 *Physical Review* **59** 960–981
[10] Slaney M and Kak A 1988 *Principles of computerized tomographic imaging* (IEEE Press)
[11] Schulz M 2010 *forthcoming PhD thesis, TU München*
[12] URL `http://www.volumegraphics.com/products/vgstudiomax/`

Polarized Neutrons and Synchrotron X-rays for Magnetism Conference 2009 IOP Publishing
Journal of Physics: Conference Series **211** (2010) 012026 doi:10.1088/1742-6596/211/1/012026

XYZ-polarisation analysis of diffuse magnetic neutron scattering from single crystals

Werner Schweika

Institut für Festkörperforschung, Forschungszentrum Jülich, D-52425 Jülich, Germany

E-mail: w.schweika@fz-juelich.de

Abstract. Studies of diffuse magnetic scattering largely benefit from the use of a multi-detector covering wide scattering angles. Therefore, the different contributions to the diffuse scattering that originate from magnetic, nuclear coherent, and nuclear spin-incoherent scattering can be separated by the so-called XYZ-polarization analysis. In the past this method has been successfully applied to the analysis of diffuse scattering by polycrystalline samples of magnetic disordered materials. Single crystal studies that exploit the vector properties of spin correlations are of particular interest for furthering our understanding of frustration effects in magnetism. Based on the symmetry properties of polarised scattering a suitable extension of the conventional XYZ method has been derived, which allows for the complete separation and the analysis of features of diffuse magnetic scattering from single crystals.

1. Introduction

The scattering theory of polarised neutrons including magnetic interactions started with the early work of by Halpern and Johnson [1] and has been essentially completed by Blume[2] and Maleyev[3] more than fourty years ago. The major milestones in development of experimental means and applications have been set by technique of longitudinal polarisation analysis (Moon *et al.* [4]) and by spherical neutron polarimetry (Tasset and Brown) [5, 6], which nowadays provides a high precision tool to analyse the full polarisation tensor for single crystal magnetic scattering. The development for diffuse scattering utilizing a wide angle detector has been essentially driven by Schärpf, who established with *his* D7 instrument at the ILL and the so-called XYZ method[7, 8] efficient means for measuring and analysing spin-correlations in powder samples. With recent instrumental progress, the D7 at ILL [9] and the DNS at JCNS [10] have gained the required efficiency to explore more routinely the diffuse magnetic scattering from single crystals. To overcome the pending problem that the XYZ-method is invalid for the separation of the more complex scattering from single crystals, here an appropriate extension of the XYZ-method is derived using the symmetry properties of the polarised scattering.

2. Polarised neutron scattering and symmetry

According to Blume [2] and Maleyev [3] the neutron scattering process including magnetic interactions can be completely described by two master equations, Eq.(1) for the scattering cross-section, here for brevity denoted by the intensity I, and Eq.(2) for $\mathbf{P}'I$, where $\mathbf{P}'$ denotes the final polarization:

$$I = N^\dagger N + I_{si} + \mathbf{M}_\perp^\dagger \mathbf{M}_\perp + \mathbf{P}\cdot\mathbf{M}_\perp^\dagger N + \mathbf{P}\cdot\mathbf{M}_\perp N^\dagger + \mathrm{i}\mathbf{P}(\mathbf{M}_\perp^\dagger \times \mathbf{M}_\perp) \qquad (1)$$

$$\begin{aligned}\mathbf{P}'I &= \mathbf{P}(N^\dagger N - \frac{1}{3}I_{si}) + (\mathbf{P}\cdot\mathbf{M}_\perp^\dagger)\mathbf{M}_\perp + (\mathbf{P}\cdot\mathbf{M}_\perp)\mathbf{M}_\perp^\dagger - \mathbf{P}(\mathbf{M}_\perp^\dagger\mathbf{M}_\perp) \\ &\quad +\mathrm{i}N(\mathbf{P}\times\mathbf{M}_\perp^\dagger) - \mathrm{i}N^\dagger(\mathbf{P}\times\mathbf{M}_\perp) + N\mathbf{M}_\perp^\dagger + N^\dagger\mathbf{M}_\perp - \mathrm{i}(\mathbf{M}_\perp^\dagger\times\mathbf{M}_\perp)\,,\end{aligned} \tag{2}$$

where $N(\mathbf{Q}) = \sum_n b_n \exp(\mathrm{i}\mathbf{Q}\cdot\mathbf{R_n})$ is the nuclear structure factor and $\mathbf{M}_\perp = \mathbf{e_Q}\times\mathbf{M}(\mathbf{Q})\times\mathbf{e_Q}$, is the vector of dipolar magnetic interaction of magnetic moments with the neutron spin, and $\mathbf{e_Q}$ is the unit vector along the direction of the scattering vector $\mathbf{Q}$. Accordingly, the parallel components to the scattering vector $\mathbf{Q}$ do not contribute to magnetic scattering. Here, $\mathbf{M}(\mathbf{Q}) = \sum_n \mathbf{M}_n \exp(\mathrm{i}\mathbf{Q}\cdot\mathbf{R_n})$ denotes the Fourier transform of the magnetic moments. I_{si} denotes the nuclear spin-incoherent scattering, assuming that the nuclear spins are randomly oriented. This diffuse background, which is usually small compared to magnetic and nuclear Bragg peaks, can be relatively large when considering diffuse magnetic and diffuse nuclear scattering.

The full information about the scattering terms in Eqs. (1,2) can be retrieved by spherical neutron polarimetry [6]. Therefore, the most convenient and standard choice of a x,y,z-coordinate system for the setting of the polarisation $\mathbf{P}$ is to have one axis x parallel to $\mathbf{Q}$, with the axes y and z perpendicular to $\mathbf{Q}$ pointing in- and out-of the scattering plane respectively.

However, when using multi-detectors $\mathbf{P}$ can be set ideally parallel to $\mathbf{Q}$ only for a single detector. Therefore, the XYZ-polarisation analysis for multi-detectors the axes x and y are chosen to be in the (horizontal) scattering plane and z to point (vertically) out of the scattering plane. In an excellent contribution to the present subject, the polarised scattering has been revisited and treated within the density matrix formalism by Schärpf [11]. Following Ref.[11] we consider first the general case of an arbitrarily rotated x,y,z-system and the polarised intensities that can be measured in spin-flip and non-spin flip modes and their relation to the above equations. With the definition of polarised intensities and polarisation

$$I = I_{\nu\nu} + I_{\nu\bar{\nu}} \quad , \quad P'_\nu I = I_{\nu\nu} - I_{\nu\bar{\nu}} \tag{3}$$

for any cartesian coordinate $\nu = x, y, z$, the non-spin flip (nsf) and spin-flip intensities (sf), $I_{\nu\nu}$ and $I_{\nu\bar{\nu}}$ respectively, have been derived [11] from Eqs. (1) to (3):

$$\begin{aligned}I_{\nu\nu} &= N^\dagger N + NM_{\perp\nu}^\dagger + N^\dagger M_{\perp\nu} + M_{\perp\nu}^\dagger M_{\perp\nu} + \frac{1}{3}I_{si} \\ I_{\nu\bar{\nu}} &= \mathbf{M}_\perp^\dagger\mathbf{M}_\perp - M_{\perp\nu}^\dagger M_{\perp\nu} + \mathrm{i}(\mathbf{M}_\perp^\dagger\times\mathbf{M}_\perp)_\nu + \frac{2}{3}I_{si}\,.\end{aligned} \tag{4}$$

Eq. 4 shows a well known result namely that only components of $\mathbf{M}_\perp$ parallel to the neutron spin appear in the non-spin flip scattering and only components of $\mathbf{M}_\perp$ perpendicular to the neutron spin can contribute to the spin-flip scattering and is exemplified for $\nu = x$

$$I_{x\bar{x}} = M_{\perp y}^\dagger M_{\perp y} + M_{\perp z}^\dagger M_{\perp z} + \mathrm{i}(M_{\perp y}^\dagger M_{\perp z} - M_{\perp z}^\dagger M_{\perp y}) + \frac{2}{3}I_{si}\,.$$

It is further worthwhile to note that with any specific choice of the coordinate system the average of non-spin flip and spin-flip intensities differ for the different directions

$$I_x = \frac{1}{2}(I_{xx} + I_{x\bar{x}}) \neq I_y \neq I_z.$$

So far we followed the analysis of Schärpf. For a further analysis and separation of terms in the polarised neutron scattering, here, we continue with separating intensities in symmetric and antisymmetric parts. Therefore, we consider scattering for reversed polarisation by the intensities $I_{\bar{\nu}\bar{\nu}}$ and $I_{\bar{\nu}\nu}$ which are experimentally likewise accessible, and we obtain the analogue to Eq. (4) from Eqs. (1-3):

$$\begin{aligned} I_{\bar{\nu}\bar{\nu}} &= N^\dagger N - NM^\dagger_{\perp\nu} - N^\dagger M_{\perp\nu} + M^\dagger_{\perp\nu} M_{\perp\nu} + \frac{1}{3} I_{si} \\ I_{\bar{\nu}\nu} &= \mathbf{M}^\dagger_\perp \mathbf{M}_\perp - M^\dagger_{\perp\nu} M_{\perp\nu} - \mathrm{i}(\mathbf{M}^\dagger_\perp \times \mathbf{M}_\perp)_\nu + \frac{2}{3} I_{si} \,. \end{aligned} \tag{5}$$

In comparison to Eq. 4 and 5, a further simplification is achieved by considering the average non-spin-flip and spin flip intensities, Σ^{nsf}_ν and Σ^{sf}_ν respectively:

$$\begin{aligned} \Sigma^{nsf}_\nu = \frac{1}{2}(I_{\nu\nu} + I_{\bar{\nu}\bar{\nu}}) &= N^\dagger N + M^\dagger_{\perp\nu} M_{\perp\nu} + \frac{1}{3} I_{si} \\ \Sigma^{sf}_\nu = \frac{1}{2}(I_{\nu\bar{\nu}} + I_{\bar{\nu}\nu}) &= \mathbf{M}^\dagger_\perp \mathbf{M}_\perp - M^\dagger_{\perp\nu} M_{\perp\nu} + \frac{2}{3} I_{si} \end{aligned} \tag{6}$$

showing the expected property of the unpolarised intensity I

$$I = \Sigma_x = \Sigma_y = \Sigma_z = \Sigma^{nsf}_\nu + \Sigma^{sf}_\nu \;=\; N^\dagger N + \mathbf{M}^\dagger_\perp \mathbf{M}_\perp + I_{si} \,.$$

By this we have eliminated the interference terms related to nuclear-magnetic correlations (spin-orbit coupling) and vector chirality terms (cross products) that now appear in the corresponding averaged deviations Δ^{nsf}_ν and Δ^{sf}_ν respectively, :

$$\begin{aligned} \Delta^{nsf}_\nu = \frac{1}{2}(I_{\nu\nu} - I_{\bar{\nu}\bar{\nu}}) &= NM^\dagger_{\perp\nu} + N^\dagger M_{\perp\nu} = 2\Re(NM^\dagger_{\perp\nu}) \\ \Delta^{sf}_\nu = \frac{1}{2}(I_{\nu\bar{\nu}} - I_{\bar{\nu}\nu}) &= 2\mathrm{i}(\mathbf{M}^\dagger_\perp \times \mathbf{M}_\perp)_\nu \,. \end{aligned} \tag{7}$$

Eq. (7) reveals a well known result that is the basis of experiments with polarized neutrons using polarisation reversal without polarisation analysis. For the purpose of separating the interference terms due to chirality or due to nuclear-magnetic correlations, polarisation analysis is actually not required and can be achieved by setting ν, with polarisation reversal, either parallel (for chirality) or perpendicular (for nuclear-magnetic interference) to the scattering vector $\mathbf{Q}$. However, as can be seen, Eq. (7) also provides a complete separation of these terms for any choice of the cartesian coordinate system if polarisation analysis is applied.

Finally, one may note that it is also straightforward to derive this symmetry decomposition from Brown's tensor representation of polarised scattering [6] for the case of the standard coordinate system ($x||\mathbf{Q}$) and also to include spin-incoherent scattering.

3. XYZ-polarisation analysis, powders and single crystals

The standard XYZ-method has been successfully applied in many studies of magnetic correlations using powder samples. From the directional dependence of the polarised scattering, the purely magnetic scattering contribution can be separated by experimental means of polarisation analysis. For further interest the reader is referred to a recent and detailed review [9] and a recent own study [12]. Here we shall briefly recall the fundamental equations (Eq. 8), and discuss the additional potential of instruments like DNS and D7 for studies of powder samples with polarisation analysis. An interesting consequence is that the interference of nuclear and induced magnetic scattering could be observed even for powders if chiral domains relate to satellites with different Q moduli or in presence of a symmetry breaking external field, whereas usually, chiral scattering is found in single crystals with a preferred domain orientation. In the following (3.2) the central issue will be to separate and analyse the different contributions to the polarised scattering from single crystals based on the conventional tools of the standard XYZ method. Hence, the separation will be achieved without approximations only from diagonal intensity elements.

3.1. The application to powders

In case that orientational averaging applies, e. g. for powder samples, we do not distinguish the components $M_{\perp\nu}$, and the magnetic scattering can be separated via the XYZ polarisation analysis [7, 8] from the diagonal intensities. In case of only weak guide fields one does not need to distinguish the intensities with respect to the sign of polarisation as proposed in the previous section. It is a particular virtue of the XYZ-method that this method applies even if the Cartesian coordination system is arbitrarily rotated in the scattering plane, so that the new coordinates are x', y', z (see figure). In case of multi-detectors covering a larger Q-range it is trivially impossible to have the polarisation simultaneously parallel to all different Q vectors; the difference between the polarisation axis x' and the actual $\mathbf{Q}$ vector is the *Schärpf angle* α.

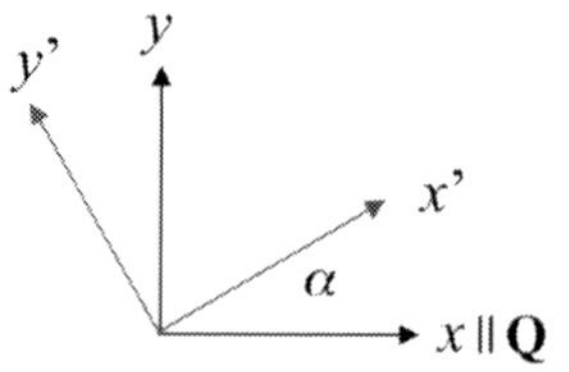

The information about the in-plane magnetic intensities, the square of magnetic amplitudes $\mathbf{M}_{\perp y}$, is recovered from $\mathbf{M}^2_{\perp y} = \mathbf{M}^2_{\perp x'} + \mathbf{M}^2_{\perp y'} = \mathbf{M}^2_{\perp y}(\sin^2\alpha + \cos^2\alpha)$. Hence, the magnetic, nuclear coherent and spin-incoherent scattering can be separated by combinations of polarised intensities [7, 8, 9]:

$$\begin{aligned} |\mathbf{M}_\perp|^2 &= 2(I_{x'\bar{x}'} + I_{y'\bar{y}'} - 2I_{z\bar{z}})^{sf} = -2(I_{x'x'} + I_{y'y'} - 2I_{zz})^{nsf} \\ I_{si} &= \frac{3}{2}\left(-I_{x'\bar{x}'} - I_{y'\bar{y}'} + 3I_{z\bar{z}}\right) \\ N^2 &= I_{zz} - \frac{1}{2}|\mathbf{M}_\perp|^2 - \frac{1}{3}I_{si}. \end{aligned} \tag{8}$$

Here, we have assumed that the intensities do not change for a common polarisation reversal of initial and final polarisation. More carefully, for the application of this analysis we have to consider whether the antisymmetric part actually vanishes. This is fulfilled only if there is no distinguished direction by any external field. Guide fields applied in the XZY method are typically in the order of magnitude of ten Gauss and they are too weak to induce any significant sample magnetization in paramagnetic or antiferromagnetic samples.

Strong fields invalidate the standard XYZ separation method, however, this allows for the study of the correlations and interference terms between nuclear and magnetic scattering amplitudes by determining the antisymmetric part of the polarised intensity, *id est* measuring intensity differences for polarisation reversal. A unique response of chiral terms to an applied horizontal external field is less likely, although not impossible. A well-known strategy for single crystals used in a "half-polarised" setup [13] can also be applied to powder diffraction. Such a determination of correlations between nuclear and magnetic scattering amplitudes from powders looks promising particularly for detectors covering a large solid angle without polarisation analysis. With respect to polarisation analysis of scattering it may be advantageous that only the difference of the *nsf*-intensities contributes to the signal, and the background from *sf*-scattering could be eliminated.

3.2. The application to single crystals

For a single crystal, even without applied external field, its intrinsic anisotropy and possible polarity may give rise to antisymmetric scattering contributions. Therefore, the symmetry decomposition resulting in Eqs. (6,7) is important and helpful to analyse the polarised scattering and its magnetic contributions. For a Cartesian coordinate system (x', y', z) rotated by the angle α around the vertical z-axis with respect to x parallel to $\mathbf{Q}$, the following relations can be derived from Eqs. (6) and (7) providing a complete separation

$$N^\dagger N = \frac{1}{2}\left(\Sigma_x^{nsf} + \Sigma_y^{nsf} - \Sigma_z^{sf}\right) = \frac{1}{2}(\Sigma_{x'}^{nsf} + \Sigma_{y'}^{nsf} - \Sigma_z^{sf}) \quad (9)$$

$$I_{si} = \frac{3}{2}(\Sigma_x^{nsf} - \Sigma_y^{nsf} + \Sigma_z^{sf}) = \frac{3}{2}\frac{\Sigma_{x'}^{nsf} - \Sigma_{y'}^{nsf}}{\cos^2\alpha - \sin^2\alpha} + \frac{3}{2}\Sigma_z^{sf} \quad (10)$$

$$M_{\perp y}^\dagger M_{\perp y} = \Sigma_z^{sf} - \frac{2}{3}I_{si} \quad (11)$$

$$M_{\perp z}^\dagger M_{\perp z} = \Sigma_z^{nsf} - \frac{1}{3}I_{si} - N^\dagger N \quad (12)$$

$$I_{chiral,x'} = 2\mathrm{i}(\mathbf{M}_\perp^\dagger \times \mathbf{M}_\perp)_{x'} = \Delta_{x'}^{sf} = \cos\alpha\,\Delta_x^{sf} \quad (13)$$

$$I_{NM,y'} = 2\Re(NQ_{y'}^\dagger) = \Delta_{y'}^{nsf} = \cos\alpha\,\Delta_y^{nsf} \quad (14)$$

$$I_{NM,z'} = 2\Re(NQ_{z'}^\dagger) = \Delta_{z'}^{nsf} = \Delta_z^{nsf}\,, \quad (15)$$

where the geometrical correction factors in Eqs. (10,13,14) stem from the projections $M_{\perp x'} = \sin\alpha\, M_{\perp y}$ and $M_{\perp y'} = \cos\alpha\, M_{\perp y}$ to the scattering vector[1]. Compared to the information content of powder experiments, the study of diffuse polarised scattering from single crystals yields additionally the anisotropy of magnetic correlation functions, possible vector chirality and all nuclear- magnetic correlations, essential information about complex magnetic materials of high current interest. It is worthwhile to note that a separation of the chiral term has already been given earlier by Eq. 50 in Ref.[8].

For diffuse scattering the typical experimental strategy is to map out scattering planes by stepwise rotations of the crystal around the vertical axis z. Note, the separation by Eqs. (9) to (15) is independent of the sample rotation ω and the convenient choice is a fixed coordinate system x', y', z.

In order to interpret the magnetic scattering amplitudes in terms of the crystal lattice, the in-plane component $M_{\perp y}$ has to be decomposed further into independent lattice components in the plane (ab) and is given by the projections of M_a and M_b, while the vertical component is simply identified by M_c. For orthogonal components, only the squares of M_a and M_b determine $M_{\perp y}$, see Fig. 1. Note, this example demonstrates that the considerations here also apply to inelastic scattering. The disentanglement of M_a and M_b is, analog to structure determination

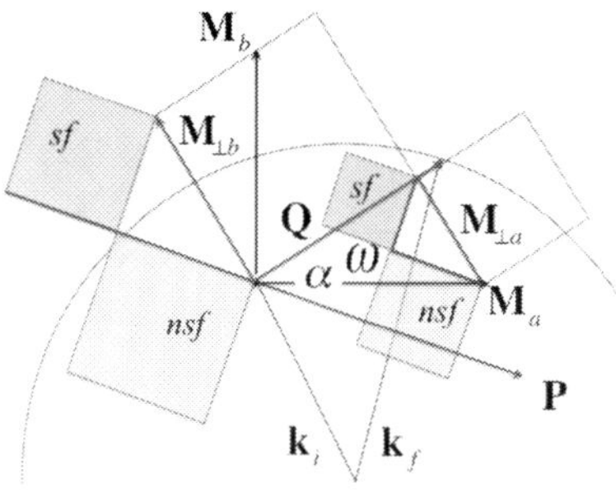

Figure 1. (color online) Polarisation **P** and magnetic (inelastic) scattering $M_{\perp\nu}^2$ (white squares) for the case of a *Schärpf angle* α between scattering vector **Q** and polarisation **P**; here ω denotes the angle between **Q** and reciprocal lattice vector $\mathbf{a}^*$ with $\mathbf{M}_a||\mathbf{a}^*$. Shaded squares represent sf- and nsf-intensities related to the projections of $\mathbf{M}_{\perp a}$ and $\mathbf{M}_{\perp b}$ $\perp$ and $||\mathbf{P}$ respectively. The vertical component ($||z$ and $\perp$ to **P**, sf-intensity) is not shown.

with short-range order and displacements, relying on sufficient information in extended Brillouin

[1] In view of the singularity at $\alpha = \pi/4$ in Eq.(10) and for smallest possible corrections, the polarisation axis x' is best set parallel to the **Q** vector for the central part of a multidetector. For a multidetector accepting scattering within angles $\Delta\vartheta = 120^o$, the *Schärpf angles* α (between **Q** and $\mathbf{P}_x$) will be equal to the deviations from the average Bragg angle $-30^o \le \alpha < 30^o$; hence, correction factors vary between 1 and 0.5 for the spin-incoherent scattering, between 1 and $\sqrt{3}/2$ for the chiral contribution (Eq. 13) and likewise for the nuclear magnetic interference (Eq. 14) resulting in small average corrections $3\sqrt{3}\,/\,2\pi \approx 0.827$ and $\pi/3 \approx 0.955$ respectively.

zones of reciprocal space. Diffuse scattering related to disorder phenomena typically implies an exponential decay of correlations in real space. Hence a Fourier analysis will efficiently reveal finite real space pair correlation functions, an approach that is a linear least squares problem having a unique solution. Therefore, the determination of three dimensional pair-correlations functions requires accordingly measurements of more than a single scattering plane.

4. Final remarks

With the substantial gain in efficiency seen with the developments on instruments in the recent years, particularly on the DNS and D7, single crystal studies with polarisation analysis became feasible and produced already first outstanding results [14]; a further example can also be found here in the proceedings of the PNSXM 2009 [15]. For a systematic and thorough understanding the separation of the different components is essential, a task that to date has been believed to be incapable by the conventional means used in XYZ polarisation analysis. The solution here, utilizing the symmetry properties, also places the emphasis on additional measurements with polarisation reversal. It is worthwhile to note that this request does not demand for more beamtime to achieve equal statistical weights.

So far we have not considered the off-diagonal terms. These terms also represent the interference terms, Eqs. (13-15), due to chiral correlations and due to coupling of spin-space and real-space variables. Spherical neutron polarimetry of off-diagonal terms distinguishes these effects irrespectively of a possible depolarisation from different magnetic domains in the crystal. Obviously, the diagonal antisymmetric polarised intensities Eq. (7) also reveal the mentioned interference terms. It should be noted that the same XYZ-setup for multi-detectors can be used to access the complete polarisation tensor and off-diagonal terms can be determined with the same accuracy by spherical neutron polarimetry with precessing incident polarization[16, 17]. However, the here proposed XYZ-polarisation analysis, a method based on only diagonal terms and applicable for multi-detectors, will certainly provide a much more convenient and efficient separation for the diffuse magnetic neutron scattering from single crystals.

Acknowledgments

Otto Schärpf is gratefully acknowlegded for discussions and valuable comments.

References

[1] Halpern O and Johnson M H 1939 *Phys. Rev.* **55** 898
[2] Blume M 1963 *Phys. Rev.* **130** 1670
[3] Maleyev S V, Baryakhtar V G and Suris P A 1962 *Fiz. Tv. Tela* **4** 3461; 1963 *Soviet Phys. Solid State* **4** 2533
[4] Moon R M, Riste T and Koehler W C 1969 *Phys. Rev.* **181** 920
[5] Tasset F 1989 *Physica* B **157** 627
[6] Brown P J 2001 *Physica* B **297** 198
[7] Brückel Th 1988 *Dissertation* Tübingen
[8] Schärpf O and Capellmann H 1993 *Phys. Stat. Sol.* (a) **135** 359
[9] Stewart J R *et al* 2009 *J. Appl. Cryst.* **42** 69
[10] Schweika W and Böni P 2001 *Physica* B **297** 155
[11] Schärpf O 2002 in *Polarized Neutron Scattering* (Series Matter and Materials vol 12) ed Th Brückel and W Schweika (Jülich: Forschungszentrum Jülich) p 89
[12] Schweika W, Valldor M and Lemmens P 2007 *Phys. Rev. Lett.* **98** 067201
[13] Schweitzer J 2006 *Neutron scattering from magnetic materials* ed T Chatterji (Amsterdam: Elsevier)
[14] Fennell T, Deen P P, Wildes A R, Schmalzl K, Prabhakaran D, Boothroyd A T, Aldus R J, McMorrow D F, Bramwell S T 2009 *Science* **326** 415
[15] Chang L J, Su Y, Schneider H, Mittal R, Lees M R, Balakrishan G and Brückel Th 2010 *J. Phys. Conf. Ser.* **211** 012013
[16] Schweika W 2003 *Physica* B **335** 157
[17] Schweika W, Easton S and Neumann K U 2005 *Neutron News* **16** 14

Journal of Physics: Conference Series **211** (2010) 012027 doi:10.1088/1742-6596/211/1/012027

Amorphous soft-magnetic ribbons studied by ultra-small-angle polarized neutron scattering

G Badurek[1], E Jericha[1], R Grössinger[2] and R Sato-Turtelli[2]
[1]Atominstitut, Vienna University of Technology, Stadionallee 2, A-1020 Wien, Austria
[2]Institut of Solid-State Physics, Vienna University of Technology, Wiedner Hauptstrasse 8, A-1040 Wien, Austria
E-mail: `badurek@ati.ac.at`

Abstract. When we investigated the magnetic structure of a variety of soft-magnetic amorphous ribbons by means of ultra-small-angle neutron scattering (USANSPOL) we were confronted with one particularly interesting $Fe_{65.7}Co_{18}Si_{0.8}B_{15.5}$ ribbon, provided by VAC Hanau. Due to a special thermal treatment during production a field- and stress-induced transverse domain texture was expected. Although the USANSPOL technique encountered its resolution limits during the investigation of this specific sample ribbon, such a texture could indeed be verified.

1. Introduction
Amorphous soft-magnetic materials represent an attractive source material for novel highly efficient magnetic shielding. Fabricated as ribbons under various heat treatment and applied stress conditions these materials may exhibit anisotropic orientation of the magnetic domains with respect to the ribbon axis. Such samples are relatively homogeneous with respect to nuclear scattering but residual material inhomogeneities may be identified by the then arising interference of nuclear and magnetic scattering contributions. Here we report on the investigation of an extremely soft-magnetic amorphous $Fe_{65.7}Co_{18}Si_{0.8}B_{15.5}$ ribbon provided by VAC Hanau, which due to its specific production conditions was expected to exhibit a field- and stress-induced transverse domain texture. According to the size ranges involved we used the novel ultra-small angle scattering technique with polarized neutrons (USANSPOL), which in combination with strong birefringent static magnetic fields exploits the extremely narrow width of Bragg reflections at perfect silicon crystals to resolve the tiny angular differences between spin-*up* spin-*down* neutrons and hence to discern magnetic from nonmagnetic scattering [1, 2, 3]. Since scattering from this sample was quite weak and partly hidden behind the resolution function, our investigations can also be understood as a demonstration of the actual capabilities of USANSPOL when it encounters its resolution limit.

2. Experimental setup
The measurements were performed at the dual purpose, so-called CRG-B instrument S18 of the Institut Laue-Langevin (ILL) in its ultra-small-angle scattering (USANS) configuration [4].

There two monolithic channel-cut perfect silicon crystals are mounted on an optical bench in non-dispersive antiparallel orientation of their (220) lattice planes (Fig. 1). Due to the narrow angular width of perfect crystal Bragg reflections scattering from the sample can be resolved with extreme angular resolution by sweeping the analyzer crystal in tiny precisely controlled rotational steps over the antiparallel orientation, thereby keeping the monochromator at a fixed angle with respect to the incident beam. Triple reflections at both the monochromator and the analyzer improve the angular resolution by causing a drastic reduction of the side wing intensity if no sample is present. To achieve optimal performance possible single-bounce particle trajectories are suppressed by cutting a vertical slit into one of the crystal plates of the analyzer that is then filled by a Cd sheet. When a monochromatic neutron beam is refracted at both

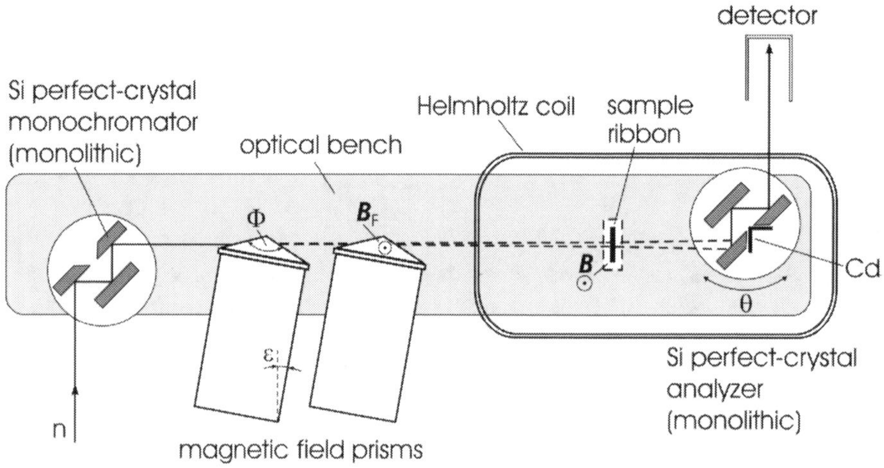

Figure 1. Schematic sketch of the experimental setup for ultra-small-angle scattering of polarized neutrons. The monochromator crystal is placed behind massive radiation shielding (not shown) in an interruption section of the thermal neutron guide H-25 of the high-flux reactor of the ILL Grenoble.

the entrance and the exit boundary of a prismatically shaped static magnetic field region the spin-*up* (↑)and spin-*down* (↓) neutrons leave this region along two different trajectories, since the field acts as a birefringent medium with an index of refraction

$$n_{\uparrow\downarrow} = 1 + \frac{\Delta k_{\uparrow\downarrow}}{k} \simeq 1 \pm \frac{m\mu B}{\hbar^2 k^2} = 1 \pm \frac{\mu B}{2E} \,. \tag{1}$$

There E is the energy, k the wave number, m the mass, and μ the magnetic moment of the neutrons. Although for thermal neutrons the angular separation of these trajectories [5]

$$\delta(E, B, \Phi, \varepsilon) = \frac{2\mu B}{E} \frac{\sin\Phi}{\cos\Phi + \cos(2\varepsilon)} \,, \tag{2}$$

where Φ is the apex angle of this 'field prism' and ε the deviation angle from symmetric passage through the prismatic field region, is typically only of the order of a few μrad, the extremely narrow resolution function $R(\theta)$ (usually called '*rocking curve*') of a USANS facility can be exploited for a clear separation of the two spin states, as illustrated in Fig. 2. This clearly twin-peaked rocking curve was obtained by means of two NdFeB-based magnet yokes, establishing a homogeneous field $B_p \simeq 0.9\,\mathrm{T}$ within their prismatically shaped ($\Phi = 116°$) air gaps of 1 cm

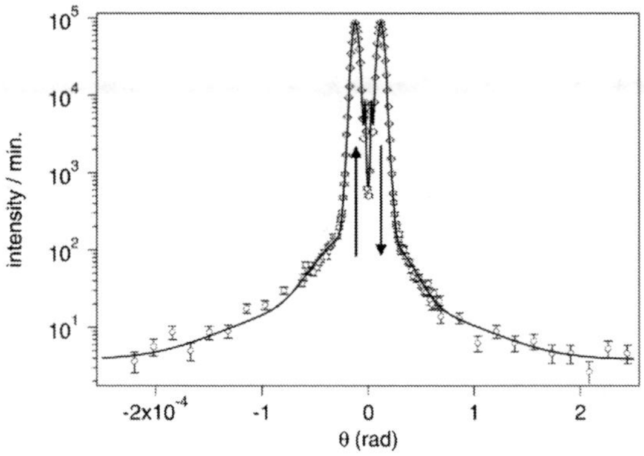

Figure 2. Rocking curve measured with empty sample holder. The arrows indicate the spin orientation with respect to the field direction. Due to the logarithmic vertical scale the peaks caused by $\lambda/2$ neutrons appear quite exaggerated. They have, however, negligible influence on the data evaluation which is restricted to the outward shoulders and wings of the measured intensity distributions.

height, at a neutron wavelength of 1.9 Åand an asymmetric passage angle $\varepsilon \simeq 25°$ (which turned out to be the maximally possible value under the conditions given). The $21.4 \pm 0.5\,\mu$m thin soft-magnetic amorphous $Fe_{65.7}Co_{18}Si_{0.8}B_{15.5}$ sample ribbon was provided by VAC, Hanau. It had a width of 25 mm and was cut to a length of 250 mm. Due to a specific thermal treatment (20 s at $T = 370°$) under simultaneously applied longitudinal tensile stress ($\sim$50 MPa) and transverse magnetic field the ribbon was expected to exhibit some transverse magnetic anisotropy with both stress- and field-induced contributions. Hysteresis measurements provided by VAC revealed a highly linear zero remanence magnetization curve with a relative permeability H_r $\sim$5000, an extremely low coercive field $H_c \simeq 0.01$ A/cm, a saturation induction $B_s = 1.737$ T and a magnetostriction constant $\lambda_s = +4.2 \times 10^{-5}$. A dedicated sample holder allowed to orient the ribbon either horizontally or vertically. In the latter case a variable tensile stress could be applied to the otherwise freely hanging ribbon by clamping Pb weights on its bottom end. A large Helmholtz-coil surrounding both the sample region and the analyzer crystal allowed to apply a small homogeneous magnetic field to the ribbon which was sufficiently large, however, to achieve magnetic saturation. During each individual measurement this field was kept precisely at its respective preset value by means of a flux-gate sensor whose output was used to control the DC current that was fed into the coils. Disturbing influences of large magnet facilities installed in the vicinity of the instrument S18 could thus be eliminated completely.

3. Results of measurements

The differential cross-section for elastic scattering of a fully polarized incident beam can formally be written as the sum of a purely nuclear and a purely magnetic contribution and a corresponding interference term:

$$\left(\frac{\mathrm{d}\sigma}{\mathrm{d}\Omega}\right)_{\uparrow\downarrow} = F_\mathrm{n}^2 + F_\mathrm{m}^{\perp 2} \pm 2F_\mathrm{n}F_\mathrm{m}^{\perp} \,. \qquad (3)$$

The subscript $^\perp$ indicates, however, that magnetic scattering is caused only by magnetization components with orthogonal orientation relative to the scattering plane. Fig. 3 shows the measured intensity distributions both for horizontal and vertical alignment of the ribbon, when the field at the sample position is set to zero as close as possible and no external mechanical stress is applied. Fig. 4 shows the analogous scattering patterns if a magnetic field of 10.9 G is applied. According to Eq. (3) the obvious lack of scattering asymmetry for the two spin states implies that there is no interference between nuclear and magnetic scattering. However, since scattering is largely reduced if the ribbon is saturated magnetically, it is of purely magnetic origin and contributions from structural or chemical inhomogeneities are negligible. Since in the case of $B = 0$ G the measured scattering distribution is narrower for vertical than for horizontal

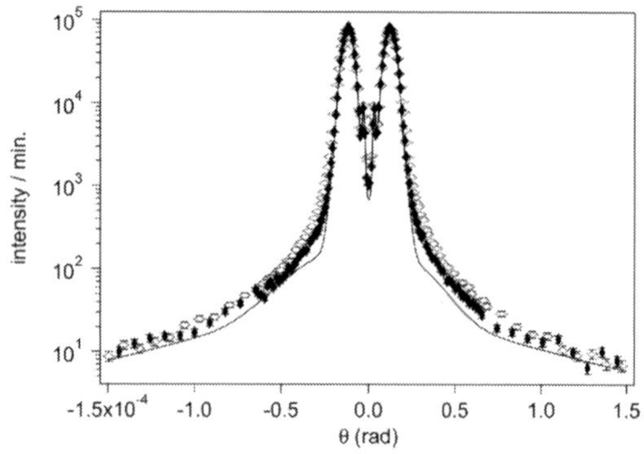

Figure 3. Diffraction patterns obtained with $B = 0\,\mathrm{G}$ for vertical (♦) and horizontal (◊) orientation of the amorphous sample ribbon. The full line represents the rocking curve.

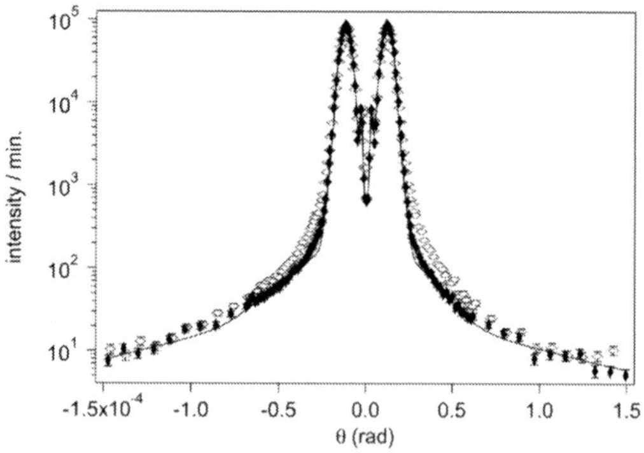

Figure 4. Analogous intensity distributions measured when this extremely soft-magnetic sample is magnetically saturated at $B = 10.9\,\mathrm{G}$.

orientation of the ribbon, the ribbon possesses a transverse magnetic texture, indeed.

On the other hand, Fig. 5 shows that a significant reduction of scattered intensity for the vertically aligned ribbon upon applying a longitudinal external tensile stress of 82 MPa, well above the (~50 MPa) which were applied during the texture-forming thermal treatment of the sample. This means that because of the positive sign of the magnetostriction constant the transverse magnetic texture vanishes or may even be transformed into a longitudinal one. However, to unambiguously verify such a transformation a comparison with the results of an additional measurement with horizontally aligned ribbon under under external stress would be required.

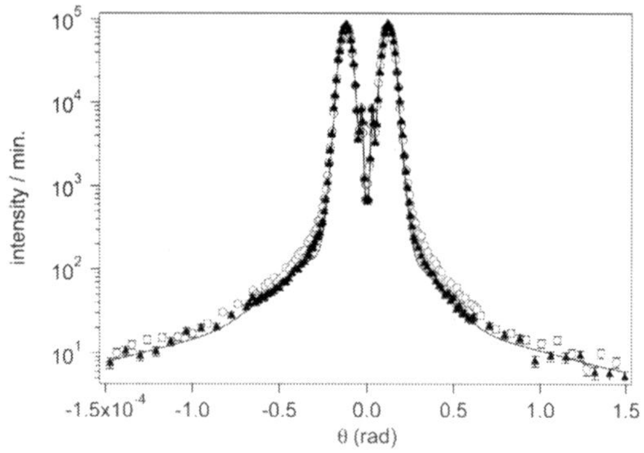

Figure 5. Influence of an externally applied tensile stress on the measured intensity distributions for vertical orientation of the sample ribbon: (○) $\sigma = 0$ Mpa, (▲) $\sigma = 82$ MPa. Again, the full line represents the rocking curve obtained without sample, i.e. the instrumental resolution function.

As mentioned already in the introduction, scattering from this sample ribbon is generally quite low and partly hidden behind the instrumental resolution function. Therefore we did not aim on a detailed determination of the shape of the magnetic inhomogeneities, which would require better statistics at the wings, but just tried to extract reasonable information about their overall size. We started by subtracting an arbitrary percentage $R(\theta) \times (1 - P_{\mathrm{scat}})$ of the rocking curve $R(\theta)$ from the measured raw data, where P_{scat} denotes the scattering probability. Subsequently a least square fit to the shoulders of the resulting 'pure' scattering distribution was performed according to a Guinier approximation of the form $I(\theta) \propto \exp(-kL\sin\theta)^2$, with L as fit parameter and θ the scattering angle. (The proportionality factor is governed by the number of scatterers, and the squares of their volumes and mean scattering length density differences.)

For each of the measured raw intensity distributions this procedure was repeated about 10^3 times with different values of $P_{\rm scat}$ and L. From the fit yielding the smallest χ^2 we inferred both the most likely scattering probability, typically $P_{\rm scat}$ of the order 10–20%, and the size parameter L. Assuming domains of approximately rectangular cross-section, the horizontal dimension of the mean effective rectangle is then given by $a = L\sqrt{12}$. Fig. 6 and Fig. 7 illustrate the results of this procedure for the field dependence of the scattered intensity with vertically aligned ribbon.

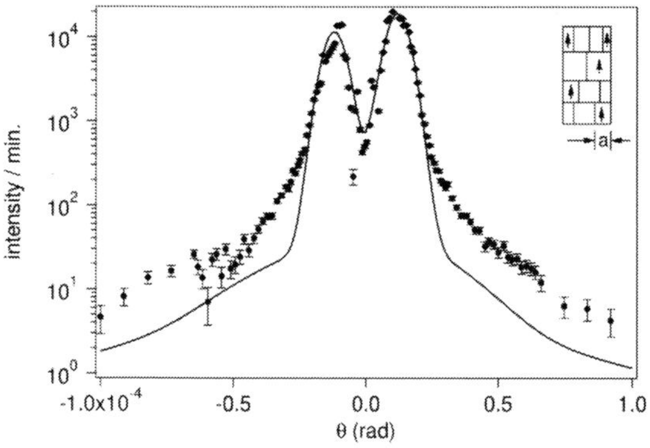

Figure 6. 'Pure' scattered intensity for vertically oriented ribbon (●) at $B = 0$ G and $\sigma = 0$ MPa after subtraction of the most likely non-scattered contribution. The solid line corresponds to a Guinier fit to the peaks of this intensity distribution, convolved with the measured resolution function of Fig. 2 (for details of the underlying procedure see the text).

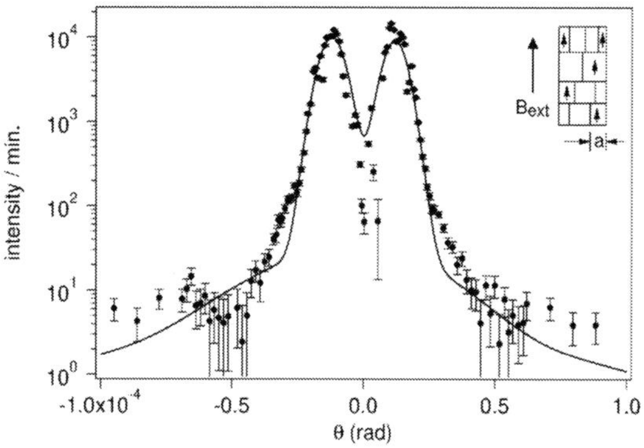

Figure 7. The analogous distribution (●) obtained at $B = 10.9$ G and $\sigma = 0$ MPa and the corresponding Guinier fit to the peaks (——), again convolved with the resolution curve. The inset illustrates the orientation of the ribbon and the applied magnetic field as well as the meaning of the size parameter a.)

The influence of an externally applied tensile stress on the scattered intensity distribution is seen from a comparison of Figs. 6 and 8.

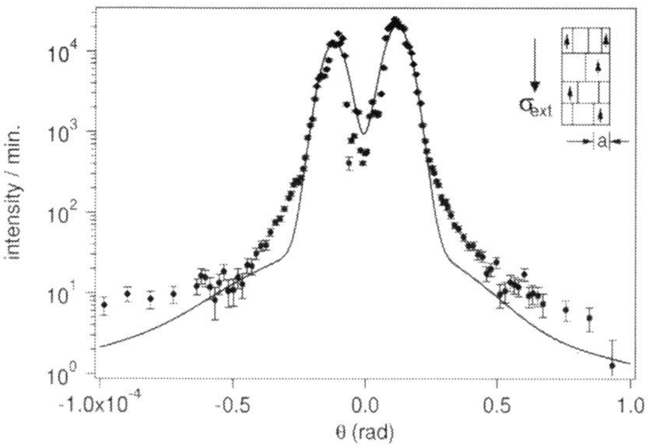

Figure 8. The 'pure' scattered intensity distribution (●) measured at $B = 0$ G when an external tensile stress $\sigma = 82$ MPa is applied to the ribbon, which is appreciably larger than the stress of 50 MPa which was applied during the fabrication process of the ribbon in order to induce a transverse magnetic anisotropy. The meaning of the inset is analogous to that of the previous figures.

Finally, in Fig. 9 the field and the stress dependence of the size parameter a is shown. It shall be noticed that the statistical errors are about as large as the symbols. Systematic errors of the absolute values of the size of the magnetic inhomogeneities are difficult to estimate, however. But the overall tendency upon variation of field and tensile stress seems to be quite reliable. At a very small external field the transverse magnetic texture of the ribbon is obviously removed and the magnetic domains have roughly the same size in vertical and horizontal direction. With increasing field the size of the now vertically aligned domains starts to increase as well (naturally on the cost of their number). In the case of horizontal ribbon alignment, where the field is orthogonal to the ribbon axis, the size of the transversely oriented domains is slightly decreasing with increasing field. It is interesting to see that an increasing external tensile stress along the ribbon axis leads to a similar behavior than that obtained by applying a magnetic field. Up to a tensile stress of $\sigma \sim 60$ MPa the size of the transverse domains is decreasing. Above this value, which is in good agreement with the expected limit due to the thermal treatment, the stress-induced transverse magnetic anisotropy is completely overcome and growing of the domains - now aligned along the ribbon axis because of positive magnetostriction - starts again.

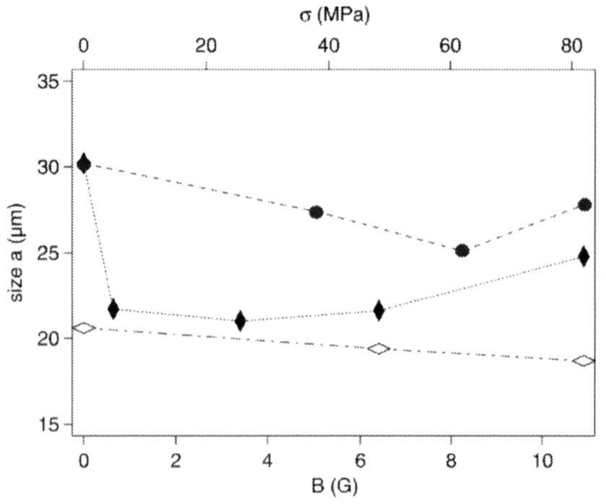

Figure 9. Field dependence of the domain size parameter a (see inset of Fig. 6) for vertical (♦) and horizontal (◊) orientation of the sample ribbon. Furthermore the variation of a with increasing externally applied tensile stress is shown for vertical ribbon orientation in the absence of any magnetic field (•). The statistical errors correspond roughly to the sizes of the symbols.

4. Concluding comment

It is clear that the USANSPOL technique came close or even went beyond its limits upon encountering the very large magnetic domains of this specific soft-magnetic amorphous ribbon, which turned out to be of the order of 20-30 μm. But it is remarkable that nevertheless it could provide valuable information about the internal magnetic structure of the sample. Although the scattering probabilities are relatively large compared to a typical SANS experiment, we did not need to concern ourselves with multiple neutron scattering since on their passage through the ribbon the neutrons encounter a single scatterer and its surroundings only. This is concluded from the fact that the thickness of the ribbon of $\sim$25 μm also corresponds to the domain sizes determined by our measurements.

Acknowledgments

The authors wish to thank H. Lemmel and R. Loidl for their invaluable support. This project was supported by Hochschuljubiläumsstiftung der Stadt Wien (Proj. No. H-1851/2006).

References

[1] Littrell K and Lee W 2004 *Physica* B **345** 246
[2] Wagh A G, Rakhecha V, Strobl M and Treimer W 2004 *Pramana J. Phys.* **63** 369
[3] Jericha E, Badurek G and Trinker M 2007 *Physica* B **397** 88
[4] Jericha E, Baron M, Hainbuchner M, Loidl R, Villa M and Rauch H 2003 *J. Appl. Cryst.* **36** 778
[5] Badurek G, Buchelt R J, Kroupa G, Baron M and Villa M 2000 *Physica* B **283** 389

Neutron Resonance Spin Echo Spectroscopy on Split Modes

K Habicht[1], M Enderle[2], B Fåk[3], K Hradil[4], P Böni[5] and T Keller[6,7]

[1]Helmholtz-Zentrum Berlin für Materialien und Energie, Lise-Meitner-Campus, Hahn-Meitner Platz 1, D-14109 Berlin, Germany
[2]Institut Laue-Langevin, BP 156-38042 Grenoble Cedex 9, France
[3]Commissariat à l'Energie Atomique, INAC, SPSMS, 38054 Grenoble, France
[4]Institut für Physikalische Chemie, Georg-August-Universität Göttingen, Aussenstelle am FRM II, Lichtenbergstraße 1, D-85747 Garching, Germany
[5]Physik-Department E21, Technische Universität München, D-85747 Garching, Germany
[6]ZWE FRMII, TU Munich, Lichtenbergstraße 1, D-85747 Garching, Germany
[7]Max-Planck-Institut für Festkörperforschung, Heisenbergstraße 1, D-70569 Stuttgart, Germany

Email: [1]habicht@helmholtz-berlin.de

Abstract. We present an analytical formalism investigating effects of multiple excitations with different dispersion parameters in the context of neutron resonance spin echo spectroscopy. This allows identifying conditions to obtain high-resolution information on the energy split of two dispersive modes which may differ or not in their local energy gradient. Comparison to experimental data is provided by recent measurements on magnon excitations in $RbMnF_3$. Potential future applications are investigations of hybridized phonon or magnon modes or mixed excitations.

1. Introduction

Neutron resonance spin-echo (NRSE) spectroscopy is a unique experimental tool to measure life times of dispersive elementary excitations over extended regions in the Brillouin zone with resolution in the μeV range. It is thus natural to explore the application of the technique for resolving modes with a separation in energy which is less than the resolution of standard spectrometers, e.g. triple-axis spectrometers (TAS). However, in this case, the spin-echo conditions cannot be satisfied for both modes simultaneously and it is important to understand the consequence for the echo-amplitude on a quantitative level. Based on an extension of Popovici's matrix approach for the TAS resolution function [1], we have developed the formalism which can be used for data correction for NRSE-TAS experiments [2, 3]. In this paper we report on the generalization of our earlier treatment now including instrumental parameters not matching the dispersion properties, allowing the local gradient $\nabla_q\omega$ to have an arbitrary direction in wave vector-energy space and apply it to scattering from a second mode contributing to the echo amplitude.

The significance of this work is to identify conditions which allow obtaining high-resolution information on the energy split of two overlapping dispersive modes which may differ or not in their local energy gradient. As a special case mixtures of spin-flip and non-spin-flip scattering and their spin-echo time dependent contribution to the echo signal are included. This is important in life time studies of magnons.

We first consider a simplified model for the echo amplitude in section 2 while the complete model is discussed in section 3. A comparison to NRSE experimental data is presented in section 4.

2. Simplified model for the time-dependence of the echo amplitude

In this section we introduce a simplified model for the echo amplitude as a function of spin echo time which assumes that the spin echo conditions are met for both modes simultaneously. Although this is strictly not possible, the model can be useful in practical cases when the violation of the echo condition is very small or one deals with relatively small spin echo times (compared to quasielastic NSE) which is actually often the case in NRSE spectroscopy leading to negligible depolarization of the detuned mode.

In a minimal model the echo amplitude A_E may be described in terms of the Fourier transform of two Lorentzians. Explicitly we can write:

$$A_E = \left| \frac{1}{N} \int R(\omega-\omega_T) S(\omega-\omega_S) e^{-i(\omega-\omega_N)\tau} d\omega \right|. \tag{1}$$

Here N is a normalization factor which ensures that the echo amplitude A_E=1 at spin echo time τ=0. We allow for the energy resolution of the background TAS as an additional weighting function

$$R(\omega-\omega_T) = \exp\left(-(4\ln 2) \frac{\hbar^2 (\omega-\omega_T)^2}{\Delta E^2{}_{TAS}} \right), \tag{2}$$

where ΔE_{TAS} is the FWHM of the TAS energy resolution function which is centered at energy $\hbar\omega_T$ and therefore has its maximum at energy $\hbar\omega_T$.

Assuming that the structure factor does not vary within the resolution ellipsoid we can factorize the scattering function S(**Q**,ω), i.e. the **Q**-dependent part will only contribute to absolute intensity and can be neglected for our purposes. In this minimal model there is no provision for dispersion at all and neither TAS resolution in Q-space nor spin-echo related resolution effects are considered. We do however allow for two excitations in that we parameterize

$$S(\omega-\omega_S) = A_1 \frac{\Gamma_1}{\Gamma_1^2 + (\omega-\omega_S)^2} + A_2 \frac{\Gamma_2}{\Gamma_2^2 + (\omega-\omega_S-\Delta\Omega)^2}, \tag{3}$$

Here $A_{1,2}$ are amplitudes which determine the relative intensity weight of the two Lorentzians with line widths given as HWHM $\Gamma_{1,2}$. $\Delta\Omega$ is the energy separation of the two modes. We have explicitly introduced ω_S expressing the fact that we consider the scattering function relative to the first excitation mode which has the energy $\hbar\omega_S$. Strictly speaking Eq. (3) violates the detailed balance condition for S(**Q**,ω). Here we implicitly assume ω>0 and zero temperature which allows neglecting the anti-Stokes term.

For inelastic scattering the spin echo phase (total Larmor precession phase ϕ at the detector) will have a constant term $\tau\omega_N$. This term depends on instrument parameters and is fixed by the spin echo condition. As a matter of fact $\hbar\omega_N$ is just the mode energy $\hbar\omega_S$ if the apparatus is tuned to this mode. With the variable transform $\omega = \omega_S + \Delta\omega$, where $\Delta\omega$ is the energy deviation from the mode energy of the first Lorentzian in Eq. (3), the echo amplitude reduces to

$$A_E = \left| \frac{1}{N} \int R(\Delta\omega-\omega_T+\omega_S) S(\Delta\omega) e^{-i\Delta\omega\tau} d\Delta\omega \right|. \tag{4}$$

Fig. 1 shows the scattering function (Eq.(3)), the scattering function multiplied and convoluted with the TAS energy resolution (Eq.(2)) as well as the corresponding Fourier transforms. An approximate analytic expression of the echo amplitude in case the line widths are identical is given by an exponential decay modulated by a cosine term

$$A_E = \left\{ |A_M| + \left(1 - |A_M|\right)\cos\left(2\pi \frac{\tau}{T}\right)\right\} \exp\left(-\frac{\tau\Gamma}{\hbar}\right), \tag{5}$$

where the period T of the modulation is inversely proportional to the energy difference

$$T = \frac{4\pi\hbar}{\Delta\Omega} \tag{6}$$

and the amplitude of the modulation is determined by the intensity ratio of the modes A_1/A_2

$$A_M = \frac{A_1' - A_2'}{A_1' + A_2'}, \qquad A_1' = A_1 R(-\omega_T), \quad A_2' = A_2 R(\Delta\Omega - \omega_T). \tag{7}$$

We note that in this minimal NRSE model the amplitude parameters should not only reflect the scattering probability but also contain the information about the fraction of the signal which gives rise to a polarized signal at the detector. Two limiting cases are immediately obvious: If the spin-echo conditions are strongly violated for the second mode there will be an increase of depolarized background independent of spin-echo time and A_2 can be set to zero. Provided that the signal from the second mode is completely depolarized at the smallest experimentally accessible spin-echo time [4] the only effect to be considered is that the initial echo amplitude at τ=0 will be less than 1. The other limiting case is when the spin-echo conditions are perfectly satisfied for both modes and thus the modulation of the signal extends to large spin-echo times. In the latter case the amplitudes $A_{1,2}$ are determined by the structure factor and the energy resolution function only. Any intermediate case requires going beyond the simple model.

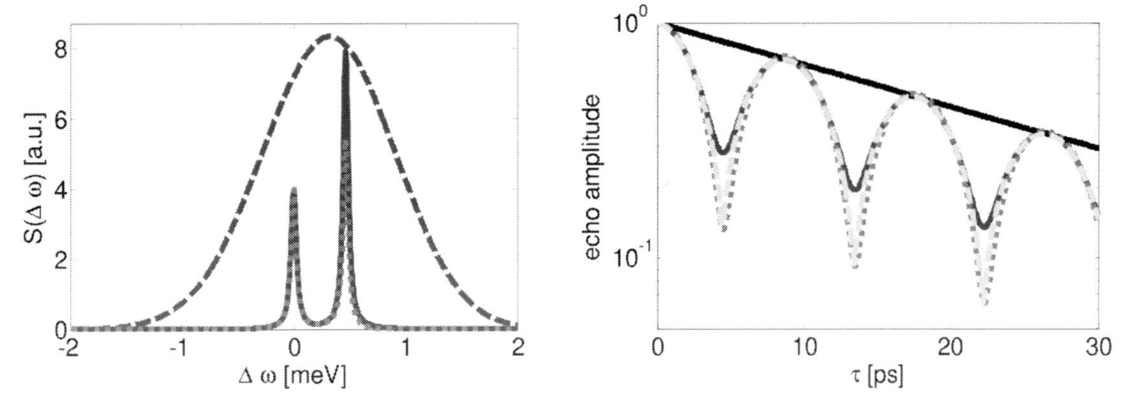

Fig. 1: Left: Blue solid: Scattering function S(Δω) assuming $\Gamma = \Gamma_1 = \Gamma_2 =$ 27 μeV, A_1=1, A_2=2, ΔΩ= 0.464 meV. Red dotted: scattering function multiplied with the TAS energy resolution R S(Δω) assuming ΔE_{TAS}=1.25 meV. Grey dashed: scattering function convoluted with the TAS energy resolution. Right: Black solid: Single exponential decay corresponding to Γ. Blue solid: Fourier transform of S(Δω). Red dashed: Fourier transform of R S(Δω). Green dashed: The approximation of Eq. (5).

3. Extended model for the time-dependence of the echo amplitude

Next we outline the main generalizations required to attain a full model which explicitly takes into account the decaying contribution of the detuned mode. An earlier treatment of the resolution function [3] assumed that the spin-echo conditions are perfectly satisfied and that the TAS resolution ellipsoid coincides with the centre of the dispersion. We have now derived corresponding analytical expressions giving up this restriction. In addition we allow for instrumental parameters not satisfying

Journal of Physics: Conference Series **211** (2010) 012028 doi:10.1088/1742-6596/211/1/012028

the spin-echo conditions for inelastic, dispersive NSE, and two dispersion surfaces with different local gradients in wave-vector-energy space. As a starting point we always express the echo amplitude using the Larmor phase ϕ as a function of the incident and final wave vectors $\mathbf{k}_{i,f}$.

$$A_E = \left| \frac{1}{N} \int R(\mathbf{k}_i, \mathbf{k}_f) S(\mathbf{Q}, \omega) e^{i\phi(\mathbf{k}_i, \mathbf{k}_f)} d^3k_i d^3k_f \right|. \tag{8}$$

Larmor phase, dispersion relation and TAS resolution function are expanded to second order which allows using a covariance matrix formulation of the resolution problem.

3.1. Violated spin-echo conditions

For brevity we outline the result for the functional form of the echo amplitude assuming a single dispersion branch but violated spin-echo conditions. A full derivation will be given elsewhere [5]. We find that instrumental parameters introduce terms in the Larmor precession phase which are linear in the wave-vector variables $\mathbf{k}_i$, $\mathbf{k}_f$. This implies a Gaussian decay of the echo amplitude on top of the usual second-order resolution contribution

$$A_E = \left| \sqrt{\frac{\det \tilde{L}(\tau_2''=0)}{\det \tilde{L}(\tau_2'')}} \exp\left(-\frac{1}{2}\kappa\tau_2''^2\right) \right|. \tag{9}$$

Here τ_2'' is a reduced spin-echo time, i.e. the usual spin-echo time but modified by an additional factor introduced by the detuning. and κ is essentially a linear transform of the inverse of the complex resolution matrix $\tilde{L}$ [2] providing an effective width for the Gaussian. Detuning curves of individual instrument parameters (Fig. 2) are readily obtained from the present formalism and provide an estimate on the sensitivity of the experiment with respect to instrumental detuning which may be due to limited knowledge of the dispersion parameters.

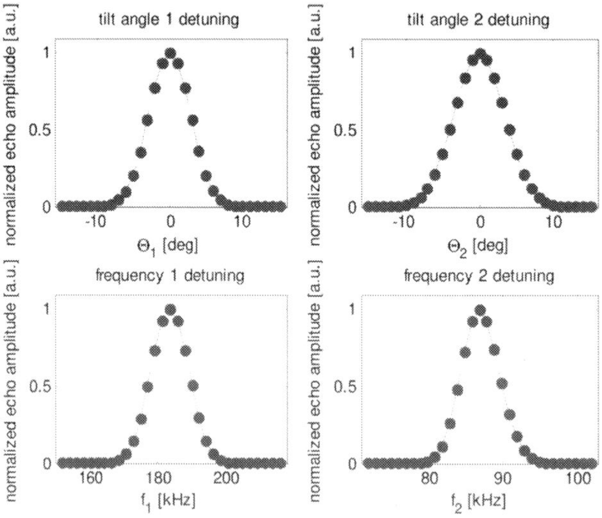

Fig. 2: Detuning of the tilt angles of the RF coils $\theta_{1,2}$ and the frequencies $f_{1,2}$ of the RF field in first and second spectrometer arm evaluated at spin echo time τ=15 ps.

3.2. Second dispersion surface

For the functional form of the echo amplitude assuming two dispersion branches, one perfectly tuned, the other violating the spin-echo conditions we obtain for the echo amplitude

$$A_E = \left| \frac{1}{N} \left\{ \int S_1(\Delta\omega) e^{-i\tau''_{2,1}\Delta\omega} d\Delta\omega + \exp\left(\tilde{\mathbf{T}}^T \tilde{\mathbf{L}}^{-1} \tilde{\mathbf{T}} \right) \int S_2(\Delta\omega) e^{-i\tau''_{2,2}\Delta\omega} d\Delta\omega \right\} \right| . \qquad (10)$$

This result is very similar to Eq. (4) since to good approximation $\tau_{1,2}'' \approx \tau_{2,2}''$. However the integral over the second Lorentzian has now an additional complex prefactor

$$f_p = \exp\left(\tilde{\mathbf{T}}^T \tilde{\mathbf{L}}^{-1} (\tau) \tilde{\mathbf{T}} \right) \qquad (11)$$

allowing to estimate to what extent the second dispersion branch contributes to the echo amplitude. Here the column vector **T** is linear in the spin-echo time and thus the prefactor f_p leads to a Gaussian damping of the modulation. In the limit of satisfied spin echo conditions **T** vanishes.

Fig. 3 shows the model scattering function multiplied and convoluted with the TAS energy resolution as well as the corresponding Fourier transforms now explicitly considering the spin-echo-time-dependent prefactor f_p.

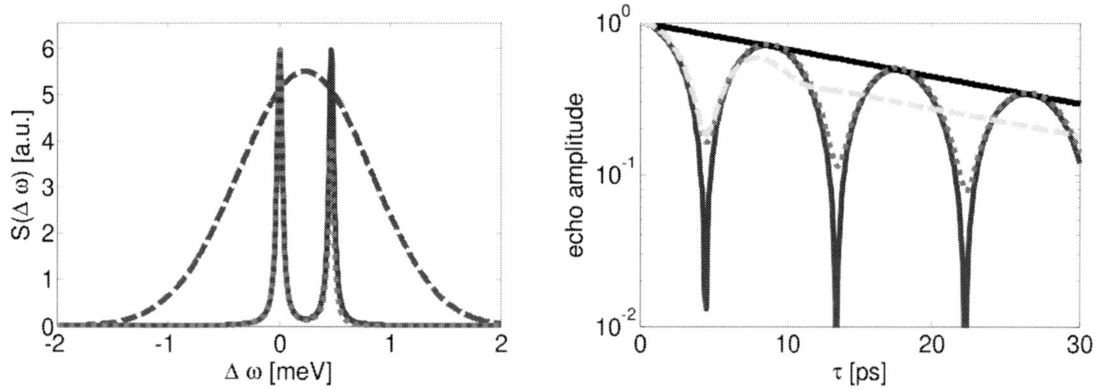

Fig. 3: Left: Blue solid: Scattering function S(Δω) assuming $\Gamma = \Gamma_1 = \Gamma_2 =$ 27 μeV, A_1=1, A_2=1, ΔΩ= 0.464 meV. Red dotted: scattering function multiplied with the TAS energy resolution R S(Δω) assuming ΔE_{TAS}=1.25 meV. Gray dashed: scattering function convoluted with the TAS energy resolution. Right: Black solid: Single exponential decay corresponding to Γ. Blue solid: Fourier transform of S(Δω). Red dotted: Fourier transform of R S(Δω). Green dashed: The full model echo amplitude of Eq. (10).

We note that a special case of the detuning problem is realized investigating magnon excitations with non-spin-flip scattering while the second precession field region has a polarity adapted to spin-flip scattering. This case is covered assuming that the RF frequency f_2 is detuned to $-f_2$. All other parameters stay tuned.

4. Comparison to experimental data

The case of a second dispersion branch within the background spectrometer resolution has been experimentally realized in recently performed NRSE measurements at TRISP, FRM II, with a particular sample of the simple cubic Heisenberg antiferromagnet $RbMnF_3$ consisting of two grains of comparable size. This provides an ideal ground for checking the model against experimental data since the second dispersion surface is identical to the first and its position in wave-vector-energy-space is easily obtained from the experiment. The spin wave dispersion of magnons in $RbMnF_3$ is known in sufficient detail [6]. The relative orientation of the two crystallites has been obtained by standard procedures maximizing the intensity of two Bragg peaks for each of the two grains. Applying an extension of the original UB matrix formalism [7] to inelastic scattering following Lumsden *et al.* [8]

we obtain the peak positions of the inelastic signal in good agreement with experimental data collected on the thermal TAS PUMA, FRM II (see Fig. 4 for an example).

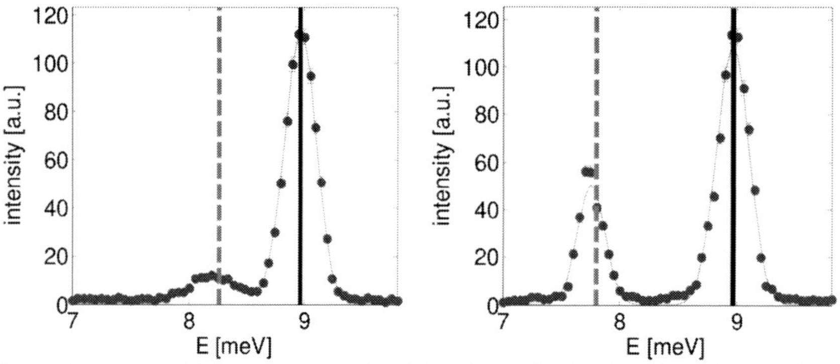

Fig. 4: Experimental peak positions (data points) arising from the double dispersion surface compared to calculated peak positions indicated by the vertical lines. Note that these two scans are at wave vectors [0.75 0.75 -0.25] (left panel) and [0.75 0.75 +0.25] (right panel) where a clear separation of the two modes is observed. There are however positions in Q-space where the split is unnoticeable within standard TAS resolution.

The experimental NRSE data for the zone boundary magnon at **Q**=[0.5 0.5 -1] and E=8.46 meV has been fit to the simple model Eq. (5) (see Fig. 5). The energy split from the fit was found to be 0.420(19) meV at a sample temperature T_S=3.18 K and 0.414(6) meV at T_S=7.75 K. This is in reasonable agreement with the calculated energy split of 0.464 meV. Also shown in Fig. 5 is the time dependence applying the full model Eq. (10) found to be in agreement with the NRSE data within statistical error. In order to fit the energy split using the full model, more data points are required than available. This inadequacy is due to the fact that the original aim of the spin-echo experiment was actually not to determine the energy split, but rather the line width of the mode.

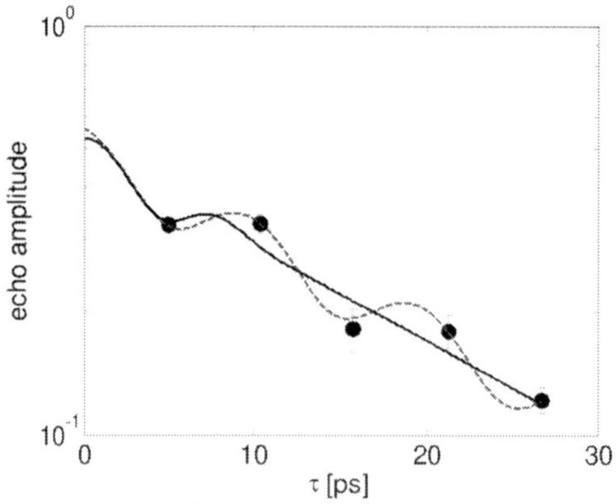

Fig. 5: Experimental NRSE data measured on the zone boundary magnon **Q**=[0.5 0.5 -1], E=8.46 meV, T_S=3.18 K. The dashed blue line is a fit of Eq. (5) to the data. The solid black line is the echo amplitude calculated according to Eq. (10).

Finally we note that in the particular experiment on $RbMnF_3$ magnons our model allows identifying those regions in the Brillouin zone where the line-width data is affected by the second grain. Two conditions must be met for spurious effects to be significant: The energy separation of the two dispersion surfaces must be less than the FWHM of the background TAS resolution and the magnitude of the complex prefactor must be less than 50% at the smallest experimentally accessible spin echo time used in the data analysis. It is found that a precise measurement of the line width is unaffected by the second grain in most of the Brillouin zone except for small regions close to the zone boundary.

5. Conclusion and Outlook

The present analytical formalism allows investigating effects of multiple excitations with different dispersion parameters in the context of NRSE spectroscopy. The significance of this work is to identify conditions which allow obtaining high-resolution information on the energy split of two overlapping dispersive modes which in general may differ in their local energy gradient. Potential future applications may be seen for investigations of hybridized phonon or magnon modes or mixed excitations. More experimental work is clearly needed to establish the method when applied to split modes.

References

[1] Popovici M 1975 *Acta Cryst.* A **31** 507
[2] Habicht K, Keller T and Golub R 2003 *J. Appl. Cryst.* **36** 1307
[3] Habicht K, Keller T and Golub R 2004 *Physica* B **350** E803
[4] The smallest accessible spin echo time is assumed to be finite as in any typical NRSE experiment unless special instrumental measures are taken to avoid reducing the frequency of the resonance coils to less than 50 kHz.
[5] Habicht K *et al.* to be published
[6] Windsor C G, Stevenson R W 1966 *Proc. Roy. Soc.* **87** 501
[7] Busing W R and Levy H A 1967 *Acta Cryst.* A **22** 457
[8] Lumsden M D, Robertson J L and Yethiraj M 2005 *J. Appl. Cryst.* **38** 405

Isotope-periodic multilayer method for short self-diffusion paths – a comparative neutron and synchrotron Mössbauer reflectometric study of FePd alloys

D G Merkel[1], Sz Sajti[1], Cs Fetzer[1], J Major[3], M Major[3†], R Rüffer[2], A Rühm[3], S Stankov[2‡], F Tanczikó[1] and L Bottyán[1]

[1]KFKI Research Institute for Particle and Nuclear Physics, P.O.B 49, H-1525, Budapest, Hungary
[2]European Synchrotron Radiation Facility, BP 220, 38043 Grenoble CEDEX 9, France
[3]Max-Planck-Institut für Metallforschung, Heisenbergstr. 3, D-70569 Stuttgart, Germany

E-mail: merkel@rmki.kfki.hu

Abstract. FePt, FePd, CoPt, and CoPd in equilibrium exhibit the $L1_0$ structure with high perpendicular magnetic anisotropy making them attractive candidates for high-density magnetic recording. Magnetic properties of these films depend on the distribution and orientation of the $L1_0$ fraction controlled by diffusion on atomic scale. Epitaxial isotope-periodic ^{nat}FePd/^{57}FePd alloy films were prepared by molecular beam epitaxy and heat treated at 500°C for various retention times. Isotope-sensitive non-destructive methods, neutron reflectometry and synchrotron Mössbauer reflectometry were applied to follow very short diffusion paths normal to the film plane. Squared diffusion lengths and diffusion profiles were obtained from the fitting of experimental reflectivity curves for each annealing treatment steps. The somewhat different diffusion lengths obtained for the neutron and synchrotronMössbauer reflectograms of the same samples are explained by the larger footprint of the sample in the neutron experiment for which interface inhomogeneities are to be averaged. Diffusion in the microscopically different local environments were modelled by piecewise constant diffusion coefficients in the regions identified as different species by conversion electron Mössbauer spectroscopy.

1. Introduction

Due to their high perpendicular magnetic anisotropy [1,2,3], CoPt, CoPd, FePt as well as FePd are candidate materials for future ultra-high density magnetic recording [4,5,6]. In the composition range between $0.5 < x < 0.6$, in equilibrium $Fe_{1-x}Pd_x$ exhibits $L1_0$ (CuAu(I)-type) order with alternating Fe and Pd planes along the crystallographic [001] direction. In this ordered structure the face centred cubic (fcc) unit cell with random Fe and Pd occupation is distorted to tetragonal, the lattice parameter

† Present address: Institute of Materials Science, TU Darmstadt, Petersenstr. 23, D-64287 Darmstadt, Germany On leave from KFKI Research Institute for Particle and Nuclear Physics, P.O.B. 49, H-1525 Budapest, Hungary

‡ Present address: Institute for Synchrotron Radiation, Karlsruher Institute of Technology Campus Nord, Hermann-von-Helmholtz-Platz 1, D-76344 Eggenstein-Leopoldshafen, Germany

ratio (*c/a*) varying between 0.96 and 0.97 [7]. The changes of magnetic and structural properties in this alloy are associated with the atomic-scale diffusion processes governing the local structure.

There are several methods to study chemical diffusion processes but the spatial resolution of most of them is far above the atomic scale. By sectioning and profiling techniques [8,9,10] it is possible to measure very low diffusion paths but they are not suitable to follow self-diffusion processes in chemically homogeneous materials since the solely electronic interaction with the x-ray photons produces no contrast between the adjacent chemically identical layers.

In order to study self diffusion, an isotope-sensitive technique is needed. Due to the difference in the nuclear cross section of different isotopes of the same element, neutron reflectometry (NR) is a suitable non-destructive method to study self diffusion in a chemically homogeneous isotope multilayer with a diffusion length in the range of a few angstroms [11,12,13]. Another isotope-sensitive non-destructive method is Synchrotron Mössbauer Reflectometry (SMR) [14,15,16] due to the nuclear hyperfine (hf) interaction between the atomic nuclei and the highly monochromatized synchrotron radiation.

In this paper we present a comparative study of Fe self diffusion in FePd by NR and SMR using an isotope-periodic ^{nat}FePd/^{57}FePd multilayer film.

2. Experimental procedure

Partially ordered (52% $L1_0$) isotope-periodic [$^{nat}Fe_{47}Pd_{53}$(3 nm)/$^{57}Fe_{47}Pd_{53}$(2 nm)]$_{10}$ multilayers were evaporated onto MgO(001) substrates by molecular beam epitaxy (MBE). To improve epitaxial growing, Cr seed and Pd buffer layers (of 3 and 15 nm, resp.) were applied. To prevent oxidation, a 1 nm Pd capping layer was grown on top of the periodic multilayer. During evaporation the substrate temperature was held at 350°C.

For NR experiments, the originally 20×20×2 mm^3 size sample was cut to four equal (10×10×2 mm^3) pieces. One of them was left in the as-deposited state, the others were annealed at 500°C for 90, 360, and 1800 minutes. The heat treatments were carried out under UHV condition. The NR experiments were performed using a 2D ^{3}He detector and the non-polarized monochromatic beam (λ=5 Å) at the NREX$^+$ reflectometer (operated by MPI-MF Stuttgart at FRM II, Garching) [17]. A supermirror filter was used to suppress higher harmonics.

The SMR experiments were performed at the ID18 beam line [18] of the European Synchrotron Radiation Facility (ESRF, Grenoble) in 16-bunch mode. The samples previously measured by NR were then cut into half (5×10×2 mm^3). The beam was sequentially monochromatized by a Si(111), then a Si(4.2.2)/Si(12.2.2) double channel cut monochromator, respectively, to the vicinity of the 14.4 keV Mössbauer transition (λ=0.86025 Å) of ^{57}Fe. The lateral beam size was restricted to 0.2 mm. Nuclear resonant (delayed) time-integrated and electronic (prompt) reflectograms were simultaneously recorded by avalanche photo diode (APD) detectors.

Conversion electron Mössbauer spectroscopy (CEMS) was performed using a ^{57}Co(Rh) source and a home-made gas-flow single-wire proportional counter operating with ^{4}He and additional 4.7% CH_4 extinction gas at a bias voltage of 830(±10) V.

3. Results and discussion

In former studies [16,19,20] SMR spectra were evaluated by normalizing the Bragg-peak height to the totalreflection peak height, but, as Andreeva showed, the peak height ratio is very sensitive to the value (and distribution) of the hf fields [21]. Furthermore, the actual peak heights may be modified by overlapping with the Kiessig beats, a condition that can hardly be avoided if, like in the present case, variable non-periodic (buffer and capping layer) contributions are present in the reflectivity. For the above reasons the hf fields (actually, the hf field distributions) were independently determined from CEMS spectra and fed into the robust full reflectivity curve fit [22] to extract the extent of the diffusion layer mixing. For the proper evaluation of the layer parameters, simultaneous fits were performed to the prompt (non-resonant) and delayed (nuclear resonant) reflectivity curves. Of course, in case of NR measurement, the latter is not possible. Substrate and interface roughness is always present in a multilayer. Individual rms roughnesses were taken into account except for the multilayer stack, where a common roughness for all periods was fitted. With the above-described procedures and constraints both the NR and the SMR data were fitted by the FitSuite program [22]. The program accounts for the diffusive interface roughness in the form of a 1D diffusion profile characterized by $D{\cdot}t$ (squared diffusion length, D and t being the diffusion coefficient and the time of diffusion, respectively) and adjusts this profile through $D{\cdot}t$ to fit the measured reflectivity curves. Since the roughness and diffusion both have error-function profiles, the contributions are separated simply by assuming the non-diffusive interface roughness to be fully present in the as prepared sample and to remain constant during heat treatments.

Figure 1 shows the experimental and fitted SMR (a) and NR (b) curves of the Pd(3nm)/[$^{nat}Fe_{47}Pd_{53}$(3 nm)/$^{57}Fe_{47}Pd_{53}$(2 nm)]$_{10}$/Pd(15nm)/Cr(3nm)/MgO(001) alloy multilayer along with the corresponding isotope bilayer diffusion profiles adjusted in the fit.

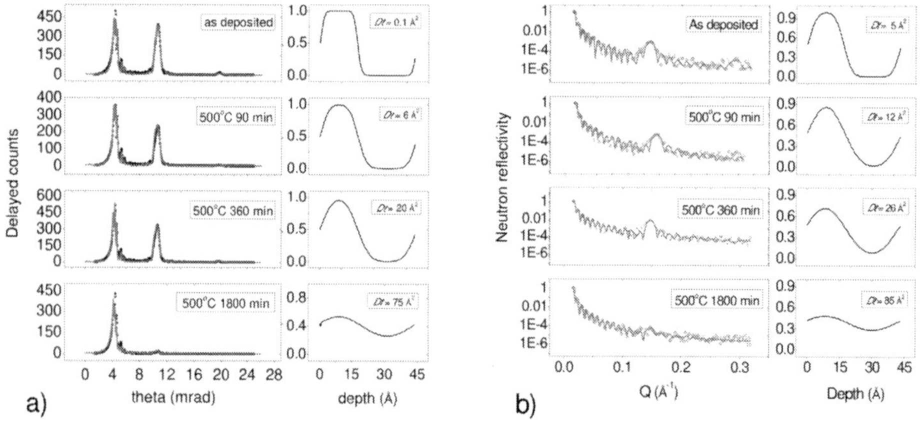

Figure 1 SMR (a) and NR (b) curves taken on the isotope-periodic FePd multilayer annealed at 500°C for 0, 90, 360 and 1800 minutes, resp. In the right columns the corresponding bi-layer diffusion profiles are shown.

It is clearly seen that, with increasing retention time, the diffusion profile flattens indicating the intermixing of the adjacent isotope layers. However, the extracted $D{\cdot}t$ values do not completely agree in the SMR and NR experiments (see Fig. 2a). The difference is already noticeable in the as deposited state. The diffusion profile in the SMR spectrum is more square-like than it is in the NR spectrum. The $D{\cdot}t$ values in the as deposited state extracted from SMR and NR are 0.1 and 5 Å^2, respectively.

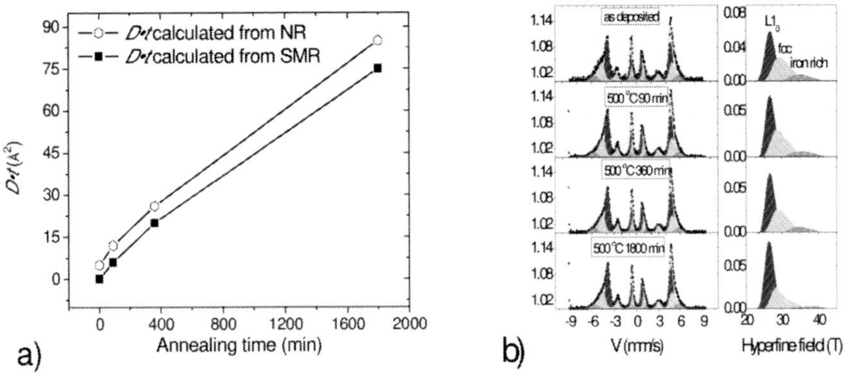

Figure 2 Variation of the squared diffusion length measured by NR and SMR. (panel *a)*, full lines for guiding the eye). Measured and fitted conversion electron Mössbauer spectra (taken after the annealing steps) along with the corresponding hyperfine field distributions (panel b).

This kind of deviation can be observed in all samples. The squared diffusion lengths increased from 0.1 and 5, 6 and 12, 20 and 26, 75 and 85 Å^2 in the 90, 360 and 1800 min heat treated samples obtained from SMR and NR spectra, respectively (Figure 2a). Since the measured samples were the same in the NR and SMR experiments, the difference cannot originate from different structural properties. Ageing of the samples between SMR and NR beam sessions can also be excluded since we re-measured several samples by CEMS at later dates and found no difference in the spectra. Moreover, in a later SMR session several re-measured samples gave unchanged fitted $D{\cdot}t$ values. However, such deviation may well originate from the difference of the footprint of the beams in the two experiments. In the case of NR, the reflected beam comes from a much wider area and therefore averaging the lateral thickness and interface roughness variation in the sample are reflected in the spectrum as an initial diffusive roughness causing an overall decrease of the Bragg peak intensity and a corresponding artificially higher $D{\cdot}t$. This effect is much smaller and will be neglected in case of SMR because the x-ray beam was 0.2 mm as compared to the 10 mm width of the neutron beam. (Further lateral restriction of the neutron beam was not possible for reasons of intensity and beam time constraints.) One can also see that the difference of the squared diffusion length reaches 10 Å^2 for the longest annealing time (1800 min) as compared to ~6 Å^2 for shorter annealing times. A possible explanation of this fact may be that, when the adjacent layers are almost completely inter-diffused, the error of the fitting can be as high as a few Å^2. Another possible explanation may be that new sources of roughness emerge at very long annealing times. It is also observable that the tendency of $D{\cdot}t$ as a function of annealing time is not linear as it should be for a structurally homogeneous material [23]. This deviation from the linear tendency indicates that the diffusion coefficient varies during the heat treatments. Therefore, one may suspect that the structure of the layers in the sample, changes. Indeed, from CEMS experiments three distinct environments, namely a low-hf field, a large-hf-field, and an intermediate-hf-field species were identified with wide distribution and average hf fields of ~27, ~35 and ~30 T and assigned to ordered $L1_0$, disordered fcc and a (here non-specified) iron-rich structural unit, respectively. These are visualized by the hf field distributions in Figure 2b. For details of the assignment see ref. [24]. The CEMS spectral fraction of the ordered $L1_0$ phase raises from 52% to 66% while the fraction of the disordered fcc unit decreases from 24% to 22% and the fraction of the iron-rich unit decreases from 24% to 12% with increasing annealing time from 0 to 1800 min. The significant increase of the $L1_0$ fraction and decrease of iron-rich regions indicate that the deviation from linear tendency in the $D{\cdot}t$ originates from the different diffusion properties of these environments, namely, the iron self diffusion in the crystallographic *c*-direction is considerably slower in the $L1_0$ phase than that in the iron-rich phase. With the plausible assumption that the diffusion constant takes an arbitrary positive value in volumes of distinct environments like in a random alloy;

the total $D{\cdot}t$ of the system can be described as a weighed sums of the individual squared diffusion lengths [25]. Since these annealing experiments were performed so far only at one temperature, the determination of the activation energies and pre-exponental factors of the distinct species is not yet possible. Further experiments on FePd films of different initial structures annealed at various temperatures are in progress.

4. Summary

In order to follow short-range diffusion processes in an isotope-periodic multilayer, a comparative neutron and synchrotron Mössbauer reflectometric study was performed on isotope-periodic, partially ordered FePd samples. Three distinct local structural units were identified and quantified by conversion-electron Mössbauer spectroscopy. The diffusion profiles and hence the squared diffusion lengths ($D{\cdot}t$) extracted from full-curve fits of neutron reflectivities show a positive and somewhat increasing difference of about 6 to 10 Å^2 for diffusion lengths 0 to 70 Å^2 for SMR. This deviation is attributed to the difference in the footprints of the beams and the correspondingly different lateral averages of the layer thickness and/or interface roughness variations in the NR and SMR experiments, resulting in an artificial extra flattening of the diffusion profile in the NR curve. The observed slight deviation from linear dependence in the $D{\cdot}t$ *vs* t curve is attributed to the variation of the fractions of the different structural units with different diffusion coefficients.

Acknowledgements

This work was partially supported by the Hungarian National Fund (OTKA) and the National Office for Research and Technology of Hungary under contracts K 62272 and the NAP-VENEUS and by the European Commission under the 6th and 7th Framework Programmes through the Key Action: Strengthening the European Research Area, Research Infrastructures, contract no.: 226507 (NMI3). Careful reading of and valuable corrections to the manuscript by Prof. Dénes L. Nagy are gratefully acknowledged. The authors would like to thank the European Synchrotron Radiation Facility and the Forschungsneutronenquelle Heinz Maier-Leibnitz (FRM II) for providing the synchrotron and neutron beams.

References

[1] Visokay M R and Sinclair R 1995 *Appl. Phys. Lett.* **66** 1692
[2] Shima H, Oikawa K, Fujita A, Fukamichi K, Ishida K and Sakuma A 2004 *Phys. Rev.* B **70** 224408
[3] Ersen O, Parasote V, Pierron-Bohnes V, Cadeville M C and Ulhaq-Bouillet C 2003 *J. Appl. Phys* **93** 5
[4] Devolder T, Bernas H, Ravelosona D, Chappert C, Pizzini S, Vogel J, Ferré J, Jamet J P, Chen J and Mathet V 2001 *Nucl. Instr. and Meth.* B **175-177** 375
[5] Piao K, Li D J and Wei D 2006 *J. Magn. Magn. Mater* **303** e39
[6] Suzuki T, Harada K, Honda N and Ouchi K 1999 *J. Magn. Magn. Mater* **193** 85
[7] Pearson W 1958 *Handbook of Lattice Spacings and Structures of Metals, Pergamon*, New York
[8] Drinklage J and Frerichs R 1963 *J. Appl. Phys.* **34** 2633
[9] Cook H E and Hilliard J E 1969 *J. Appl. Phys.* **40** 2191
[10] Rosenblum M P and Turnbull D 1980 *Appl. Phys. Lett.* **37** 184
[11] Gupta M, Gupta A, Chakravarty S, Gupta R and Gutberlet T 2006 *Phys. Rev.* B **74** 104203
[12] Gupta M, Gupta A, Stahn J and Gutberlet T 2008 *New J. Phys.* **10**, 053031
[13] Greer A L 1993 *J Magn. Magn. Mater.* **126** 89
[14] Nagy D L, Bottyán L, Deák L, Szilágyi E, Spiering H, Dekoster J and Langouche G 2000 *Hyperfine Interact.* **126** 353

[15] Deák L, Bottyán L, Major M, Nagy D L, Spiering H, Szilágyi E and Tanczikó F 2002 *Hyperfine Interact.* **144/145** 45
[16] Gupta A, Gupta M, Dasannacharya B A, Kikuta S, Yoda Y and Seto M J 2004 *Phys. Soc. Jpn.* **73** 423
[17] Major J, Vorobiev A, Rühm A, Maier R, Major M, Mezger M, Nülle M and Dosch H *Rev. Sci. Inst.* in print; www.mf.mpg.de/en/abteilungen/dosch/frmII/frmIIintro_en.shtml
[18] Gerdau E, Rüffer R, Winkler H, Tolksdorf W, Klages C P and Hannon J P 1985 *Phys. Rev. Lett.* **54** 835-838
[19] Gupta A, Gupta M, Chakravarty S, Rüffer R, Wille H and Leupold O 2005 *Phys. Rev.* B **72** 014207
[20] Rennhofer M, Sepiol B, Sladecek M, Kmiec D, Stankov S and Vogl G 2006 *Phys. Rev.* B **74** 104301
[21] Andreeva M A, Monina N G and Stankov S 2008 *Moscow University Physics Bulletin* **63** No. 2 132-136
[22] Sajti Sz, Deák L and Bottyán L FitSuite 1.0.4 , arXiv:0907.2805v1 [cond-mat.other]; www.fs.kfki.hu
[23] Matteson S, Paine B M, Grimaldi M G, Mezey G and Nicolet M A 1981 *Nucl. Instr. Meth.* **182/183**, 43
[24] Merkel D G, Bottyán L, Tanczikó F, Sajti Sz, Major M, Németh A, Horváth Z E, Waizinger J and S. Stankov 2008 *J. Appl. Phys.* **104**, 013901
[25] Revathi S and Balakrishnan V 1993 *J. Phys.* A: *Math. Gen.* **26** 5661

Polarized Neutrons and Synchrotron X-rays for Magnetism Conference 2009 IOP Publishing
Journal of Physics: Conference Series **211** (2010) 012030 doi:10.1088/1742-6596/211/1/012030

Description of the new I1011 beamline for magnetic measurements using synchrotron radiation at MAX-lab

I A Kowalik[1], G Öhrwall[1], B N Jensen[1], R Sankari[1], E Wallén[1], U Johansson[1], O Karis[2] and D Arvanitis[2]

[1] MAX-lab, Lund University, Box 118, 22100 Lund, Sweden

[2] Department of Physics, Uppsala University, Box 530, 75121 Uppsala, Sweden

E-mail: iwona.kowalik@maxlab.lu.se

Abstract. We report on the characterization of the new I1011 beamline at the MAX-II storage ring, in the MAX-lab synchrotron radiation laboratory and give examples of first results. This beamline is using an Elliptically Polarizing Undulator source, producing soft x-rays of a variable polarization state. It delivers high flux and high brightness circularly polarized x-rays in the energy range 0.2 to 1.7 keV, covering the L-edges of the late 3d elements. The new beamline will operate with an octupole magnet endstation. It is specially engineered to solve the problem of the limited optical access typically associated with magnetic fields and synchrotron radiation endstations. Eight water-cooled magnets allow the application of the magnetic field of up to 1 T in any direction. X-ray absorption spectroscopy, X-ray resonant reflectivity and the corresponding magnetic variants, i.e., XMCD, XMLD and XRMS experiments are possible also under an applied magnetic field. The high flux allows working with dilute magnetic systems such as ultra-thin films and nano structures.

1. Introduction

The I1011 beamline at MAX-lab, located at the MAX II ring [1], uses an Elliptically Polarizing Undulator (EPU) light source, which produces soft x-rays of a variable polarization state. This beamline delivers, in particular, high flux and high brightness x-ray light covering the L-edges of the 3d elements and is therefore useful for elementally resolved magnetic investigations of technologically relevant samples. Also, the M edges of the rare earth elements can be reached in the spectral region between 1 and 2 keV. The EPU source can be used to produce linearly polarized light in the vertical and horizontal planes, as well as left and right handed circularly polarized x-rays. Comparing the spectral areas of core level spectra while varying the x-ray light polarization allows in particular for X-ray Magnetic Linear Dichroism (XMLD) and X-ray Magnetic Circular Dichroism (XMCD) spectroscopy. Both XMCD and XMLD make possible the element specific characterization of the magnetic properties of both ferromagnetic and anti-ferromagnetic materials.

An important issue for magnetic measurements by means of the XMCD technique using synchrotron radiation is, among other things, the full control of the degree of polarisation over the whole energy range covering the absorption edges of all relevant elements, in combination with good energy resolution and sufficient photon flux. For the endstation a variable direction of the applied magnetic field vector is very useful, allowing for measurements in both the easy as well as the hard direction, without a sample rotation or any other type of sample motion. These aspects concerning new I1011 beamline will be described below.

Journal of Physics: Conference Series **211** (2010) 012030 doi:10.1088/1742-6596/211/1/012030

2. A linearly and circularly polarised x-ray light source: the EPU

To reveal the spin-dependence of the x-ray absorption process, best is to use an x-ray light source capable of delivering variable polarisation. In particular, the XMCD technique requires a well defined degree of circular polarisation. Advanced Design Consulting USA Inc. (ADC) together with the MAX-lab machine group staff has optimized the design and manufactured the variable polarization EPU x-ray source, with four moveable sub-girders, for the MAX II (1.5 GeV) storage ring. The total length of the EPU is 2.14 m, with a period length of 46.6 mm. It produces soft x-rays with sufficient flux and brilliance over the important energy range, 100 to 2000 eV, with variable polarization: linear (vertical and horizontal) to circular, depending on the phase of the undulator magnet sub-girders. A user-friendly interface provided by ADC allows for control of the undulator gap and the EPU phase with almost no impact on orbit of the electron beam and the stored beam life-time. The high photon flux makes possible to measure even dilute magnetic samples and nanostructures. As seen in Fig. 1, the third harmonic of the undulator radiation covers the full range of the L-edges of the 3d transition metal series, allowing for element specific investigations of most technologically important materials. This situation can still be improved by increasing the energy of the stored electron beam, to 3 GeV, such as in future the MAX IV storage ring, where one will be able to reach the L-edge 3d transition metal series within the 1st undulator harmonic.

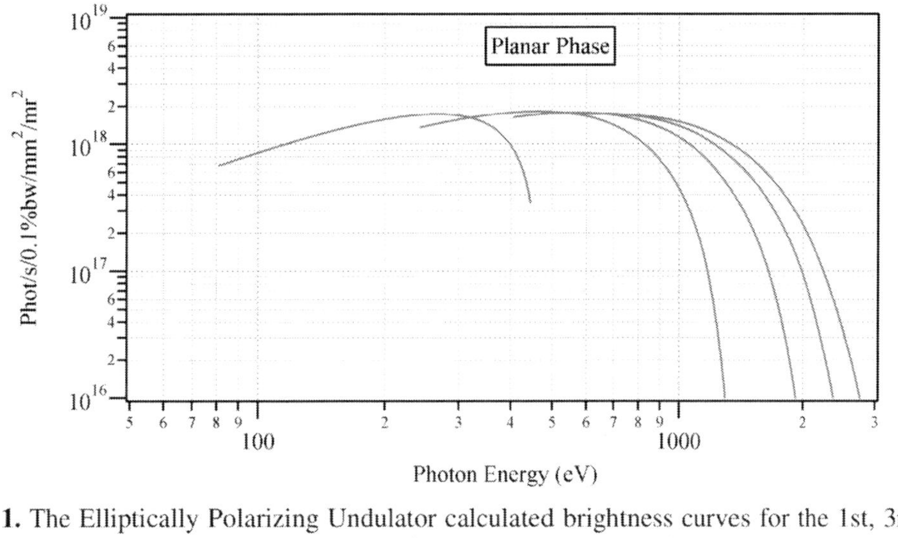

Figure 1. The Elliptically Polarizing Undulator calculated brightness curves for the 1st, 3rd, 5th, 7th and 9th harmonics for linearly polarized light in the horizontal plane light.

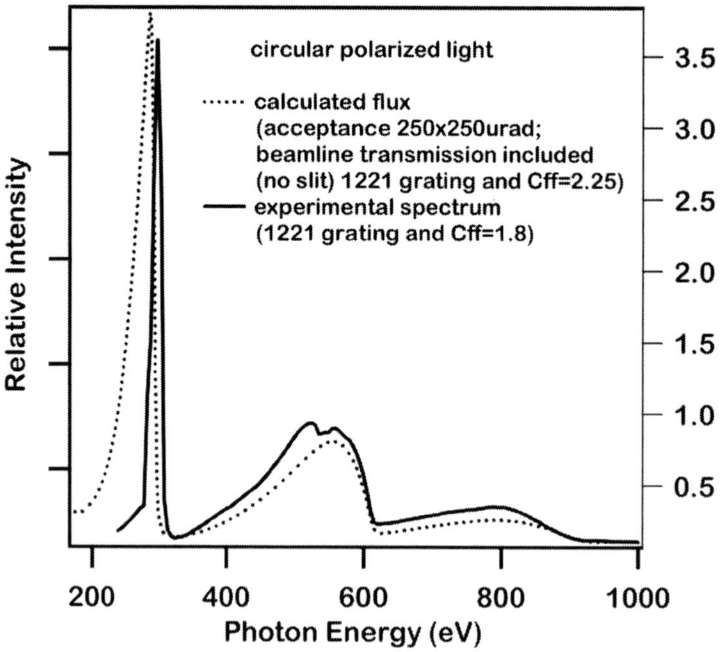

Figure 2. The I_0 photocurrent measured from Au mesh on grating 1221 l/mm compared with theoretical predictions for circular polarized light. 1st and 2nd harmonic clearly visible. The minimum in the measured spectrum around 530 eV, is due to oxygen contamination of the optical elements of the beamline. The carbon mimima can also be observed, here not so clearly, as they are located in the low energy side of the first harmonic peak. These features can be normalized out in the measured spectra using the I_0 flux reference monitor of the beamline.

3. The beamline: a cPGM for optimal photon flux and energy resolution

The beamline is based on a collimated Plane Grating Monochromator (cPGM) which covers under high photon flux and good energy resolution, the energy range of 0.1 -1.7 keV. The cPGM and several other beamline components were provided by the BESTEC company (Berlin, Germany). The cPGM contains three blazed gratings, which can be exchanged in vacuum: 336, 1221, or 1400 l/mm. A plane grating solution is used for the x-ray monochromatization, which maintains the polarization state over the whole energy range covered by the monochromator. The possibility of selecting the so-called c_{ff}-parameter allows balancing between energy resolution, photon flux or higher order suppression [2,3,4]. The energy resolution was characterized using a gas cell, by measuring the photo ion yield. Various gases were used to cover the energy range of about 0.2 to 1 keV. Here we show the case of Ne gas using the 1221 l/mm grating. The energy range probed using the K-edge of Ne is close to the Co L-edge and characteristic of the performance of the I1011 BL for the late 3d series ferromagnets. The cPGM is designed to deliver a resolution of 10 000 over the whole energy range of operation but even with rather relaxed conditions (c_{ff} = 4.5, slit 30 μm), the resolving power which we reached is more than 9000 for N_2 and 6000 for Ne gas, while the flux at the sample is about 10^{11} photons/sec. The flux, within an order of magnitude, is estimated using the photocurrent from a fine electroformed Au mesh in the beam path. The quantum efficiency of gold is obtained from Ref. [5]. The transmission of the gold mesh, is reported by the manufacturer. The geometry of the mesh was checked using an optical miscroscope. This mesh is also used as an I_0 reference monitor. The confirmed high energy resolution and flux means that the alignment of the optical elements is satisfactory. The resolving power and flux at I1011 compares favourably to the other beamlines at MAX-lab [6]. To maximize energy resolution the beam is focused onto the exit slit. The spot size of the diverging beam, after the exit slit, into the experimental station at the sample position is about 1 mm both in the vertical and horizontal directions.

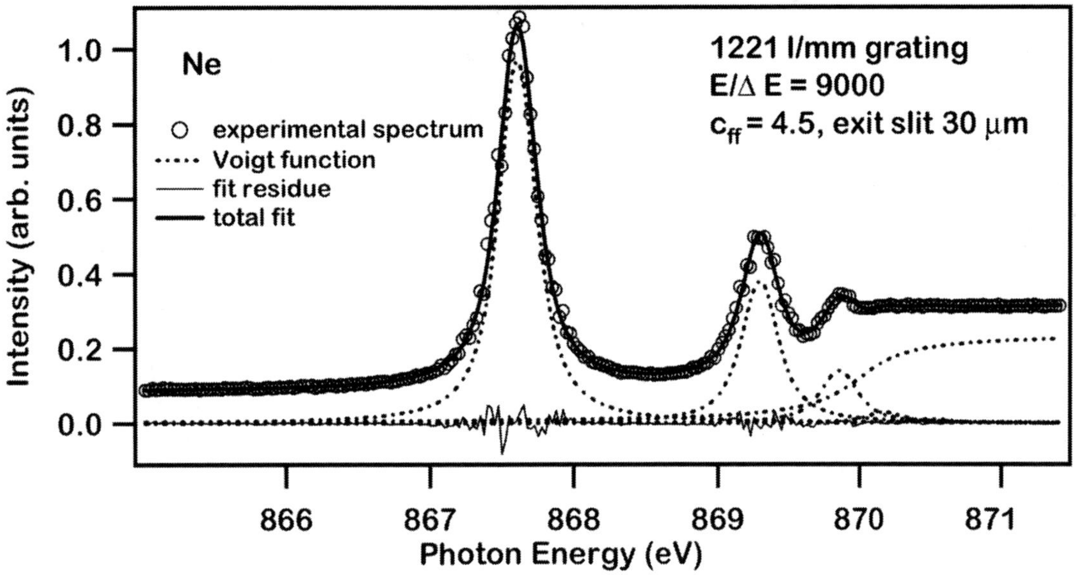

Figure 3. Photo ion yield spectrum recorded at Ne 1s ionization threshold. The spectrum is dominated by the 1s to 3p absorption peak, between 867 and 868 eV. A Voigt profile analysis is shown; the instrumental width corresponds to the resolving power of 6000.

4. Element specific magnetic measurements using an "octupole" magnet endstation

We present the first results obtained with the endstation we designate as the "octupole" magnet endstation. The "octupole" endstation has been specially engineered to solve the problem of the limited optical access typically arising from magnetic measurements using synchrotron radiation [7]. An octupole Ultra High Vacuum chamber contains eight water-cooled electromagnets, spaced equidistantly over the surface of a sphere, which allows the application of the field in any direction [8,9]. This solution together with the possibility to rotate the whole chamber allows measuring the magnetic properties with a variable geometry of applied magnetic field, electromagnetic radiation and sample orientation. These high current magnets generate a field in the sample space of up to 1T. Such possibilities allow to optimize the measuring conditions for various magnetic systems studied by the XMCD method [10]. XMCD relies on spin selective photo excitation of the $2p_{3/2}$, $2p_{1/2}$ electrons to the empty 3d band, in case of 3d transition metals. Circularly polarised light transfers its angular momentum ħ for right helicity (σ+) or -ħ for left helicity (σ-) to a spin-orbit-split core electron, from the $2p_{3/2}$ or $2p_{1/2}$ level, allowing for spin selectivity in the XAS process. By applying the magneto optical sum rules the orbital (m_l) and spin (m_s) moments can be determined directly from the experimental XMCD signal [11]. We assume that the amount of depolarized x-ray light is negligible. The experiments can be carried out both in remanence and under magnetic field, in particular with a constant applied magnetic field (dc mode). Another method available at the "octupole" endstation is x-ray resonant magnetic scattering (XRMS) known also as resonant x-ray reflectivity. A photodiode detector travels along a +/- 90 deg arc perpendicular to the beam axis. This motion, coupled with the entire system rotation of +/- 90 deg and incoming light polarization allows for the diode to be placed anywhere on a hemisphere perpendicular to the beam for any light polarization. The XMCD or XRMS experiments can be combined with element specific magnetic hysteresis loops taken in total electron yield mode or using the photodiode. This allows for a full magnetic characterization of the specimen (measurements of saturation moment, coercive force, magnetic anisotropy) in an elementally resolved manner. A cryogenically cooled rod allows the XMCD and XRMS signals to be measured as a function of temperature - down to 20 K.

We now turn to measurements of magnetic standards and the use of the XMCD technique as a means to perform a characterization of the degree of circular polarization of the EPU x-rays. The x-ray absorption spectroscopy can be measured in different ways, using various secondary channels. Of importance is to stay in such conditions where the signal chosen is proportional to the absorption coefficient. Here we use the photocurrent of the sample and work in the Total Electron Yield (TEY) mode, avoiding grazing incidence angles to avoid saturation effects in the TEY mode. We measure the difference in absorption of circularly polarized x-rays between parallel and anti-parallel orientation of the photon spin in the TEY mode. After proper normalization the XMCD signal is obtained. We give as an example data obtained for a thick bcc Fe film. The Fe film is protected from oxidation by means of a protective cap layer. We cover the photon energies around the Fe L-edges with the 1221 l/mm grating. The data are shown in Fig. 4. Here elliptical x-rays are used using an energy range close to the 3rd EPU harmonic maximum. The angle of x-ray incidence is 45 degrees to avoid saturation effects in the secondary channel. The data are taken in the remanant state after a 0.2T magnetic field pulse was applied along the magnetic easy direction. Clear XMCD contrast is seen.

The XMCD in the XAS mode is taken by measuring the sample photocurrent with the sample in the remanent state. The data can be analyzed using the magneto optic sum rules, which allows in particular to determine the orbital and spin magnetic moments, on a per an atom basis. Here we focus on the spin moment by assuming the known number of holes for bulk Fe (3.8) and neglecting the dipole term, which is small for the Fe bulk. Knowing the experimental geometry and assuming that the remanent magnetization is to a good approximation equal to the full saturation magnetization, the degree of circular polarization can be deduced: we obtain a value of about 0.75(10). The error of order 10-15% is due in part to the imperfect knowledge of the background function introducing an uncertainty in the determination of the spin moment by means of the XMCD sum rules. This is close to the predicted value for the 3rd EPU harmonic for the chosen beamline acceptance.

5. Summary

Beamline I1011 at MAX-lab is dedicated for magnetism related investigations. With an EPU as the source of linear and circular polarized x-ray light this beamline delivers high flux and high brightness light covering the L-edges of the late 3d elements and M edges of the rare earths. The so-called "octupole" endstation has been specially engineered to solve the problem of the limited optical access typically associated with magnetic fields and synchrotron radiation, in a UHV environment, allowing to fully exploit the vector character of the XMCD technique. The electron yield measurement with a diode detector makes at present possible the element specific characterization of the magnetic properties of various types of dilute magnetic samples using XMLD, XMCD and X- XRMS.

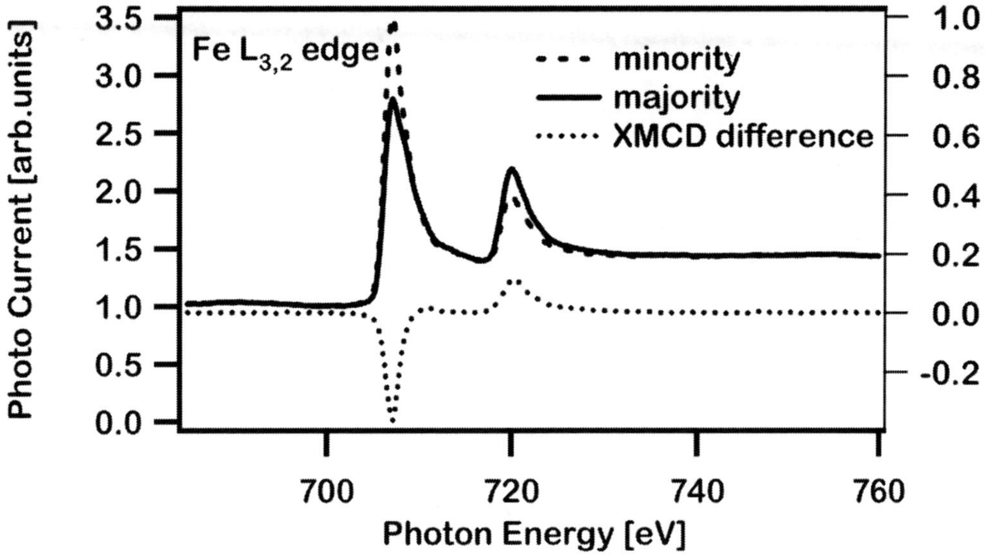

Figure 4. The photocurrent of a thick bcc Fe film is shown (left scale), using the 1221 l/mm cPGM grating. Here elliptical x-rays are used, see text. The XMCD difference is clearly seen (right axis). The angle of x-ray incidence is 45 degrees. The data are taken in the remanent state.

References

[1] Hunter Dunn J, Hahlin A, Karis O, Arvanitis D, LeBlanc G, Andersson Å and Lindgren L-J 2004 *AIP Conf. Proc.* **705** 65

[2] Follath R 2001 *Nucl. Instrum. Meth.* A **467-468** 418

[3] Follath R and Senf F 1997 *Nucl. Instrum. Meth.* A **390** 388

[4] Kaznatcheev K V, Karunakaran Ch, Lanke U D, Urquhart S G, Obst M and Hitchcock A P 2007 *Nucl. Instrum. Meth.* A **582** 96

[5] Krumrey M and Tegeler E 1992 *Rev. Sci. Instr.* **63** 797

[6] Bässler M, Ausmees A, Jurvansuu M, Feifel R, Forsell J O, de Tarso Fonseca P, Kivimäki A, Sundin S, Sorensen S L, Nyholm R, Björneholm O, Aksela S and Svensson S 2001 *Nucl. Instrum. Meth.* A **469** 382

[7] Kulesza J, Johnson E, Lyndaker A, Deyhim A, Jayne R and Hunter Dunn J 2007 *AIP Conf. Proc.* **879** 531

[8] http://www.adc9001.com/index.php?src=01-softxray-scattering

[9] Arenholz E and Prestemon S O 2004 *AIP Conf. Proc.* **705** 1170

[10] Chen C T, Sette F, Ma Y and Modesti S 1990 *Phys. Rev.* B **42** 7262

[11] Carra P, Thole B T, Altarelli M and Wang X 1993 *Phys. Rev. Lett.* **70** 694

Polarized Neutrons and Synchrotron X-rays for Magnetism Conference 2009 IOP Publishing
Journal of Physics: Conference Series **211** (2010) 012031 doi:10.1088/1742-6596/211/1/012031

Polarized neutron scattering on the triple-axis spectrometer PANDA: first results

Enrico Faulhaber[1,2], Astrid Schneidewind[1,2], Fei Tang[3], Peter Link[4], Dirk Etzdorf[4] and Michael Loewenhaupt[1,3]

[1]Gemeinsame Forschergruppe Helmholtz-Zentrum Berlin - Technische Universität Dresden
[2]Helmholtz-Zentrum Berlin für Materialien und Energie, D-14109 Berlin, Germany
[3]Technische Universität Dresden, Institut für Festkörperphysik, D-01062 Dresden, Germany
[4]Forschungsneutronenquelle Heinz-Maier-Leibnitz (FRM II), D-85747 Garching, Germany

E-mail: enrico.faulhaber@frm2.tum.de

Abstract. PANDA is a triple-axis spectrometer utilizing the cold neutron source of the German research reactor FRM II in Garching. Its typical applications comprise studies of spin dynamics and lattice dynamics, magnetic excitations and magnetic structures. In the past years, PANDA has earned its reputation by excellent performance in these fields operating in conventional, unpolarized mode. To complement this normal mode of operation, a full polarization analysis of the scattered neutrons was highly desired and, therefore, was recently implemented. First tests to characterize the different spectrometer components for the polarization analysis mode were performed with neutron spin directions perpendicular to the scattering plane, utilizing static vertical guide fields between sample, Heusler monochromator, spin flipper and Heusler analyzer. Finally, a fully automated setup of split coils around the sample space was implemented to allow for longitudinal polarization analysis for arbitrary configurations. Different neutron spin polarization directions were successfully investigated, showing polarizations of 92 % or higher.

1. Introduction

Because of its high flux, unique large dynamic range and optional flexibility PANDA acquired a position among the leading cold TAS instruments. Typical applications comprise studies of magnetic excitations, spin and lattice dynamics as well as the investigation of magnetic structures. Up to now, PG monochromator and analyzer were used for Beryllium filter wavelengths $k_f \leq 1.55\,\text{Å}^{-1}$ or for PG filter wavelengths at $k_f = 2.557\,\text{Å}^{-1}$ or $k_f = 2.662\,\text{Å}^{-1}$. The best energy resolution observed in an user experiment is $\Delta E = 37\,\mu\text{eV}$ at $k_f < 1.078\,\text{Å}^{-1}$. The momentum transfer can vary from 0.1 up to $6\,\text{Å}^{-1}$, with energy transfers up to 23 meV.

The sample environment available on PANDA allows temperatures from from 50 mK up to 2100 K and for temperatures below 100 K magnetic fields fields up to 14.5 T. The combination of very low temperatures, high magnetic fields and cold neutrons matches perfectly the energy scales in typical magnetic samples.

To complement this normal, unpolarized mode of operation of PANDA, the instrument was equipped with Heusler type monochromator and analyzer. As the polarized operation was planned from the beginning, some guide-field elements were already included during the construction of the instrument.

2. Mode of Operation

The neutrons leave the Heusler type monochromator (variable vertical, fixed horizontal focusing) with a horizontal polarization. A vertical guide field then turns the spins adiabatically to a vertical polarization. A spin flipper (sf1) can be activated to flip the spin orientation around. A vertical magnetic field at the sample position (basic setup) or a magnetic vector field around the sample (3-directional setup) might either keep the vertical direction of the spins ($P \parallel z$) or turn them into the scattering plane parallel ($P \parallel Q$) or perpendicular ($P \perp Q$) to the scattering vector Q. The polarization of the scattered neutrons is then turned vertical again. A second spin flipper (sf2) can invert the polarization direction. The Heusler analyzer (fixed vertical and variable horizontal focusing) finally accepts only neutrons with a vertical polarization pointing up and allows the energy selection required for operation as triple axis instrument. All field gradients of the guide fields along the neutron path are low enough to allow for adiabatic changes of spatial directions of the neutron polarization.

3. Basic setup

As a first step an instrumental setup with a vertical guide field around the sample position was tested and found operational [1]. The elements providing the static magnetic guide field were arranged close to the sample, avoiding a depolarization of the beam near the sample. This setup exclusively provides a vertical guide field and therefore supports only vertical polarizations. Utilizing some attenuation to safeguard the detector, only minor depolarization of the primary beam was found, as illustrated in figure 1 showing the observed neutron counts versus the flipping current at optimal field compensation. Using a single crystal of Lead, a flipping ratio of up to 40 was found in a rocking scan across the (2 0 0) Bragg reflection, as shown in figure 2(a). An energy scan across the elastic incoherent line of Vanadium shows the expected intensity ratio of 1:2 for non spin flip (nsf) vs. spin flip (sf) scattering (see figure 2(b)). Compared to the completely unpolarized setup with PG monochromator/analyser, the polarized mode provides a factor of 20 less neutrons in the detector.

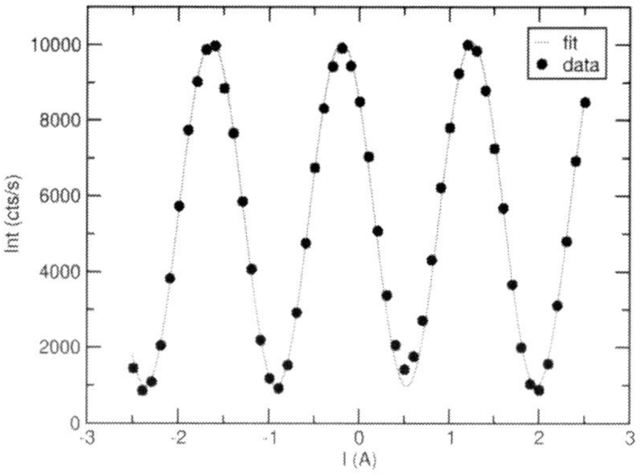

Figure 1. Direct beam intensity, varying with the spin flippers flip field current at optimized compensation current, showing only minor depolarization.

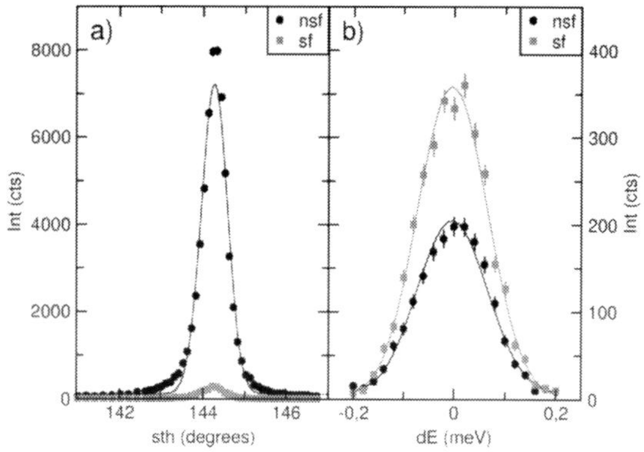

Figure 2. Scans with polarized neutrons ($P \parallel z$) on the (2 0 0) Bragg reflection of Pb (a) and on the elastic line of Vanadium (b). (Red: spin flip (sf), black: non spin flip (nsf).)

To test the setup under realistic conditions, an experiment on a single crystal of Er_2PdSi_3 was performed in the basic setup at low temperatures. The hexagonal compound Er_2PdSi_3 shows a strong crystal electric field (CEF) transition from the ground state doublet to the first excited state (also a doublet) around an energy transfer of 3.5 meV. Despite good crystal quality and heating treatment, superstructure peaks are observable in the hole series of R_2PdSi_3. The corresponding superstructure is formed by an ordering of the Pd and Si atoms [2, 3, 4] which leads to distinct Er site environments. This gives raise to (slightly) different CEF schemes. Calculations indicate the occurrence of two distinct CEF excitation energies [2]: One resulting from a Er site with local hexagonal symmetry and one with reduced local symmetry. Since the matrix elements for inelastic neutron scattering do not vanish, both transitions should be observable. In fact, in an unpolarized inelastic neutron scattering experiment instead of one transition at around 3.5 meV, two transitions at 3.3 and at 3.6 meV, were observed. In addition, both transitions differ in their directional dependence of the CEF matrix elements and hence in their intensity [4]. To proof the magnetic origin of these two signals, we performed the first polarized experiment on PANDA.

Since Er_2PdSi_3 displays an antiferromagnetic order below $T_N \approx 7\,K$ [5], the measurements were done at 10 K to avoid the magnetic order modifying the CEF scheme. The sample was mounted in a (H K 0) scattering plane with a vertical polarization of the incoming neutrons. After careful alignment of both the instrument and the sample, the overall polarization was determined on the (0 1 0) Bragg reflection. This is illustrated in figure 3, where the intensity of the Bragg peak versus the flipping current of the spin flipper at optimal compensation current is displayed. From the raw data one finds a flipping ratio of $R = 35$, a sine fit to the data points results in a more realistic value of $R = 31$ yielding an overall polarization of $P \geq 93.75\,\%$.

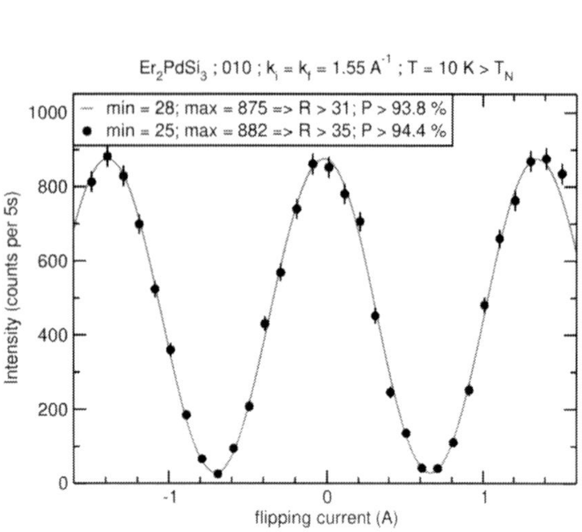

Figure 3. Intensity of the (0 1 0) Bragg reflection of Er_2PdSi_3 versus flipping current at optimum compensation field. (raw data: black, sine fit: red)

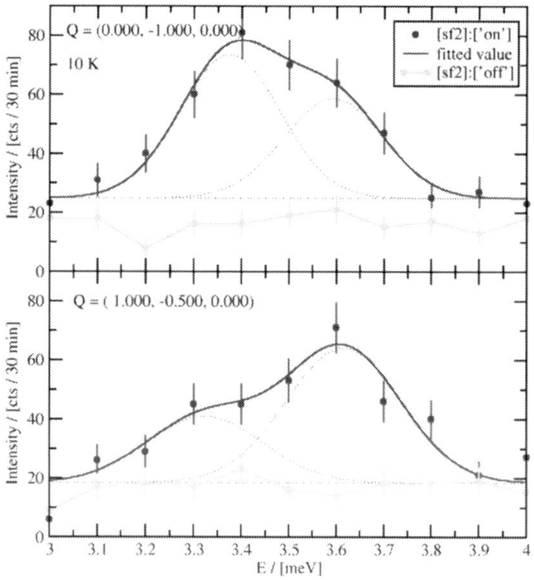

Figure 4. Inelastic signal in Er_2PdSi_3 at two perpendicular Q positions $(0\,\bar{1}\,0)$ and $(1\,\overline{0.5}\,0)$. (sf channel: blue, nsf channel: green) Blue lines indicate two-peak Gaussian fits to the data.

After this encouraging result, inelastic scans were measured at two perpendicular Q-positions $(0\,\bar{1}\,0)$ and $(1\,\overline{0.5}\,0)$, having a comparable $|Q|$ value. Figure 4 shows the obtained neutron counts for sf (blue curve, sf2:on) and nsf channels (green curve, sf2:off), respectively. At both Q positions the two inelastic CEF excitations were found in the sf channel, whereas (within the error bars) no signal was found in the nsf channels. The latter finding indicates a negligible J_z matrix element for both CEF transitions. Furthermore, the intensity of the low-energy transition at

3.3 meV shows a strong directional dependence while the behavior of the high-energy transition at 3.6 meV is isotropic in the basal plane (see upper and lower panel in figure 4). These findings correlate well with the unpolarized results [4]. From these facts we can conclude that the isotropic (i.e. 3.6 meV) excitation belongs to Er-sites with higher symmetry, whereas the Er-sites with lower symmetry show a slightly smaller excitation energy. A site occurrence of 1:1 is calculated from the averaged intensities, fitting very well to recent calculations [2].

4. 3-directional Polarization Analysis

To give access to additional polarization channels, a coil system around the sample position was implemented[1]. It allows to use any field direction for the neutron guide field at the sample position. In the experiment, the field is applied by these coils along the $P \parallel Q$ and $P \perp Q$ directions with Q within the scattering plane. Figure 5 shows a view of the setup as mounted on the sample table of PANDA. The construction allows to use most of the sample environment, except the 14.5 T cryomagnet. Due to the split design, no beam blockage occurs. Nevertheless, this design can be improved further: (i) At some point between the upper and lower coils the dipole field vanishes leading to a strong beam depolarization. The problem is fixed by an additional static vertical field next to the coils which shifts this point out of the scattering plane. (ii) The outside stray field of the vector coils influences the compensation current of the spin flippers. Therefore, the flippers need to have a certain distance from the coils.

[1] The original design of the coil system was made by Peter Böni. It was manufactured by Swiss Neutronics, Switzerland.

Figure 5. View of the coil setup as mounted on PANDA.

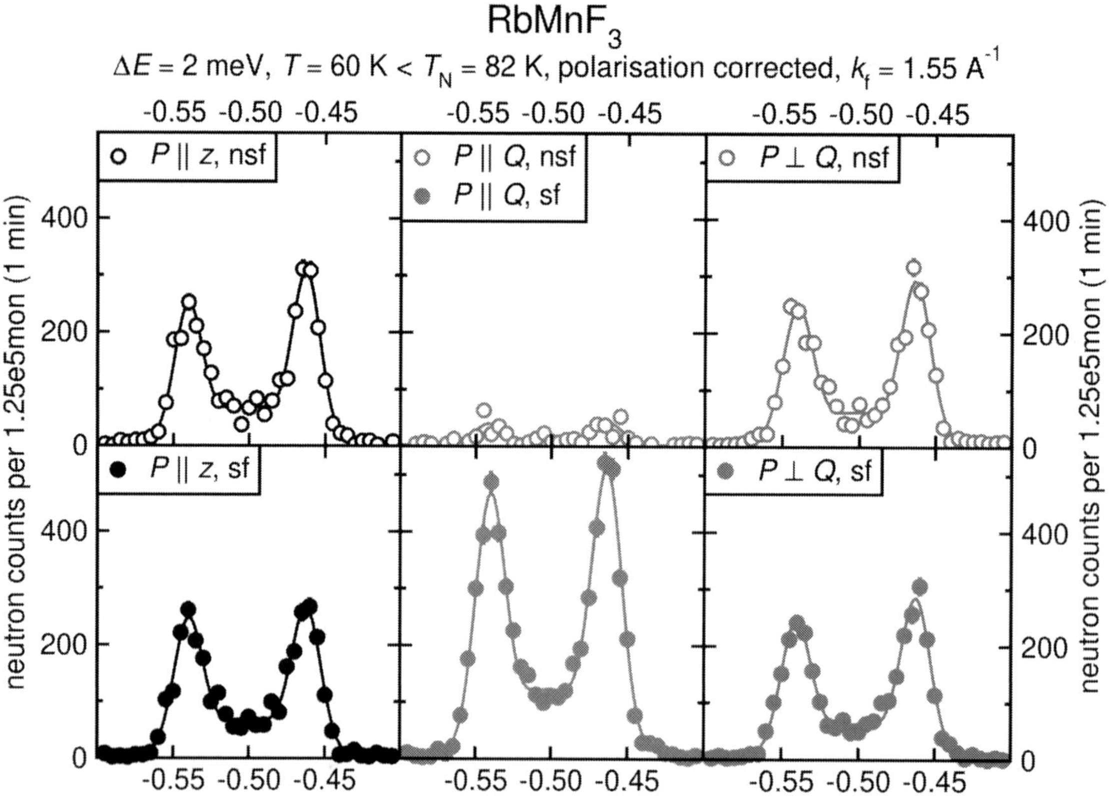

Figure 6. $RbMnF_3$: Q-scans at an energy transfer of $\Delta E = 2\,\text{meV}$ for all accessible scattering channels. The antiferromagnetic Bragg reflection is observed at (-0.5 -0.5 -0.5). Intensities were corrected to account for reduced polarization (see text).

To test the device and the controlling software, we performed an inelastic neutron scattering experiment on the magnons of a single crystal of the magnetic salt $RbMnF_3$ [6]. Although the instrument was aligned very well, we found a flipping ratio of only[2] $R = 3.5$ corresponding to a polarization of $P = (R-1)/(R+1) = 55.6\,\%$. We found that magnetic parts of the closed-cycle refrigerator add an additional z-component to the guide field at the sample position, thus effectively depolarizing the beam. In this setup we performed longitudinal Q-scans at selected energies between 1 and 5 meV across the antiferromagnetic Bragg peak at $(-0.5\ -0.5\ -0.5)$ at $T = 60\,\text{K} < T_\text{N} = 83\,\text{K}$. Figure 6 shows the polarization corrected intensities[3]for an energy transfer of 2 meV for all six accessible scattering channels. As expected, the sum of nsf and sf channels is the same for the polarization directions $P \parallel z$, $P \parallel Q$ and $P \perp Q$. The sum of all nsf channels is half of all sf channels. And, most important, there is (almost) no signal in the nsf channel for $P \parallel Q$, as expected from scattering theory (see for example [7]). The coupling and magnon stiffness constants obtained from our dataset are comparable to the results given in [6] (see formulas and text therein). The finite intensity at positions between the magnon peaks can be attributed to longitudinal moment fluctuations occurring close to T_N [8, 9]. The occurrence of domains results in an even distribution of the scattering intensity over the allowed channels.

[2] Values without this cryostat are above 25.

[3] The polarisation was determined in the elastic condition on the Bragg Peak and this value was used to correct the measured inelastic signal according to: $\left(\frac{I_{sf}}{I_{nsf}}\right)_{corr.} = \frac{1}{2P} \cdot \begin{vmatrix} P+1 & P-1 \\ P-1 & P+1 \end{vmatrix} \cdot \left(\frac{I_{sf}}{I_{nsf}}\right)_{meas.}$.

This experiment shows that despite unexpected problems a successful measurement could be performed.

5. Conclusion

The polarization option of the cold-triple axis spectrometer PANDA has been successfully commissioned. The specifically crafted vector guide field allows to access almost all possible spin related scattering channels. Due to the chosen construction, no beam blockade occurs.

Acknowledgments

We greatfully thank Thomas Keller from MPI Stuttgart for helpful discussions, references and the kind transfer of experiences.

Our special thanks go to Christian Baar for calculating the field distribution in different setups and for the calculation and realization of most of the guide field pieces. He also did a great deal of work getting the setup up and running.

References

[1] Link P 2007 *Forschungsneutronenquelle Heinz Maier-Leibnitz (FRM II), Annual Report* pp 46-47
[2] Tang F, Link P, Frontzek M, Mignot J M, Hoffmann J U, Löser W and Loewenhaupt M 2009 *Proc. Int. Conf. on Neutron Scattering* (Knoxville) accepted for publication in Journal of Physics: Conference Series
[3] Frontzek M, Kreyssig A, Doerr M, Hoffmann J U, Hohlwein D, Bitterlich H, Behr G and Loewenhaupt M 2004 *Physica* B **350** E187
[4] Tang F, Link P, Frontzek M, Schneidewind A, Löser W and Loewenhaupt M 2009 *Proc. Int. Conf. on Neutron Scattering* (Knoxville) accepted for publication in Journal of Physics: Conference Series
[5] Frontzek M, Tang F, Link P, Schneidewind A, Mignot J M, Hoffmann J U and Loewenhaupt M 2009 *Proc. Int. Conf. on Neutron Scattering* (Knoxville) accepted for publication in Journal of Physics: Conference Series
[6] Windsor C G and Stevenson R W H 1966 *Proc. Phys. Soc.* **87** 501-4
[7] Shirane G, Shapiro S M and Tranquada J M 2002 *Neutron Scattering with a Triple-Axis Spectrometer* (Cambridge University press)
[8] Cox U J, Cowley R A, Bates S and Cussen L D 1989 *J. Phys.: Condens. Matter* **1** 3031-5
[9] Coldea R, Cowley R A, Perring T G, McMorrow D F and Roessli B 1998 *Phys. Rev.* B **57** 5281-90

Beam transport and polarization at TOPAS, the thermal time-of-flight spectrometer with polarization analysis

J Voigt[1,2], E Babcock[1] and Th Brückel[2]

[1] IFF-JCNS-FRM II, Forschungszentrum Jülich, 85747 Garching, Germany
[2] IFF-4, Forschungszentrum Jülich, 52425 Jülich, Germany

E-mail: `j.voigt@fz-juelich.de`

Abstract. We present the design for the polarization analysis of the future thermal time-of-flight spectrometer at the Juelich Centre for Neutron Science (JCNS) at the FRM II. TOPAS is a time-of-flight spectrometer covering a range of incident energies 20 meV< E_i <160 meV and an angular range $-3° < 2\theta < 150°$. A set of Fermi choppers selects the incoming energy E_i with a resolution up to 3 %. The instrument is optimized for a high flux on small samples using an elliptical neutron guide. The special feature of TOPAS is the polarization analysis. The incident polarization will be realized by means of a ^{3}He continuously pumped polarizer, which is a downscaled version of the device developed for small angle applications at JCNS. The polarization analysis over a wide angular range demands either short distances between the sample and the analyzer or a large volume of polarized ^{3}He. Here we propose the latter alternative to allow the study of magnetic samples and modest magnetic fields at the sample position.

1. Introduction

Time-of-flight spectrometers have traditionally been used to study incoherent and/or dispersionless scattering processes, such as vibrational density of states (VDOS), crystal field excitations or molecular dynamics. However, with the advent of Megawatt spallation sources and large pixelized detectors, they provide an excellent tool to map coherent excitation in novel single crystalline materials with a modest momentum resolution. TOPAS will be a reactor-based TOF spectrometer that runs at high repetition rates to compensate for the lower pulsed flux at continuous sources compared to the high peak flux of spallation instruments. Particularly for thermal neutrons with comparatively large velocities, the time frames can be rather short. Accordingly all distances have been chosen: The sample to detector distance will be to 2.5m. A system of two Fermi Choppers FC1 and FC2, placed at a distance of 5 m, will select the incoming energies. Unwanted energies will be removed by a higher order removal chopper. The chopper FC2 will spin with up to 36000 revolutions per minute (rpm). Due to a larger neutron window, the chopper FC1 will be limited to a speed of 27000 rpm which determines the maximum repetition rate $\nu_{max} = 900$ Hz. Frame overlap that might be significant particularly for low incoming energies can be avoided running FC1 at half the frequency of FC2. The rather short distances between the monochromatizing choppers and the detector bank yield an energy resolution that can be tuned from 3 % to 10%. The distance between the sample position and

Journal of Physics: Conference Series **211** (2010) 012032 doi:10.1088/1742-6596/211/1/012032

the biological shielding of the FRM II and hence the neutron guide length will be aproximately 49 m. An elliptical neutron guide will focus neutrons into a small beam spot, giving a high flux over an area 1x4 cm^2.

So far polarization analysis has been applied in inelastic experiments nearly exclusively on triple axis spectrometers. Thinking of coherent magnetic excitations in particular in correlated electron systems, polarization analysis comes also in the focus of ToF spectrometers. The application requires once more a high flux, as the polarization devices reduce the flux at least by a factor 2.

2. Neutron guide design

The neutron guide design is guided by three requirements

(i) Focus a high intensity onto a small sample spot.
(ii) Generate a uniform divergence distribution.
(iii) Prevent the direct sight onto the reactor core.

Following the proposal of Schanzer *et al.* [1], we have investigated different guide geometries. It turns out, that an elliptic guide becomes a better choice than parabolic or linearly tapered guides, when the length of the neutron guide becomes larger than 40 m. This can be understood considering that in geometric optics the ellipse images the source into the image by only one reflection. However, in the case of a neutron guide, one does not image a point source. Therefore the elliptical guide shape is not obvious. Numerical optimization of different guide shapes (including a free variation of guide segments optimizing the flux) yielded the highest flux for the 'free' guide. However, the elliptical shape with virtual foci that are slightly different from the flux maximum and the sample position, gives nearly the same flux and a more homogeneous divergence distribution. In particular, closing the entrance and exit windows of the neutron guide the improves the divergence distribution. The shape of the resulting guide is given in Fig. 1 showing the horizontal and vertical direction, respectively. The entrance and exit window cover only $\frac{1}{9}$ of the area in the center of the guide. Hence the direct sight onto the reactor core can be prevented easily by placing a beamstop there as indicated in Fig. 1. Fig. 2 a) shows, that the flux at the sample position is reduced by 5% only with the addition of the beamstop. The design of such a beamstop is guided by the requirements to moderate fast neutrons, to absorb the neutrons and to absorb the gamma radiation. Finally we discuss the divergence distribution at the sample position. Fig. 2 b) shows the homogeneous distribution. It has a single maximum and a trapezoidal shape. The width of 1.0 to 1.5° (FWHM) depending on the wavelength matches the contributions from the sample size and detector pixel size to the angular resolution. It provides also short opening and closing times of the wavelength selecting Fermi choppers.

3. Polarizer

To polarize the thermal neutrons we have chosen polarized ^{3}He neutron spin filter (NSF) cells. The use of 140 meV neutrons ($\lambda = 0.76$ Å) along with other factors including beam divergence demands the utilization of NSF over supermirror polarizers or Heussler crystals especially given the current atainable levels of ^{3}He polarization of 70% or more. A discussion of the factors involved in such a comparison are given in ref [2].

The polarizer concept for TOPAS is based on the polarization analyzers for the small angle applications at instruments KWS1 and MARIA from JCNS. The demands for the different applications share major components and designs. The analyzer applications on MARIA and KWS1 will require very large cells optimized for comparatively low neutron energy from the cold moderator. For TOPAS the beam dimension at the end of the elliptical neutron guide requires an 8 cm diameter cell, whereas the analyzer systems are being designed to accept cells of 14-16

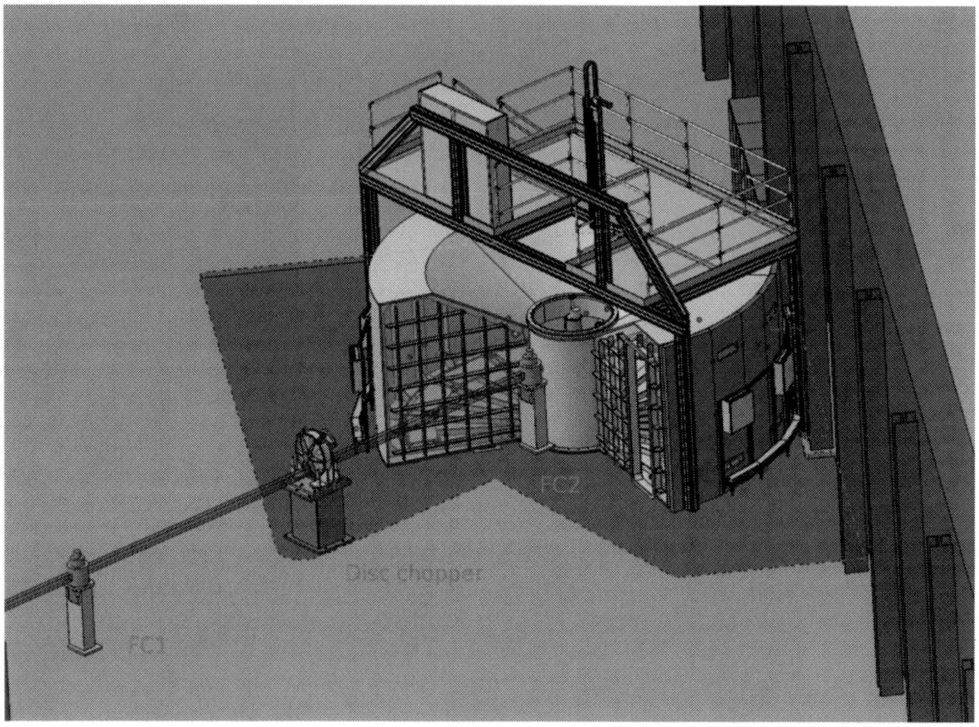

Figure 1. 3d model of the primary and secondary spectrometer. The neutron energy is selected by the phase between the Fermichoppers FC1 and FC2. The disc chopper suppresses neutron velocities that could also pass through the Fermi choppers.

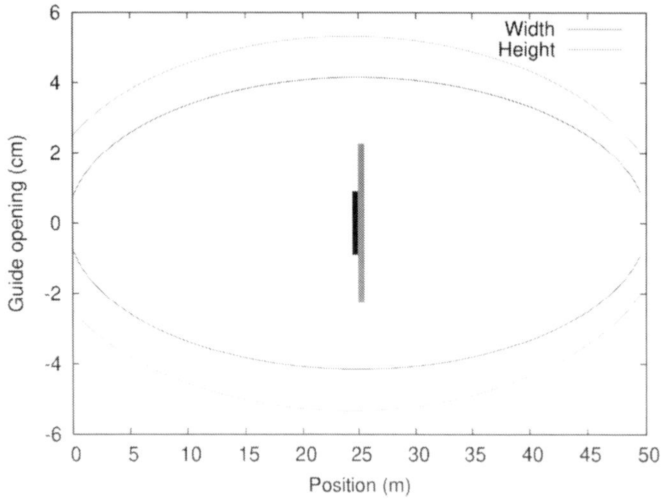

Figure 2. Horizontal and vertical guide shape. At the centre of the ellipse (Position = 25 m) an 1.8x4.5 cm^2 beamstop will block the direct sight.

cm in total height. Therefore TOPAS will use a downsized version of the continuously optically pumped analyzer being prototyped and tested for those applications. It will have a very similar format to the polarizer of [3, 4] with somewhat smaller dimensions. A scheme of the polarizer concept is shown in Fig. 4. However the high energy of the TOPAS beam will require the cells to contain 20-30 bar cm of ^{3}He leading to a cell of 10 to 15 cm long at 2 bar (absolute) pressure.

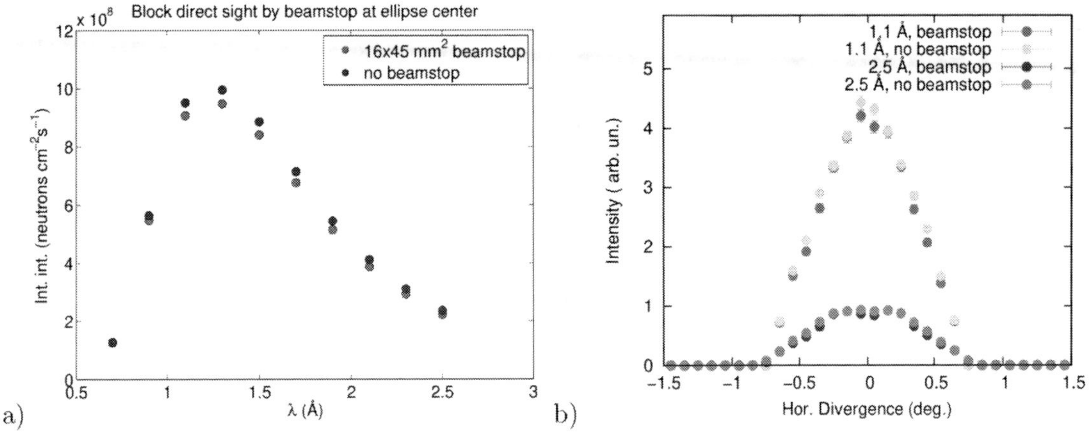

Figure 3. Comparison of flux a) and divergence b) at sample position with and without beamstop at the centre of the double elliptic neutron guide.

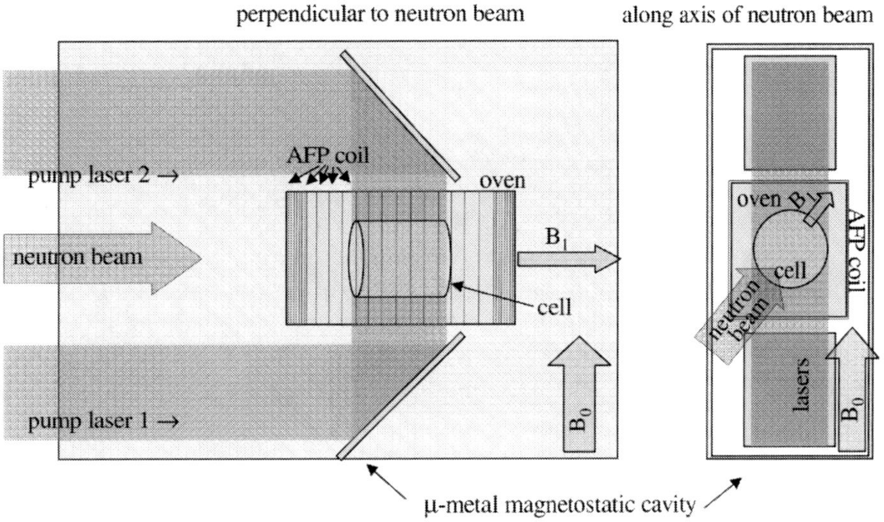

Figure 4. a)Block diagram of configuration of the SEOP system for online polarization.

We note these estimates of required cell opacity (*i.e.* ^{3}He pressure times neutron wavelength times cell length) are based on the analysis given in [5].

The cell will be polarized *in situ* because this will provide a higher time averaged figure of merit compared to ^{3}He gas polarized remotely. Off-line polarized gas undergoes T_1 nuclear spin relaxation, and thus loss of polarizing efficiency and neutron transmission over time. Consequently the cell must be replaced leading to an interruption of measurement time for the procedure and the required instrument calibration for the new cell filling. Fig. 5 shows $\Theta(t)$, the square root of the time integrated figure of merit, or P^2T for three conditions, steady ^{3}He polarization of 70%, a cell with 70% initial polarization decaying with a 100 hour polarization decay time constant, and a cell with 70% initial polarization and a 100 hour polarization decay time constant refilled or replaced each day. $\Theta(t)$ is the obtained data quality where the figure of merit can be thought of as the number of usable neutrons, thus $\Theta(t)^{-1}$ would be proportional to the measurement error assuming counting statistics. To obtain a given level of data quality for

Journal of Physics: Conference Series **211** (2010) 012032 doi:10.1088/1742-6596/211/1/012032

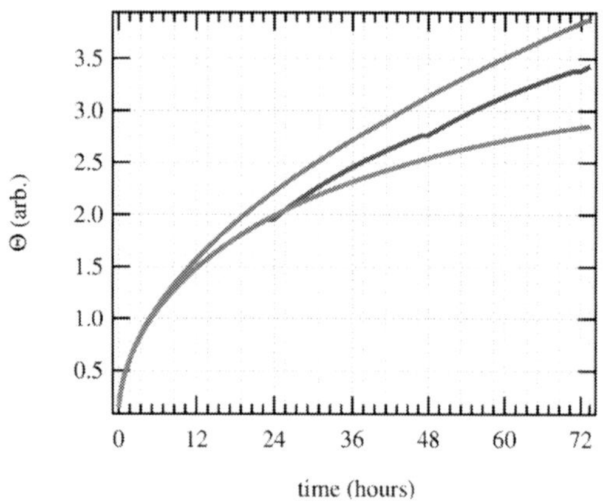

Figure 5. Comparison of three cases for use of ^{3}He NSF as a polarizer. The value plotted is the data quality $\Theta(t)$. We defined $\Theta(t)$ as the square root of the time integrated figure-of-merit or P^2T of the ^{3}He polarizer cell (P: Polarisation, T: transmission of the cell). $\Theta(t)^{-1}$ is hence proportional to the measurement error assuming counting statistics. The grey line indicates the obtained quality vs time for a cell initially polarized to 70% and allowed to decay assuming a typical 100 hour on-beam relaxation time for the NSF. The blue line shows the improvement obtained by refreshing the polarization daily accounting for a pause of one hour to change the cell and perform a calibration. The red line indicates the performance for constant ^{3}He polarization of 70% maintained by in situ polarization.

a device using an in-situ polarized NSF with equal initial values of ^{3}He polarization, 70% in this case, about 20-25% less counting time is required for a continuously polarized NSF compared to an off line polarized cell undergoing T_1 polarization decay for an experiment with a 1-2 day measuring time where we have assumed an on beam $T_1 = 100$ h.

The device will be placed just upstream the second Fermi chopper; as it is expected polarization can be maintained through the Fermi chopper if care is taken to insure an adequate guide field inside the device and eliminate sources of neutron depolarizing eddy currents [6]. Another feature of NSF is the ability to reverse the ^{3}He neutron polarizing power through adiabatic fast passage (AFP), thus no extra flipper will be required [7].

4. Analyzer

The polarization analyzer will have much more rigorous demands as it will require a similar thickness of ^{3}He, but over at least a significant fraction of the TOPAS detectors. The ^{3}He is foreseen to be polarized off-line because of these challenges. Novel techniques for *in-situ* polarization are being considered but will not be discussed here. It is also foreseen that prototyping/testing of systems similar in philosophy to the "PASTIS" concept [8] will be conducted at DNS at the JCNS [9] as an intermediate stage for a full scale TOPAS analyzer.

Here we will consider simply the 'brute force' method to obtain polarization analysis for TOPAS. For the brute-force method to cover the entire detector with a large container, or array of containers is not out of the question. This container would need approximately 100 bar L of polarized ^{3}He per day. While this number is beyond the production capacity of systems currently available to neutron research centers, researchers are prototyping systems with this daily capacity [10]. Further single MEOP systems have capacities of 20 bar L per day [11] and single SEOP systems similar to the ones we are using in the in-situ polarizers, have been shown to produce over 2.5 bar liter per day when used in the laboratory [12]. Thus it is possible to obtain partial coverage starting with one or several of these smaller systems and scale up to full detector coverage via additional systems. AFP inversion of the ^{3}He gas can be performed in the polarizers to reverse the analyzing efficiency of the NSF eliminating the need to develop a wide angle, large area spin flipper.

5. Conclusion

We have presented the neutron guide and polarization analysis concept for TOPAS. A high flux, which is a pre-requisite for single crystal studies or application of polarization analysis, is realized by a focusing neutron guide. A continuously pumped ^{3}He polarizer provides the best performance for experiment lasting for several days, as it is anticipated for inelastic experiments. For the polarization analysis it is crucial to have ^{3}He system with a large capacity $\approx$ 100 bar L per day, if the polarization analysis has to be spatially displaced from the sample position and covers still a large fraction of the solid angle.

References

[1] Schanzer C, Böni P, Filges U and Hils Th 2004 *Nucl. Instr. Meth. Phys. Res.* A **529** 63
[2] Cussen L D, Goossens D J and Hicks T J 2000 *Nucl. Instr. Meth. Phys. Res.* A **440** 409
[3] Babcock E, Boag S, Andersen K H, Becker M, Beecham C, Bordenave F, Chastagnier J, Chen W C, Chung R, Chupp T E, Elmore S, Fouilloux P, Gentile T R, Jullien D, Lelievre-Berna E, Mouveau P, Petoukhov A K, Revert M and Soldner T 2009 *Physica* B **404** 2655
[4] Boag S, Babcock E, Andersen K H, Becker M, Charlton T R, Chen W C, Dalgliesh R M, Elmore S D, Frost C D, Gentile T R, Lopez Anton R, Parnell S R, Petoukhov A K, Skoda M W A and Soldner T 2009 *Physica* B **404** 2659
[5] Gentile T R, Babcock E, Borchers J A, Chen W C, Hussey D, Jones G L, Lee W T, Majkzrak C F, O'Donovan K V, Snow W M, Tong X, Velthuis S G E T, Walker T G and Yan H 2005 *Physica* B **356** 96
[6] Stewart J R, Deen P P, Andersen K H, Schober H, Barthlmy J-F, Hillier J M, Murani A P, Hayes T and Lindenau B 2009 *J. Appl. Cryst.* **42** 69
[7] Babcock E, Petoukhov A K, Chastagnier J, Jullien D, Lelievre-Berna E, Andersen K H, Georgii R, Masalovich S, Boag S, Frost C D and Parnell SR 2007 *Physica* B **397** 172
[8] Andersen K H, Jullien D, Petoukhov A K, Mouveau P, Bordenave F, Thomas F and Babcock E 2009 *Physica* B **404** 2652
[9] Su Y, Schweika W, Mittal R, Ioffe A and Brückel T 2008 *JCNS experimental report* www.jcns.info/src/report/sr2008_DNS-1_new.pdf
[10] Hersmann F W *et al.* 2009 www.xemed.com/products/hyperpolarizers/helibox_z100
[11] Petoukhov A K, Andersen K H, Jullien D, Babcock E, Chastagnier J, Chung R, Humblot H, Lelivre-Berna E, Tasset F, Radu F, Wolff M and Zabel H 2006 *Physica* B **385-386** 1146
[12] Chen W C, Gentile T R, Walker T G, Babcock E 2007 *Phys. Rev.* A **75** 013416

AUTHOR INDEX

AUTHOR INDEX

Institute of Physics
Dirac House, Temple Back
Bristol BS1 6BE UK

CURRAN ASSOCIATES INC
proceedings.com